The Study of Dyslexia

The Study of Dyslexia

Edited by

Martin Turner
Consulting Educational Psychologist
and Head of Psychology, The Dyslexia Institute
Egham, United Kingdom (1991–2003)

and

John Rack
The Dyslexia Institute
University of York
York, United Kingdom

Kluwer Academic / Plenum Publishers
New York, Boston, Dordrecht, London, Moscow

Library of Congress Cataloging-in-Publication Data

The study of dyslexia/edited by Martin Turner and John Rack
p. cm.
Includes bibliographical references and index.
ISBN 0-306-48531-1 (hbk.) — ISBN 0-306-48535-4 (pbk.) — ISBN 0-306-48534-6 (e-book)
1. Dyslexia. I. Turner, Martin, 1948– II. Rack, John Paul.

RC394.W6S78 2004
618.9298553—dc22

2004047315

ISBN 0-306-48531-1 (hardback)
ISBN 0-306-48535-4 (paperback)
ISBN 0-306-48534-6 (e-book)

233 Spring Street, New York, New York 10013

http//www.wkap.nl/

10 9 8 7 6 5 4 3 2 1

A C.I.P. record for this book is available from the Library of Congress

Printed in the United States of America

Contributors

Dr Alan A. Beaton, Senior Lecturer, Department of Psychology, University of Wales, Swansea, Singleton Park, Swansea SA2 8PP, UK

Bruce J. W. Evans, Director of Research, Institute of Optometry, 56-62 Newington Causeway, London, SE1 6DS, UK, and Private practice, 23 Shenfield Road, Brentwood, Essex, CM15 8AG, UK

Prof. Rhona S. Johnston, Department of Psychology, University of Hull, Hull HU6 7RX, UK

Richard K. Olson, Department of Psychology, University of Colorado, Boulder, CO 80309, USA

Dr Valerie Muter, Department of Psychology, University of York, Heslington, York YO1 5DD, UK

Susan J. Pickering, University of Bristol, Department of Experimental Psychology, Bristol, BS8 1TN, UK

Dr John Rack, The Dyslexia Institute, The Henry Wellcome Building for Psychology, The University of York, Heslington, York YO10 5DD, UK

Jacky Ridsdale, The Dyslexia Institute, Broom Hall, 8-10 Broomhall Road, Sheffield S10 2DR, UK

Professor Margaret J. Snowling, Department of Psychology, University of York, Heslington, York YO1 5DD, UK

Martin Turner, Chartered Educational Psychologist, Brocksett Cottage, Kennel Lane, Windlesham, Surrey GU20 6AA, UK

Joyce Watson, School of Psychology, University of St Andrews, St Andrews KY16 9JU, UK

Preface

In long-ago 1999, the Dyslexia Institute and Plenum Press conceived a plan for *two* books which would gather the best of current knowledge and practice in dyslexia studies. This would benefit those—but not only those—many individuals who train with us, acquiring a postgraduate certificate and diploma with our higher education partner, the University of York.

Since then, the century changed, the hinge of history creaked and Plenum was taken over by Kluwer Academic Publishers, but the first of the pair, *Dyslexia in Practice*, emerged quickly and on schedule (Townend and Turner, 2000). Written by staff and close associates of the Institute, its chapters were produced under close scrutiny and with the expedition of a command economy. To our delight, the book has seen a success which went beyond the dreams of its editors: it has been adopted by other courses similar to our own and is widely referred to.

The same was never likely to be true of *The Study of Dyslexia*, which was envisaged as a theoretical companion volume written by authors and researchers of international repute. Nearly five years after the idea first took shape, this second volume now arrives to complete the enterprise, but it has been a very different project. The evolution of specialist dyslexia tuition—the so-called multisensory, cumulative method—proceeds with graceful but gradual elaboration, contingent neither upon fundamental research nor upon direct evaluation (but see Chapter 8 by Rack, this volume); more like a craft skills tradition, or apostolic transmission, this teaching is informed to a greater extent by the grain of the language itself, by close-up familiarity with the idiosyncratic byways—the lanes and hedgerows—of written English.

Research into dyslexia, by contrast, proceeds at an exciting pace, marked not, as in former times, by the dignified appearance of soberly considered books, but by e-papers and press conferences. Each research "breakthrough" that makes the front page of the *New York Times* sets the faxes whirring this side of the Atlantic and arms with timely questions trainees in the front row of next day's training course. While this maintains a high level of intellectual excitement, what scope does it afford reviewers who wish to capture, within the confines of a book, a stable synopsis that will not too quickly date or fade?

We may have been too ambitious for our own good. There is now too much material, as Maggie Snowling says, to be contained within a single book (though her own seems the most satisfying single volume survey of essentials; Snowling, 2000). New facets emerge or re-emerge into prominence (naming speed deficit, magnocellular theory, self-esteem) as older ones cease to interest researchers (laterality, motor skills). The main scientific consensus of the 1980s—the phonological deficit hypothesis—continues to survive vigorous challenges from 'visual' and rapid processing theories, but is, in turn, stimulated to greater explanatory productivity.

This is perhaps the most scientifically attractive state for any field to be in: settled enough to provide students with coherence, lively enough to allow change and even upsets. In the last decade, we have seen "dual route" theories of reading accommodate the rise of phonological processing by defining the "direct lexical" (visual word recognition) route in more and more morphological terms. Challenged by Seymour's format distortion experiments, Ehri's and (independently) Stuart's findings of alphabetic advantage in whole-word-taught infants, non-phonological "routes" have ceded ground that was once their sole preserve; simultaneously, in British schools, phonic teaching methods have been reintroduced officially and promoted by inspection, while the defense of Whole Language has been reduced to "whole words" (analytic phonics) and something called reading-by-analogy (onset and rime). But even here, Morag Stuart's East London and Rhona Johnston's Clackmannanshire studies have shown a truly striking advantage for children taught by means of "synthetic phonics" (see Chapter 7 by Johnston's, this volume).

During this decade, too, we have seen the rise of behavioural genetics. In the Preface to *Dyslexia in Practice,* I quoted the March 1996 Dyslexia Institute definition of dyslexia. In July 2002, this was revised once more, as follows:

> Dyslexia is a developmental disorder which results in difficulties in learning to read, write and spell. Short-term memory, mathematics, concentration, personal organisation and sequencing may also be affected.
>
> Dyslexia usually arises from a weakness in the processing of language-based information. Biological in origin, it tends to run in families, but environmental factors also contribute.
>
> Dyslexia can occur at any level of intellectual ability. It is not the result of poor motivation, emotional disturbance, sensory impairment or lack of opportunities, but may occur alongside any of these.
>
> The effects of dyslexia can be largely overcome by skilled specialist teaching and the use of compensatory strategies.

Though spurred by the desire to be more accessible, this further simplification casually embodied a new assumption—that of "biological" causation. This has long been accepted as a truism in the research community but is a relative novelty in education. Since Rack, Hulme, and Snowling (1993) first welcomed the appearance of a cross-disciplinary "synthesis" in research studies, the integration of molecular genetics, behavioral genetics, neurobiology, cognitive psychology, and behavioral teaching has continued rapidly; though many of these "levels" are now wrapped up in the term "neuroscience," it is perhaps the neurology that still seems the least mature area, with new theories taking little account of previous ones (see Chapter 3 by Beaton, this volume).

Where a straight test of phonological, magnocellular and cerebellar theories has been attempted, the phonological core deficit emerges the victor (Ramus *et al.*, 2003), but clearly a broad rather than a narrow version is needed to accommodate the abundance of diagnostic phenomena that are not easily explained by memory. By contrast, phonological theory can probably accommodate an enhanced sense of the importance of executive processes (Pickering, 2003).

This volume cannot contain, and does not attempt to encapsulate, all these strands of developing argument. Nevertheless most of the progress in current theory is reflected in some detail in its chapters, either by main players or others closely involved. This gives the collection a special authority, even while it aims to achieve, above the mêlée of current controversies, a stable perspective that a student can rely on.

Snowling's chapter sets the scene, with a historical approach to the prevailing language disorder view of dyslexia. Hypothetical, internal phonological representations—confirmed by early developmental cross-language studies—are strong in some, weak in other individuals. The purity of this explanation is defended against the common clinical observation of "comorbidities." Implications for practice are helpfully spelled out.

Muter provides much of the detail for this developmental, linguistic view of the matter with an accomplished and very full review of the research on phonemic awareness, phonological skill and verbal memory. Like Snowling, she wishes to broaden this emphasis in further linguistic directions (grammar, semantics). In order to round out an emphasis on dyslexia, this survey then focuses on the deficits identified and the instructional remedies to be applied.

Pickering seeks a theoretical rapprochement between a descriptive account of observed memory deficits and possible experimental cognitive models. This leads her to a broader discussion of working memory, taking a componential view. Such work has recently uncovered, as we have seen, the predictive importance of central executive processes and the close involvement of working memory in measured academic attainment.

Olson reviews evidence from a behavior genetic point of view, addressing not only heritability of reading disorder but also shared influence between problems of attention and learning. If determinism is now an habitual feature of media treatment of these issues, Olson strives to avoid any such simplification. Once again, phonemic awareness emerges as a central causal feature under significant genetic control.

In what must be one of the most comprehensive surveys of its kind, Beaton examines, in his chapter, the many varieties of neurological and physiological approach to describing and explaining the problems of dyslexia. Though this has long been a popular perspective with teachers, there remains, as he shows, a bewildering multiplicity both of experimental paradigm and theoretical focus. Conveniently, he takes in endocrinological, motor, sensory, and neuroanatomical topics, though it may sometimes be hard to see whether these points of view are compatible or incompatible. Some of this work has arisen in the course of technological progress and Beaton helpfully elucidates the variety of investigative techniques that bear on the "associations" of dyslexia.

Evans has done much to establish clinical optometric practice in relation to learning disability in the United Kingdom. His chapter in this volume is a useful entrée to his book (Evans, 2001) and gives an overview of the principal axes of investigation in practical optometry, while taking care not to inflate the rôle of vision factors in reading and learning.

Johnston and Watson give a clear and objective account of their recent research in that most contested of areas, the initial teaching of reading. With support for non-phonic curricula fading, they move the debate on by showing the remarkable advantage that children acquire, not just from initial phonic instruction, but especially from so-called synthetic phonics, in which "small units" (letter sounds) are taught first and then blended into word and subword units. This may indeed secure not only core literacy skills, but much-valued phonemic awareness into the bargain.

If different instructional philosophies are to be competitively evaluated, why not specialist teaching also? Until recently, this was hardly done or even talked about in polite circles. In his chapter, Rack shows that there is now a thriving and frequently inconvenient literature on the effects of specialist teaching on dyslexic individuals (e.g., Torgesen, 2001). He reviews some of the more significant studies and concludes by presenting and elucidating the findings of the Dyslexia Institute's recent SPELLIT study, which he conducted.

Turner attempts in his chapter to include most of the background that a trainee specialist teacher needs to inform his or her testing and assessment. After all, a test is often no more than a piece of paper and a result is often easily obtained. What intercedes between the selection of an appropriate test and the interpretation and communication of a pattern of such results is a technology whose complexity is habitually underestimated by many in education. Over a decade we have attempted to develop within the Dyslexia Institute, a sound body of practice and professional competence—an assessment 'model'—that trainees will find assuring as they attempt to exercise with confidence their newly expanded rôle in dyslexia testing.

Finally, Ridsdale addresses the aspects of the dyslexia condition that are frequently of most importance to parents, teachers, and peers. When children become so desperate that they harm themselves, withdraw, or show physical symptoms, their whole families suffer. Under these conditions, parents most of all desire some improvement in the "self-esteem" situation. Social and emotional dimensions of learning difficulty have always been acknowledged but less frequently investigated; still less so personality and self-esteem (but see Hales, e.g., 1994, for an honourable exception). Ridsdale introduces us to a useful theoretical framework with which to understand this complex area and then carefully addresses the many relevant links between self-esteem and academic attainment.

This, then, is our offering to those students and consumers of dyslexia research whose persistence and devotion are legendary. We hope we deserve the attention which they will undoubtedly bring to bear. We trust, in return, that they will find this evaluation of the major topics within dyslexia careful rather than tendentious, comprehensive and yet focused, open and suitably non-partisan, informative but not excessively technical. Though each of the contributions presents the view from the front line, we seek to arrive at analyses that will endure the vicissitudes of research fashion in the medium term.

The aim, if indeed it needs to be stated, is that theory and practice should continue to meet and combine in the adventure that so often astounds visitors to the dyslexia world. Theory is the means by which we see, quite literally, what is there in front of us (Popper, 1979) and understand what we see; and it remains desirable to get some idea of what we are dealing with before we deal with it.

Martin Turner

REFERENCES

Evans, B. J. W. (2001). *Dyslexia and vision*. London: Whurr.

Hales, G. 1994. The human aspects of dyslexia. In G. Hales (Ed.), *Dyslexia matters*. London: Whurr.

Pickering, S. (Ed.). (2003). Memory and education. Special issue of *Educational and Child Psychology, 20*(3). Leicester: The British Psychological Society, Division of Educational and Child Psychology (DECP).

Popper, K. R. (1979). *Objective knowledge: An evolutionary approach* (Rev. ed.). Oxford: Clarendon Press.

Rack, J. P., Hulme, C., & Snowling, M. J. (1993). Learning to read: A theoretical synthesis. In H. Reese (Ed.), *Advances in child development and behavior* (vol. 24, pp. 99–132). New York: Academic Press.

Ramus, F., Rosen, S., Dakin, S. C., Day, B. L., Castelotte, J. M., White, S. *et al.* (2003). Theories of developmental dyslexia: Insights from a multiple case study of dyslexic adults. *Brain* (vol. 126, pp. 841–865).

Snowling, M. J. (2000). *Dyslexia* (2nd ed.). Oxford: Blackwell.

Torgesen, J. K. (2001). The theory and practice of intervention: Comparing outcomes from prevention and remediation studies. In A. Fawcett (Ed.), *Dyslexia: Theory and good practice.* Whurr: London,

Townend, J. & Turner, M. (Eds.) (2000). *Dyslexia in practice: A guide for teachers.* New York: Kluwer Academic/Plenum.

Contents

1

Visual Factors in Dyslexia

Bruce J. W. Evans

INTRODUCTION

Background

Vision is a core component of the reading process, so do visual problems cause dyslexia? Despite a century of research, there is no clear answer to this question, and the lack of clarity reflects the vagueness of the terms "dyslexia" and "vision." Definition, classification, and the etiology of dyslexia are discussed elsewhere in this book.

Vision can be broadly classified into *sensory* and *motor* functions. Sensory visual function refers to the flow of information into the brain. Motor visual function refers to the control of the various muscles that move, coordinate, and focus the eyes.

The use of the word *vision* in this chapter is different to its use in psychometric testing. Cognitive tests or profiles often identify a "visual" or "visuo-spatial" anomaly, which refers to "higher level" processing of complex visual information. These types of visual deficits are distinct from the motor and low-level sensory visual pathway anomalies that are discussed below.

This chapter will concentrate on the visual problems that have been identified as correlates of dyslexia. But research suggests that the visual correlates of

Bruce J. W. Evans, Director of Research, Institute of Optometry, 56-62 Newington Causeway, London, SE1 6DS, UK and Visiting Professor, City University, Northampton Square, London, EC1V 0HB, UK.

The Study of Dyslexia, edited by Turner and Rack.
Kluwer Academic Publishers, New York, 2004.

non-dyslexic reading difficulties may be the same as the visual correlates of dyslexia. So, the battery of optometric tests that is suggested below should really be aimed at any person who underachieves at school, not just those who have been formally diagnosed as having dyslexia.

SUMMARY

- Vision can be subdivided into sensory and motor
- The "visual" or "visual-spatial" deficits that psychologists might detect are very different from the visual problems that an optometrist might diagnose
- All children with difficulties at school may benefit from an assessment with an eye care practitioner who has specialized in specific learning difficulties

THE ASSOCIATION BETWEEN VISUAL ANOMALIES AND READING SKILLS

Ocular Health

Ocular diseases are rare in childhood and do not seem to be correlated with dyslexia. Poor ocular health will not necessarily cause symptoms, and some diseases cause a loss only of peripheral vision so that a child may be able to read a letter chart normally. Interestingly, people with ocular pathology that does interfere with vision often seem to automatically adjust their reading rate so that they would be unlikely to make as many errors as a person with good ocular health and dyslexia.

Accommodation

Ocular accommodation is the process whereby the shape of the lens inside the eye is changed to increase its power. Accommodation occurs naturally during near vision and serves to keep the eye in focus, rather like a camera might be focused on a near object. Young children can usually focus clearly very close to; they have a high amplitude of accommodation. An apparent low amplitude of accommodation can result from uncorrected longsightedness, as described below, and this possibility must be excluded before poor accommodation can be diagnosed. To read comfortably for sustained periods, the amplitude of accommodation needs to be substantially greater than that predicted for clear vision at the relevant viewing distance.

Research has shown that the average amplitude of accommodation of children with dyslexic difficulties is statistically, significantly reduced compared with control good readers (Evans, 2001). However, despite this *average* difference, the vast majority of children with reading difficulties have an amplitude of accommodation that exceeds the level that is necessary for clear and comfortable reading.

The amplitude of accommodation is measured by bringing a target of small text toward patients until they report blurring. It can also be measured objectively by using reflected light (dynamic retinoscopy). Some optometrists include other tests of accommodative function in their assessment of children with reading difficulties. The two most common of these are tests of accommodative lag and facility. During near vision, the eyes are not usually precisely focused on the object of regard, but the accommodation lags a small amount behind the target. If the accommodative lag is low, then the blur it causes is insignificant; if it is high, then it can result in blurred print during reading. Accommodative lag is measured objectively with a retinoscope (see below). Accommodative facility is a measurement of the ability to change focusing. In the classroom, accommodative infacility can result in a difficulty changing focus between the board and a book, although this can also be a sign of shortsightedness.

It is rare for a patient to have very poor accommodation, although accommodative paralysis is occasionally encountered and requires glasses. A milder degree of accommodative dysfunction is more common and requires treatment if it is associated with symptoms (e.g., blurring, eyestrain, headaches). Some children may become so used to their symptoms that they fail to report them, so accommodative dysfunction may occasionally be treated in the absence of symptoms, if it is of a degree that would be expected to interfere with academic performance. Treatment often consists of eye exercises (vision therapy) or, if exercises are unsuccessful, spectacles (possibly bifocals).

SUMMARY

- Ocular accommodation refers to the ability of the eyes to focus close to
- Optometrists use subjective and objective tests to assess accommodation
- Accommodative problems can occur in dyslexia
- If an accommodative anomaly is severe enough to require intervention, then it may be treated with exercises or glasses

Visual Acuity and Refractive Error

Visual Acuity

Visual acuity refers to the angular subtense of the smallest size of detail that the eye can resolve. It is usually expressed as a fraction where the numerator refers to the testing distance in meters (feet in the United States). Six meters is the reference distance (20 feet in the United States), and the denominator changes proportionally according to the change in size of the lettering, while reading from the same reference distance. The easiest way to interpret a visual acuity is to divide out the fraction. For example, average visual acuity is 6/6 (in feet, 20/20), which divides out to 1.0, or 100%. Although this is the theoretical normal level of visual acuity, most children do better than this, typically achieving 6/4 (20/30), which divides out at 1.5 or 150%. The bottom line of many letter charts is 6/3 (half the size of 6/6 letters), and the top line of most letter charts is 6/60. A 6/60 letter is 10 times the size of a 6/6 letter and 6/60 divides out at 0.1 or 10% of "normal" visual acuity.

Most of the near print that 6- or 7-year-old children would be expected to read is equivalent to about 6/24, and typical text for a 9-year-old is equivalent to about 6/18. The smallest print that any children are expected to read (e.g., dictionaries) is equivalent to about 6/12. Although this is approximately twice the threshold acuity, this margin may be necessary for comfortable vision.

Myopia (Shortsightedness)

Shortsightedness (myopia) blurs distance vision (e.g., the board in class), and the prescribing of spectacles for myopia by eye care practitioners is usually a fairly straightforward decision. An optometrist might consider prescribing a refractive correction if the myopia reduced the binocular visual acuity to 6/9 and would almost certainly prescribe if it was reduced to 6/12. Low to moderate myopia will not impair near vision and some studies have even found a correlation between myopia and above-average reading performance (Evans, 2001).

Hypermetropia (Longsightedness)

Most young, normal, eyes have a very low degree of longsightedness (hypermetropia or hyperopia), and young eyes have an ability to compensate for low to moderate degrees of longsightedness. The eyes compensate by using the power of ocular accommodation (described above) to "over focus" the eyes and correct the longsightedness. This is a misuse of accommodation, which usually serves to keep the image clear during near vision. A very different function during near vision is convergence, which is a turning in of the two eyes so that they both point at an approaching object. Accommodation keeps near objects clear, and convergence keeps them single (if the two eyes were not pointing at the object of regard, then it would be perceived as double). In the natural world, accommodation and convergence always act together and these two functions are linked in the brain. In other words, a given amount of convergence induces an equivalent amount of accommodation and vice versa.

This relationship between accommodation and convergence causes problems in significant degrees of longsightedness. In this condition, accommodation can be used (or, really, misused) to overcome the longsightedness rather than to focus on near objects. But this use of the accommodation will usually trigger an associated convergence. Hence, when accommodation is used to compensate for excessive hypermetropia, it can induce a turning inward of the eyes and uncorrected longsightedness is a common cause of convergent strabismus (squint, see below). In this case, the correct treatment for the strabismus is to correct the longsightedness, and surgery would be an inappropriate treatment.

A constant accommodative effort to overcome excessive longsightedness may also cause eyestrain, headaches, and blurred vision. Longsightedness in children would not usually be detected by a distance letter chart test, and frequently does not even affect performance at brief near vision tests. Longsightedness in one eye will not be detected by binocular (both eye) acuity tests and, although unlikely to affect reading, can result in a permanent visual loss if not detected in the first 5 or 6 years of life. Some studies have found a weak correlation between longsightedness and reading difficulties, others have suggested that this may result from a weak negative correlation between hypermetropia and IQ (Evans, 2001).

Astigmatism

Astigmatism is the refractive error that results when an eye has a different refractive power in different meridians (e.g., in the horizontal and vertical meridians). Low astigmatism is ubiquitous and harmless; higher astigmatism will cause blurred distance and near vision and may cause eyestrain.

Assessment of Refractive Errors

Refractive errors, like most of the things that optometrists measure, can be determined objectively and subjectively. Objective tests require little or no contribution from the patient. Subjective tests require a response from the patient, but invariably include an element of "double-checking." It should be stressed that children of any age (including infants) and of any capabilities (even with severe intellectual disabilities) can undergo an eye examination with an optometrist.

The principal objective technique for determining the refractive error uses a handheld instrument, the retinoscope, to shine light through the pupil, and to neutralize (with lenses) the rays that are reflected back from the retina. Eye care practitioners rely more on objective techniques when examining younger and less reliable children; patient cooperation is not essential for such an eye examination. Sometimes, eye drops are used to relax the accommodative mechanism and thus reveal the full refractive error. Modern drugs only sting a little, and the duration of blurred vision is just a few hours (the pupils remain dilated for longer so that sunglasses may still be needed on the day after the drops are inserted).

SUMMARY

- Shortsightedness impairs the ability to read a letter chart or board in class
- A low degree of longsightedness is normal
- Moderate or high longsightedness is often not detected by screening tests, and can cause an eye to turn inward and/or blurring and eyestrain when reading
- Children do not need to be able to read or to be cooperative to have an eye examination

Binocular Vision (Orthoptics)

Strabismus

Binocular vision refers to the ability of the two eyes to work together in a coordinated way. Perfect binocular vision requires the two eyes to be aligned precisely (to within about 0.03°) in their fixation of the object of regard. Strabismus (synonyms: heterotropia, squint, cast, turning eye) describes the condition where the eyes manifestly fail to maintain alignment. Strabismus affects about 2.5% of the population. In young children, the image in the eye that turns will be at least partially suppressed to prevent double vision. This almost invariably results in a permanent visual loss (amblyopia, or "lazy eye") in that eye. This visual loss can be largely prevented or treated in young children (up to about 7 years) by patching the non-strabismic (normal) eye. The strabismus may be of a small angle and is often undetected by parents, and sometimes is not discovered by school screening tests.

Most types of strabismus are not usually thought to be associated with reading difficulties, and some authorities have suggested that a stable unilateral strabismus might actually be associated with better than average reading because the ocular dominance is well established. For this reason, it has been suggested, controversially, that excessive patching to treat amblyopia in early life could cause problems with unstable ocular dominance in the school years (Fowler & Stein, 1983). Although most cases of strabismus are unilateral, a few patients alternate freely from using one eye to the other. If this alternation occurs during reading, then this could in some cases cause a confused perception of text.

Eye care practitioners assess strabismus with a variety of tests, but the most important is the cover test (Evans, 2002). In this test, each eye is covered in turn while the patient fixes a target and the movement of the uncovered eye is observed. If, for example, the right eye is covered, and the left is seen to move in to take up fixation, then it can be deduced that there is a left divergent strabismus (the left eye is deviated outward). The cover test requires very little cooperation and no verbal response from the child.

It is perhaps surprising in most cases of strabismus, the eye muscles work perfectly. The misalignment of the eyes usually results from a problem with the brain's control of the eye muscles, rather than from a problem with the muscles themselves. Not all cases of strabismus require treatment, but those that do are variously treated with spectacles, eye exercises, or surgery, in addition to the patching described above.

Heterophoria

Most people do not have a strabismus, but instead, the eyes are aligned during everyday vision. But this alignment is not a fixed feature: the visual system is not like the two monocular tubes of binoculars, which are permanently aligned. Even when the eyes are normally aligned in everyday vision, their resting position is usually out of alignment and this is called a heterophoria. This is best demonstrated by a variation of the cover test described above. With a heterophoria, when one eye is covered, the eye behind the cover moves: the covered eye takes up its resting position, which may be diverged, converged, elevated, or depressed. Because the eyes are aligned during normal viewing, when the cover is removed, the previously covered eye is seen to move to take up fixation once more. This movement of an eye when the cover is removed diagnoses a heterophoria, which can be classified according to the direction, size, and rapidity of movement (Evans, 2002). During everyday visual tasks, a slight constant effort has to be maintained to overcome the heterophoria, but usually this effort is insignificant and causes no symptoms. Most heterophoria are horizontal in which case this effort acts to create convergence (a turning inward of the eyes) or divergence (a turning outward). The amount of convergence or divergence that is "held in reserve" to overcome a heterophoria is called the *fusional reserve*. Sometimes, the heterophoria is unusually large and/or the fusional reserves unusually low, and the heterophoria becomes *decompensated*.

Typical symptoms of decompensated heterophoria include eyestrain, headaches, blurred vision, double vision, and visual perceptual distortions. Children may accept these symptoms as the "normal" situation, and may not fully realize that these problems are present. Some people with these symptoms develop the habit of covering one eye when reading. In the classroom, children may tip their head on one side when reading so that their nose acts to occlude the view of one eye and so prevent double vision (an unusual head position may also be a sign of a weak eye muscle). Optometrists diagnose decompensated heterophoria using symptoms, the objective cover test and subjective tests (including fixation disparity and stereo-acuity tests). It is possible that some cases of decompensated heterophoria might break down to a strabismus if left untreated.

Research has found dyslexia to be correlated with low fusional reserves and an unstable heterophoria (Evans, 2001). Together, these two visual anomalies characterize a condition termed *binocular instability* (Evans, 2002), and the symptoms of this, when reported, are similar to those of decompensated heterophoria. It should be noted that some people with binocular instability or decompensated heterophoria read very well, so these conditions are not necessarily causes of dyslexia. However, in some cases, these binocular vision problems can make reading very difficult and occasionally the label of dyslexia is used inappropriately when in fact the person's problems solely stem from a binocular vision anomaly. Decompensated heterophoria and binocular instability usually respond very well to treatment, either by exercises or spectacles.

One other very common binocular vision anomaly in children is convergence insufficiency. If a child observes an object that is approaching his/her nose, then his/her eyes will usually continue to turn inward (converge) until the object is less than about 5 cm away from the eyes. A convergence insufficiency is when a person is unable to converge closer than a certain distance, usually 8–10 cm. A marked convergence insufficiency is often associated with a decompensated heterophoria at a normal reading distance and may require treatment. Most convergence insufficiencies respond well to simple exercises.

SUMMARY

- Binocular vision problems are quite common
- A strabismus, or turning eye, will not necessarily be apparent to teachers or parents, yet needs to be detected as young as possible
- A subtle weakness in the coordination between the two eyes, *binocular instability*, is particularly common in dyslexia
- Binocular vision anomalies can contribute to a reading difficulty, although the child may not report any symptoms
- Most binocular vision anomalies can be treated with exercises or glasses

Ocular Dominance

The concept of a dominant eye is simplistic and its relevance to dyslexia is probably overstated. One cause of confusion is that there are at least three different types of ocular dominance: sighting, motor, and sensory ocular dominance (Evans, 2001). In a given individual, the right eye might be the dominant eye for a sighting task and the left for a motor task; and it may even be the case that the right eye is the dominant eye for one motor task and the left for another motor task. It is not clearly known which, if any, of these types of ocular dominance is most relevant for reading.

Sighting tests of ocular dominance were once used to determine whether an individual was "crossed dominant" (sighting eye on the opposite side to the dominant hand), and crossed dominance was thought to be related to reading difficulties. However, some studies have failed to support this hypothesis and the consensus now seems to be that sighting dominance is largely irrelevant to the assessment and treatment of reading and spelling problems (Moseley, 1988).

More recently, it has been claimed that an unstable result on a test of motor ocular dominance, the Dunlop Test, could diagnose "visual dyslexia," which was treated by occluding one eye (Stein & Fowler, 1985). The use of the term "visual dyslexia" has caused some confusion since it is probably not intended to be equivalent to the visual-spatial subtype of dyslexia identified by Boder and other classifications (see later section on Higher Visual Processing). It was later asserted that an unstable result on the Dunlop Test indicated "poor visuomotor control" or "binocular instability." The Dunlop Test has been criticized on many grounds and, although preliminary research supported the use of this test, most subsequent studies have failed to agree with the original findings (Evans, 2001). There is a fairly wide agreement that the Dunlop Test is unreliable, although there is growing evidence that many dyslexic children have some form of binocular instability (the classic meaning of this term is described above). It is unlikely that any isolated test, let alone one as unreliable as the Dunlop Test, is able to diagnose binocular instability. The best approach is for eye care practitioners to apply a battery of tests to assess binocular function.

SUMMARY

- There are different types of ocular dominance, and the main ones are sighting, motor, and sensory
- Sighting dominance and crossed laterality are probably not especially relevant to dyslexia
- A recent test of motor ocular dominance, the Dunlop Test, has been largely discredited

Eye Movements

Saccadic Eye Movements in Reading

During reading, the eyes constantly make a series of "flick" (saccadic) eye movements. The eyes typically fix on a group of letters for about 1/4 of a second while the information from these is assimilated and they then make a saccadic eye movement to the next group of letters, and so on. At the end of the line, the eyes make a large backward saccade to the beginning of the next line. Occasional smaller backward saccades, regressions, occur to aid comprehension. The control of this pattern or eye movements is influenced by both the layout of the text and by the reader's comprehension.

It should be noted that this general pattern of fixations separated by saccadic eye movements is not specific to reading (Stark, Giveen, & Terdiman, 1991). It is the visual system's basic method of acquiring information from any visual scene in which the observer and target are relatively stationary. The unusual factors in reading are the sequential and predominantly horizontal nature of successive targets, and the similarity, in gross spatial terms, of these targets.

Generally speaking, with more difficult reading material or with less expert readers, the number of fixations is increased, the duration of fixations is increased, and the number of regressions increases. The reading eye movements of people with severe reading difficulties are, therefore, somewhat atypical. A key question is whether these atypical eye movements are the result of the poor reading skills or whether they are underlying causes of the poor reading. This question has been addressed in two main ways (Evans, 2001). First, many researchers have studied the sequential horizontal saccadic eye movements of dyslexic readers in nonreading tasks. The second approach has been to compare the reading eye movements of people with dyslexia with those of younger good readers.

Although a large body of research has used these paradigms, the results have been contradictory and there is still a lack of agreement in this area (e.g., Stark *et al.*, 1991; cf. Biscaldi, Fischer, & Aiple, 1994). Although case studies have identified one or two individuals who have an irrepressible tendency to read from right to left, this appears to be abnormal, even in dyslexic individuals. It would seem to be prudent to conclude that, in view of the large amount of research, abnormal eye movements are unlikely to be a major factor in dyslexia.

Pursuit Eye Movements

It is interesting to note, in passing, some research findings on pursuit eye movements. These are the following eye movements that occur in pursuit of a slowly moving target, for example, watching an aeroplane moving in the distance. If the target moves too quickly, then the smooth pursuit eye movements break down into "jerky" saccadic eye movements. Some researchers have noted that people with dyslexic difficulties often exhibit saccadation of their pursuit eye movements for target speeds where this does not usually occur. This observation has received relatively little interest because pursuit eye movements are not used during normal reading tasks. Some authorities have described pursuit eye movements as an extreme case of a fixation reflex (the reflex that occurs to keep the image of the object of regard on the fovea) and fixation instability has been described in a group of children with poor reading performance (Eden, Stein, Wood, & Wood, 1994). More research is needed to determine whether these findings might be linked to attentional deficits (see below).

Tracking Skills

The term "tracking skills" is sometimes used to describe subjective estimates of ocular motor function in relation to reading performance. Unfortunately, the term is vague and is used variously to refer to saccadic eye movements (e.g., performance when changing fixation between two pens), saccadation of pursuit (e.g., the presence of nonsmooth movements when following a slowly moving pen), or a remote near point of convergence (e.g., eyes appear unable to converge on a pen approaching the nose). These three types of eye movements are very different, but it is possible that all three types of observation may be related to lapses of concentration by the subject. Dyslexia is often associated with attention deficit disorder (with or without hyperactivity) and this might account for much of the controversy relating to the relationship between eye movement dysfunction and dyslexia (Evans, 2001). It should also be noted that even when subjects can converge to their nose, when asked to make the effort, this does not necessarily mean that they have a full range of vergence eye movements, or that their binocular vision is normal.

SUMMARY

- Saccadic eye movements, which are fundamental to the reading process, are a basic form of eye movements that are essential for normal vision in everyday life
- Although research is inconclusive, the general consensus is that saccadic dysfunction is not a major cause of dyslexia
- The term "tracking" in relation to eye movements is vague, and is variously used to describe very different types of eye movements which bear little relationship to each other
- Attentional lapses are likely to be the explanation for many observations of unstable eye movements in dyslexia

Behavioral Optometry

Behavioral optometry (BO) is a subdiscipline of optometry that has some practitioners in the United States and a few followers in the United Kingdom. The term covers a very broad range of activities, and the simplest interpretation of BO is that optometrists should take account of the whole person and his/her environment. This is unquestionably sound advice that is likely to be followed by all good eye care practitioners. Many behavioral optometrists, like other conventional optometrists, concentrate on a very thorough orthoptic investigation as outlined above, and the treatment of anomalies detected in this way is not controversial. However, most behavioral optometrists seem to investigate unconventional areas of visual function. These practitioners often prescribe exercises and/or glasses to the vast majority of children they examine, even when these patients would not be thought to be abnormal following a thorough eye examination by a conventional optometrist or ophthalmologist. For example, many behavioral optometrists regularly treat saccadic or pursuit eye movements in dyslexia, although these types of eye movement dysfunction have not been clearly established as a strong correlate of dyslexia, let alone a cause.

One of the areas that BO concentrates on is perceptuo-motor function and exercises that are reminiscent of the "patterning treatment" of Dolman and Delacato (Silver, 1986). Often BO involves the assessment of functions which might conventionally be thought to lie within the domain of psychology, such as analyzing reading errors, subtyping dyslexia, and investigating higher visual processing skills such as decoding and short-term memory skills.

In recent years, the health care sciences have been strongly influenced by the "evidence-based approach." This advocates a skeptical approach to received wisdom and new ideas alike, until they have been validated by objective research. Unfortunately, BO has not been validated by these methods (Evans, 2001; Jennings, 2000). It seems likely that at least some of the successes in improving reading that have been attributed to BO are simply manifestations of the placebo effect. These cases may have benefited more if the resources had been spent on conventional management, such as extra teaching tailored to the child's individual needs.

SUMMARY

- Behavioral optometry (BO) is a controversial approach originating in the United States
- Extreme practitioners of BO treat virtually all the children they see with either glasses or exercises
- BO has been slow to adopt an evidence-based approach
- The benefit that some children seem to derive from BO may be explained as placebo effects owing to the extra attention they receive
- Resources might be better spent on psychological assessments and extra teaching

Meares–Irlen Syndrome

Critchley (1964) cited a case study by Jansky in 1958 of a dyslexic child who was unable to read words on a white card, but could manage to read words printed on a colored card. Meares (1980) suggested that some children's perception of text and reading disabilities are influenced by print characteristics. She found that in some cases the white gaps between the words and lines masked the print and caused perceptual anomalies, such as words blurring, doubling, and jumping. She noted that this was helped by various strategies including using colored paper.

Following these early reports, Helen Irlen established a proprietary system, based on "Irlen Institutes," in several countries to detect and treat "Scotopic Sensitivity Syndrome." This term is probably etymologically inappropriate and "Meares–Irlen Syndrome" may be a suitable alternative (Evans, 1997). The condition is characterized by symptoms of eyestrain and visual perceptual distortions when reading. The eyestrain is typified by sore, tired eyes, visual discomfort, and headaches. The distortions include blurring, doubling, fading, shimmering, movement of words or letters, and seeing patterns and shapes on the page. Treatment of Meares–Irlen Syndrome usually is with colored filters, either colored sheets placed on the page (overlays) or with colored glasses. It is claimed that the required color differs from person to person and is very specific: if a person is given colored glasses of a color that is similar but slightly different to their optimal color, then they will receive less benefit than with the appropriate color (Irlen, 1991).

Proponents of the Irlen system claim that up to 60% of people with a reading problem and 10% of good readers suffer from this disorder. However, until recently, Meares–Irlen Syndrome has lacked both an adequate scientific explanation and valid research, and this has led many professionals to dismiss it as a placebo or attributional effect. For any treatment, or intervention, to be accepted as valid, it should be supported by a type of research study called a double-masked randomized placebo-controlled trial, or RCT.

Wilkins developed an instrument, the Intuitive Colorimeter, which facilitated an RCT of the beneficial effect of colored filters. Children who had the symptoms described above were tested with the Intuitive Colorimeter to determine the precise color of filter that most improved their perception of text (Wilkins *et al.*, 1994). One pair of colored glasses were made up to this "optimal tint," while a second pair were made up with a tint that was very similar, but slightly different to their optimal one (the "control tint"). The optimal and control tints were each worn for a period of four weeks, in random order. The children reported significantly fewer symptoms with their optimal than with their control tints.

This study did not investigate whether Irlen's claims about the high prevalence of Meares–Irlen Syndrome are accurate, although recent studies suggest that about 20–30% of typical (unselected) schoolchildren select and continue to use a colored overlay for a sustained period of time (Evans and Joseph, 2002). Interestingly, the condition appears to be almost as common in good readers as in dyslexia (Kriss, 2001), but it may be that the symptoms are more apparent to children who have reading problems (Evans, 2001).

Several potential mechanisms for the benefit from colored filters have been proposed, few of which seem to be able to account for the high degree of specificity of the required color (Evans, Busby, Jeanes, & Wilkins, 1995). It seems that the most likely explanation centers around "pattern glare" (Wilkins, 1995). Striped patterns (Figure 1) can be unpleasant to look at and can trigger symptoms of eyestrain and visual perceptual distortions. In fact, these symptoms are remarkably similar to those reported by patients with Meares–Irlen Syndrome. Some people with photosensitive epilepsy and some with migraine are particularly prone to these symptoms and the symptoms probably result from a hyperexcitability of the visual cortex (Wilkins, 1995). Lines of print on a page can cause pattern glare, and this mechanism is probably responsible for at least some symptoms of "visual stress" with reading. Some areas of the visual cortex are sensitive to specific colors and this could account for the benefit from specific colored filters (Wilkins, 1995). Recent research suggests that pattern glare is a correlate of Meares–Irlen Syndrome (Evans, 2001; Evans *et al.*, 1995).

The Intuitive Colorimeter was developed by Wilkins at the Medical Research Council (MRC) Applied Psychology Unit and is manufactured under license (to the MRC) by Cerium Visual Technologies. Wilkins also developed a system of Precision Tinted Lenses to facilitate the accurate reproduction of colors chosen in the Intuitive Colorimeter[1] and a system of colored plastic sheets that can be placed

Figure 1. A pattern which may cause pattern glare. DO NOT STARE AT THE PATTERN IF YOU SUFFER FROM EPILEPSY OR MIGRAINE.

[1] Available from Cerium Visual Technologies, Cerium Technology Park, Appledore Road, Tenterden, Kent, TN30 7DE; 01580 765211.

on the page to inexpensively, yet systematically, screen children for a benefit from color (Intuitive Overlays[2]). A clinical system is in operation where teachers, optometrists, and psychologists screen children with the colored overlays. Children who show a sustained benefit from an overlay, or an immediate improvement with a new Rate of Reading Test,[3] are then referred to an optometrist or hospital department who have the Intuitive Colorimeter.

Three caveats should be stressed about this system (Evans, 2001). First, about one third of children who are issued with a colored overlay return for Intuitive Colorimetry. This suggests that the system of trying overlays first is a good method of differentiating the "genuine" cases from those who may simply want to try color for a novelty. Second, for sound optical reasons, the color of a person's preferred overlay is different from the color of their preferred lens. Spectacle lenses should not be tinted on the basis of a preferred overlay color. Finally, several studies have shown a high prevalence of other eye anomalies (e.g., binocular vision problems) in patients presenting with the symptoms of Meares–Irlen Syndrome. In some cases, these symptoms resolve after eye exercises, although other cases still need colored filters. Therefore, people should be referred to an eye care practitioner who is skilled in the assessment of people with reading difficulties before providing colored overlays.

The Wilkins Intuitive Colorimeter system and the Irlen system are the two largest systems in use in the United Kingdom. Other approaches are available in some parts of the country, including the Chromagen system, which only uses a very limited range of colors, Tintavision, and Orthoscopics. Published rigorous research seems to be lacking on the last two of these systems, so they are best considered as experimental.

SUMMARY

- The symptoms of Meares–Irlen Syndrome are eyestrain, headaches, and visual distortions
- The treatment for Meares–Irlen Syndrome is individually prescribed colored filters
- Although colored overlays are used for screening, the color of glasses cannot be predicted from the color of preferred overlay
- Suspected sufferers should have a detailed eye examination before colored filters are used

Deficit of the Magno Visual Subsystem

The earlier sections of this chapter, on accommodation, binocular vision, and eye movements, predominantly refer to the control of the eyes by the brain: these are motor visual functions. The simple optical image that is formed in the eye is

[2] Available from i.o.o. Sales Ltd, 56-62 Newington Causeway, London, SE1 6DS; 0207 378 0330.

[3] See footnote 2.

heavily processed before it reaches consciousness and the various components of the processing of the visual image are described as sensory visual factors.

It is generally accepted that visual processing under daytime light levels takes place in parallel through the magno (magnocellular; transient) and parvo (parvocellular; sustained) subsystems (Lennie, Trevarthen, Van Essen, & Wassle, 1990; and see Chapter 3 by Beaton, this volume). The magno system is a sort of rapid "early warning system" that responds best to fast moving or flickering coarse objects in peripheral vision. The parvo system is slower and exists to provide a more detailed analysis of objects that are likely to be close to the center of the field of view, stationary or only slowly moving. The parvo system is good at differentiating colors, the magno system is not.

Convincing research suggests that up to 75% of people with specific reading difficulties have a deficit of the magno visual system (Evans, 2001; Lovegrove, Martin, & Slaghuis, 1986). The fact that the parvo system appears to be normal in dyslexia may explain why many researchers have concluded that there are no visual correlates of reading disability since vision tests tapping the parvo subsystem have been used most frequently. Some researchers have suggested that the magno visual system deficit may reflect a general impairment of fast systems throughout the central nervous system in dyslexia (Livingstone, Rosen, Drislane, & Galaburda, 1991). An intuitive weakness of this hypothesis is that such an impairment might be expected to cause a more general deficit than the specific difficulty of dyslexia.

Predictably, several theories have been proposed to transform this visual *correlate* of reading difficulty into a *cause* of reading problems. Although there is some evidence to suggest that a magno deficit may contribute to certain reading errors (Evans, 2001), more research is needed to evaluate these theories. In any event, there are no widely used clinical tests of magno function, and there are no known treatments for a magno deficit. Since the magno deficit has not been clearly shown to be a cause of reading difficulties, the development of treatments seems premature.

SUMMARY

- The magno visual system is an "early warning system" for detecting the coarse details of moving or changing objects in our field of view
- Up to 75% of people with specific reading difficulties have a deficit of the magno visual system
- There are no validated clinical tests or treatments for a magno deficit and it is not clear whether treatment would be helpful

Higher Visual Processing

There have been many attempts to classify dyslexia, although it should be noted that one large study failed to support the concept of clearly defined subgroups (Naidoo, 1972). Nevertheless, many authors have classified dyslexia into three subgroups and a good example is the classification of Boder (1971). Her system is based upon three reading–spelling patterns. Dysphonetic dyslexia is characterized by people who read words globally, as instantaneous visual wholes from the limited sight vocabulary, and as a result cannot cope with new or unusual words. The opposite is dyseidetic dyslexia, exemplified by the analytic reader who cannot perceive letters or words as visual wholes and who consequently reads laboriously and cannot deal with words that are irregularly spelled or pronounced. The final group, mixed dysphonetic–dyseidetic dyslexia–alexia, exhibits the combined deficits of both groups and are usually the most severely handicapped educationally.

The type of high-level cognitive visual deficit that, for example, Boder's dyseidetic subgroup manifest is very different to the visual deficits that have been described earlier in this chapter. The relationship between these two types of visual deficit has been studied very little and is far from clear. Lovegrove *et al.* (1986) argued that the visual disabilities shown by the dyseidetic (visuo-spatial) subgroup do not reflect visual processes of the type characterized by their research on a magno system deficit. There is even some evidence to suggest that the magno deficit is present in the dysphonetic, but not the dyseidetic, subgroup of dyslexia (Ridder, Borsting, Cooper, McNeel, & Huang, 1997). This has been linked to the theory described in the last section about magno visual and phonological deficits in dyslexia both representing a common impairment of fast systems in dyslexia.

A similar distinction needs to be drawn between the "visual IQ" assessed by tests such as the British Ability Scales and the type of visual deficits described in this chapter. It should not be assumed that children with lower "visual IQs" are particularly likely to have the visual anomalies described in this chapter. There is no good evidence to suggest that psychometric results obtained in a psychological evaluation can be used to decide which dyslexic children need to see an eye care practitioner.

SUMMARY

- Dyslexia is sometimes classified into visual, auditory, and mixed subtypes
- The "high level" deficit of the visual subtype is very different from the types of visual anomalies described in this chapter
- Similarly, "visual IQ" is probably not related to the visual correlates of dyslexia identified in this chapter

LINKING THE VISUAL CORRELATES OF DYSLEXIA

The visual correlates of dyslexia can be summarized as motor deficits in accommodation and vergence, the sensory magno visual system deficit, and a benefit from colored filters. Are these correlates linked to one another?

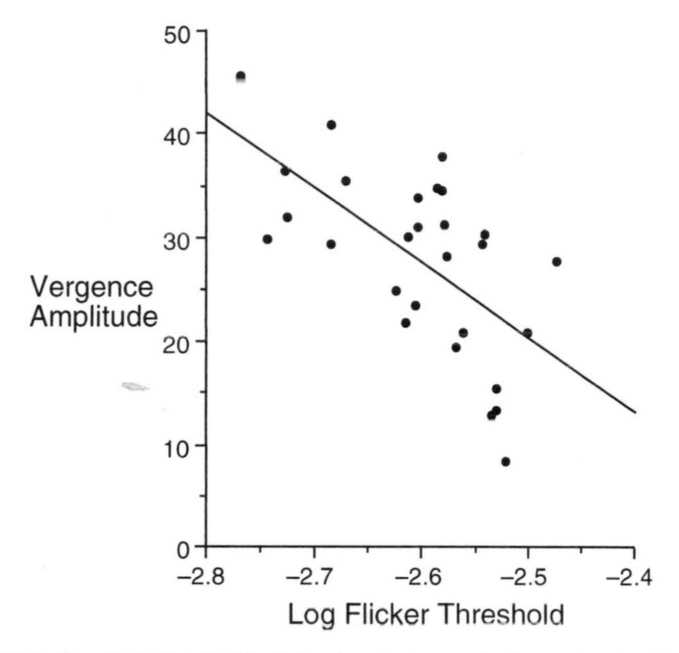

Figure 2. Regression of vergence amplitude, in prism diopters, on log temporal contrast threshold for the dyslexic group. No points are hidden by overlap.

It is well known that there is a strong crosslink between accommodation and vergence, although the role (if any) of this link in the visual correlates of dyslexia remains to be established. The hypothesis that the magno deficit in dyslexia may be linked to poor vergence control was investigated by Evans, Drasdo, and Richards (1996). They found that a group of dyslexic children manifested a reduced sensitivity to faint flicker (a sign of a magno deficit) and significantly reduced vergence amplitudes (a sign of binocular instability). These sensory and visual motor correlates were very significantly correlated in this dyslexic sample (Figure 2; $r = -0.627$).

It has been hypothezised that the magno visual system deficit may explain the benefit that some individuals experience from colored filters, although a recent study did not support this hypothesis (Evans *et al.*, 1995). This study did demonstrate an association between the motor visual correlates of dyslexia described above and a reported benefit from color, although the data suggested that it is unlikely that motor visual factors were the underlying cause of the benefit from color. The magno system is insensitive to color differences so is unlikely to be able to explain the color-specific nature of Meares–Irlen Syndrome. The search for a common cause of these visual factors will no doubt continue, and one recent analysis has suggested that a deficit of visual attention may play a key role (Evans, 2001).

SUMMARY

- A magno visual deficit has been shown to be significantly correlated with poor vergence control in a group of people with dyslexia
- Accommodative and vergence problems are common in people with Meares–Irlen Syndrome, but they are not usually the explanation for the symptoms in this condition
- A magno visual deficit is unlikely to directly explain why people need specific colors in Meares–Irlen Syndrome

CAUSALITY

While it is relatively easy to establish that visual factors are correlated with dyslexia, there is much less evidence to suggest that these factors are causes of reading difficulties. Unfortunately, there has been a tendency for many researchers to conclude that a visual *correlate* of dyslexia is necessarily a *cause* of dyslexia. Vision is clearly essential for reading, but this does not mean that all visual anomalies will interfere with reading. All of the optometric correlates that are listed in this chapter also occur in some people who do not have reading difficulties. Further, many people with reading difficulties do not have any visual problems. So the relationship between visual problems and reading performance is unlikely to be a simple causal one.

Ocular motor factors and Meares–Irlen Syndrome might, in severe cases, have a direct effect on reading through visual perceptual distortions (Evans, 2001), and could have an indirect effect if eyestrain and headaches were making a person reluctant to read. There is some evidence that ocular motor anomalies and a magno deficit might cause a greater prevalence of certain types of reading errors (Evans, 2001).

The evidence for visual factors causing reading difficulties is rather less compelling than the argument that phonological factors are major causes of reading problems. The finding that the magno deficit is often present in children with poor phonological coding suggests that the discovery of low-level visual anomalies or of phonological weaknesses in a poor reader should not preclude the investigation of both areas of function. In at least some cases, dyslexia is likely to be multifactorial: both visual and phonological factors may contribute to a given person's reading difficulties.

SUMMARY

- Phonological factors are the main causes of dyslexia
- Visual (optometric) factors can be contributory factors
- Assessment of both these areas is important in dyslexia

DETECTION

The use of a distance letter chart test for school screening would be very unlikely to detect any of the visual correlates of dyslexia described in this chapter. Some schools use screening instruments that include binocular vision tests, but these tests are generally coarse and create conditions very different from normal reading. This is in direct contrast with optometric binocular vision tests which have, in recent years, evolved so as more closely to resemble the normal situation when reading (Evans, 2002).

Not all the children who require optometric intervention have symptoms. It therefore seems most appropriate for all children with suspected learning difficulties to receive optometric investigation.

The largest eye care profession in the United Kingdom is optometry. There are approximately 7,000 optometrists who are trained to diagnose diseases, diagnose and treat refractive and orthoptic (binocular vision) anomalies, and to dispense spectacles and contact lenses. There are about 1,500 ophthalmologists who are medical specialists trained in eye examination, and medical and surgical treatment of eye diseases. There are about 1,000 orthoptists, most of whom work with ophthalmologists. Orthoptists are principally concerned with detecting and treating binocular vision anomalies. About 3,500 registered dispensing opticians supply and fit glasses and sometimes, with additional qualifications, contact lenses.

The vast majority of optometrists work in primary care, usually in city center locations, and patients do not need a referral from a GP to see an optometrist. Few ophthalmologists work in primary care and most of the patients seeing ophthalmologists have been referred by GPs or optometrists. Most orthoptists work in hospital practice and receive the majority of their patients via an ophthalmologist.

Optometrists receive a fixed fee for a child's eye examination. The value of this fee has reduced in real terms since its introduction and only the basic eye examination is usually covered by the NHS fee. Because of the various visual anomalies that need to be looked for in a person with (suspected) dyslexia, an eye examination of such a person usually takes two to three times as long as a normal eye examination. Therefore, practitioners who have specialized in investigating children with reading difficulties almost always charge a private fee for these additional tests.

Unfortunately, there are no specific qualifications that eye care practitioners can take to indicate specialist expertise in the assessment of people with dyslexia. It is the hope of this author that the College of Optometrists will create a specialist diploma to indicate expertise in this area. A code of conduct for such practitioners is currently being prepared and it is hoped that a list of practitioners who have undertaken to adhere to this will be available, which will be held on or linked to the following website: http://www.essex.ac.uk/psychology/overlays/. In the meantime, the best method of finding eye care practitioners who have specialized in this area is by personal recommendation.

When, under the 1996 Education Act, a statement of special education needs is prepared, this frequently includes, under appended "medical advice," a description of vision as being "normal." In some cases, the only vision test to have been performed is distance visual acuity, which is virtually irrelevant to reading performance. This is certainly not good practice and the opinion of appropriately specialized eye care practitioners really should be sought long before the stage of drawing up a statement of special educational needs.

SUMMARY

- School vision screening tests are unlikely to detect the subtle visual correlates of dyslexia
- Children who need optometric care will not necessarily report symptoms
- The best way of detecting optometrists who have specialized in this area is through personal recommendation
- It is unlikely that the fixed NHS fee will cover the detailed testing that is required in dyslexia
- It is of great concern that some children are "statemented" without having received a detailed visual assessment

CONCLUSIONS

Ocular pathology is rare in childhood and is not a major cause of reading difficulties. Poor visual acuities and refractive errors are only rarely severe enough to interfere with reading, although longsightedness is often missed by school vision screening and can cause visual discomfort and blurred vision with reading. People with dyslexia tend, on average, to have a slightly reduced amplitude of accommodation for their age. In a few severe cases, this may cause eyestrain and blur when reading. Similar problems can result from binocular incoordination, which also has a higher than usual prevalence in dyslexia. If severe, both of these types of visual problems can be treated with eye exercises or spectacles.

Ocular dominance is not currently thought to be one of the key issues in reading, and the Dunlop Test is widely recognized as being too unreliable for routine use. The type of binocular incoordination that this test is thought to detect may be identified by more conventional optometric techniques. There is still considerable controversy about the role of abnormal eye movements in reading disability, although the consensus seems to be that eye movements are unlikely to hold the key to dyslexia. The claims that some people with reading difficulties have symptoms of eyestrain and visual perceptual distortions that are helped by colored filters have been supported by recent research. This research suggests that the tints do need to be individually prescribed, and an instrument for this, the Intuitive Colorimeter, is in widespread use within optometry in the United Kingdom. Sometimes, people who consult practitioners for this treatment are found to have conventional optometric (orthoptic) anomalies and these should be treated in the first instance.

There is considerable evidence for the existence of a certain type of visual sensory deficit in specific reading disability: a deficit of the magno visual system. There are no widespread reliable clinical tests or therapies for this and it is unclear whether detection or treatment would be of any benefit. All of these visual correlates of reading disability are not necessarily related to the higher level visual perceptual skills that may be assessed in a typical psychometric evaluation.

Although there has been some success in linking certain of the visual correlates of dyslexia, more work needs to be done in this area, and this should also take account of the effect of attentional deficits. Some research suggests that optometric anomalies may, in certain cases, contribute to a reading difficulty, but most evidence suggests that phonological factors are in many cases more important in the etiology of dyslexia. Nevertheless, it would seem prudent for children with suspected reading difficulties to be referred to an eye care practitioner who has specialized in this subject. This applies not only to children with symptoms, but also to those without. It is a cause of concern that many children are statemented without having had a specialist detailed eye examination.

ACKNOWLEDGMENTS

I am grateful to Anne Busby for providing typical samples of text that might be read by different age groups, so that equivalent visual acuities could be calculated.

REFERENCES

Biscaldi, M., Fischer, B., & Aiple, F. (1994). Saccadic eye movements of dyslexic and normal reading children. *Perception, 23*, 45–64.

Boder, E. (1971). Developmental dyslexia: Prevailing diagnostic concepts and a new diagnostic approach. In H. R. Myklebust (Ed.), *Progress in Learning Disabilities* (pp. 293–321). New York: Grune and Stratton.

Critchley, M. (1964). *Developmental dyslexia*. London: Whitefriars.

Eden, G. F., Stein, J. F., Wood, H. M., & Wood, F. B. (1994). Differences in eye movements and reading problems in dyslexic and normal children. *Vision Research, 34*, 1345–1358.

Evans, B. J. W. (1997). Coloured filters and dyslexia: What's in a name? *Dyslexia Review, 9*(2), 18–19.

Evans, B. J. W. (2001). *Dyslexia and vision*. London: Whurr Publishers.

Evans, B. J. W. (2002). *Pickwell's binocular vision anomalies*. Fourth Edition. Oxford: Butterworth-Heineman.

Evans, B. J. W., Busby, A., Jeanes, R., & Wilkins, A. J. (1995). Optometric correlates of Meares–Irlen Syndrome: A matched group study. *Ophthalmic and Physiological Optics, 15*, 481–487.

Evans, B. J. W., Drasdo, N., & Richards, I. L. (1996). Dyslexia: The link with visual deficits. *Ophthalmic and Physiological Optics, 16*, 3–10.

Evans, B. J. W., & Joseph, F. (2004). The effect of coloured filters on the rate of reading in an adult population. *Ophthalmic and Physiological Optics, 22*(6), 535–545.

Fowler, M. S., & Stein, J. F. (1983). Consideration of ocular motor dominance as an aetiological factor in some orthoptic problems. *British Orthoptic Journal, 40*, 43–45.

Irlen, H. (1991). *Reading by the colours*. New York: Avery.

Jennings, A. (2000). Behavioural optometry: A critical review. *Optometry in Practice, 1*, 67–78.

Kriss, I. (2001). *An investigation of the effects of coloured overlays on reading in dyslexics and controls*. BSc Dissertation, Manchester Metropolitan University, Manchester, UK.

Lennie, P., Trevarthen, C., Van Essen, D., & Wassle, H. (1990). Parallel processing of visual information. In L. Spillmann & J. S. Werner (Eds.), *Visual perception: The neurophysiological foundations* (pp. 163–203). San Diego: Academic Press.

Livingstone, M. S., Rosen, G. D., Drislane, F. W., & Galaburda, A. M. (1991). Physiological and anatomical evidence for a magnocellular defect in developmental dyslexia. *Proceedings of the National Academy of Science, USA, 88*, 7943–7947.

Lovegrove, W., Martin, F., & Slaghuis, W. (1986). A theoretical and experimental case for a visual deficit in specific reading disability. *Cognitive Neuropsychology, 3*, 225–267.

Meares, O. (1980). Figure / ground, brightness contrast, and reading disabilities. *Visible Language, 14*, 13–29.

Moseley, D. (1988). Dominance, reading and spelling. *Bulletin of the Audiophonology University Franche-Comté, 4*, 443–464.

Naidoo, S. (1972). *Specific dyslexia: The research report of the ICAA word blind centre for dyslexic children*. London: Pitman.

Ridder, W. H., Borsting, E., Cooper, M., McNeel, B., & Huang, E. (1997). Not all dyslexics are created equal. *Optometry and Vision Science, 74*, 99–104.

Silver, L. B. (1986). Review: Controversial approaches to treating learning disabilities and attention deficit disorder. *American Journal of Diseases in Childhood, 140*, 1045–1052.

Stark, L. W., Giveen, S. C., & Terdiman, J. F. (1991). Specific dyslexia and eye movements. In J. Cronly-Dillon (Ed.), *Vision and visual dysfunction* (Vol. 13, pp. 202–232). London: Macmillan.

Stein, J., & Fowler, S. (1985). The effect of monocular occlusion on visuomotor perception and reading in dyslexic children. *The Lancet*, 13th July, 69–73.

Wilkins, A. J. (1995). *Visual stress*. Oxford: Oxford University Press.

Wilkins, A. J., Evans, B. J. W., Brown, J., Busby, A., Wingfield, A. E., Jeanes, R. *et al.* (1994). Double-blind placebo-controlled trial of precision spectral filters in children who use coloured overlays. *Ophthalmic and Physiological Optics, 14*, 365–370.

2

Genetic and Environmental Causes of Reading Disabilities: Results from the Colorado Learning Disabilities Research Center

Richard K. Olson

INTRODUCTION

In this chapter, I will review some recent evidence from the University of Colorado and other laboratories for significant genetic contributions to dyslexia and to related deficits in language and attention. Our research at the Colorado Learning Disabilities Research Center (CLDRC) on the genetic and environmental causes of dyslexia has been a highly collaborative effort across several laboratories in the United States and the United Kingdom and most recently in Australia and Norway. It has included investigations of children's specific deficits in component word reading and spelling skills, and of related deficits in perception, language, attention, and mathematics. Genetic and environmental causes of dyslexia and related deficits have been studied by comparing identical and same-sex fraternal twins, and by the direct examination of DNA from fraternal twins and siblings (see DeFries, Alarcon, & Olson, 1997, for a review of earlier studies).

Richard K. Olson, Department of Psychology, University of Colorado, Boulder, CO 80309, USA.

The Study of Dyslexia, edited by Turner and Rack.
Kluwer Academic Publishers, New York, 2004.

We have also conducted research on enrichment of the learning environment for children with dyslexia through different types of computer-based instruction. These studies have demonstrated that genetically influenced weaknesses in dyslexics' component reading and language skills can be significantly improved (Wise, Ring, & Olson, 1999, 2000). While the main focus of this chapter is on evidence for genetic influences, I will briefly review results from our remediation studies to emphasize the importance of extraordinary environmental support for children with dyslexia, regardless of the strong genetic influences on reading and related language deficits in many children and adults.

The chapter will conclude with an overview of new studies in progress at the CLDRC on deficits in reading and language comprehension in school-age twins, and on longitudinal studies of individual differences in reading-related skills among preschool twins, prior to any formal reading instruction.

WHY STUDY TWINS?

Identical and fraternal twins are uniquely informative about the relative contributions from genes and environment to individual differences in reading across the normal range, and to the severe reading deficits found in dyslexia. Identical twin pairs are derived from the same sperm and egg (the egg is called a "zygote," so identical twins are "Monozygotic" or MZ). Therefore, MZ twins share all their genes. Fraternal twins develop from two different sperm–egg fertilizations (i.e., they are "Dizygotic" or DZ). Therefore, they share, on average, half of their segregating genes (the small minority of genes that vary across individuals). Thus, DZ twins have the same degree of average genetic similarity as ordinary brothers and sisters, but like the MZ twins, they share their intrauterine environment, are born at the same time, and then share their family environment.

Of course growing up in the same family, schools, and community does not mean that the environments are the same for twins in a pair. Children help to create their own environments within and outside the family. DZ twins' different genes might lead them to greater differences in the way they interact with and select their environments, compared to MZ twins with identical genes. For example, one twin of a DZ pair might have normal genes related to reading development, while the other has genes that are related to dyslexia. The twin with genes related to dyslexia might have greater problems in learning to read, experience less pleasure in reading, and ultimately choose to read less than the twin who has normal or superior genes related to reading development. Thus, the hypothetical dyslexic twin's smaller amount of reading practice may contribute to their reading failure. However, less reading practice is not the only cause in most cases of dyslexia. Children with dyslexia typically require far *more* reading practice than normally developing children to reach a normal reading level.

Comparisons of large groups of MZ and DZ twins can yield quantitative estimates of the relative influence on individual differences in the population from genetic, shared environment, and nonshared environment factors. The detailed mathematical procedures are beyond the scope of this chapter, but a few points may help the reader understand the basic logic of the analyses. MZ twins share their genes and their family/school environment, so any differences within MZ twin pairs is logically due to nonshared environment influences. Such influences could include differential stress during gestation, accidents, or disease after birth, and unusual differences in educational opportunity. DZ twins within a pair will also have nonshared environmental influences, but in addition, they may have genetic differences relevant to the studied behavior. Thus, any greater behavioral differences between DZ twins compared to MZ twins may be due to genetic factors. In this chapter, we are focusing on the genetic contributions to extreme reading deficits or dyslexia. Thus, if MZ and DZ twins shared dyslexia equally often, we would conclude that there is no evidence for genetic influence in the population. In contrast, if nearly all MZ twins shared the disorder (thus with little influence from nonshared environmental factors), and DZ twins shared dyslexia in nearly half the pairs (remember that DZ twins share half their segregating genes on average), we would conclude that genes accounted for most cases of dyslexia in the population.

It is important to emphasize that estimates of genetic influence from twin studies are for the average influence in the population, and not for any individual in the study. For example, we might say that 60% of the reading deficit in a group of twins is due to genetic influence, but this does not mean that the genetic influence is 60% for each individual. For some individuals in the group, there might have been little or no genetic influence on their reading deficit, while for others, the genetic influence might have been very strong. When we are ultimately able to identify specific genes related to dyslexia, we will have more information about genetic contributions in the individual case.

A second qualification of behavioral genetic estimates of heritability or genetic influence is that the proportion of genetic influence in the population depends on the range of environmental variation in the population. If there are vast differences in schooling and cultural support for literacy in a population, environmental factors will account for most of the variation in literacy and dyslexia. In contrast, if schooling and cultural support for literacy are relatively constant across the population, genes are likely to play a proportionally greater role for dyslexia and for individual differences across the normal range.

BEHAVIOR GENETIC RESULTS FROM THE CLDRC STUDIES WITH SCHOOL-AGE TWINS

We identify twins in the third to twelfth grades with reading problems based on their school records. Then we invite them to the laboratory for further testing in specific reading skills, IQ, attention, memory, and language. We also test twins with no school history for reading problems to provide a normal-range group comparison with the dyslexic group. The total sample now includes over 2,500 twins, with slightly more males in the dyslexic group.

DeFries and Fulker (1985) developed a powerful method, now referred to as DF analysis, for assessing genetic influence on a group deficit when twin pairs are selected for one or both members' deviant position on a normally distributed dimension such as reading. DF analysis compares the average regression toward the normal population mean for MZ and DZ cotwins who do not meet the affected severity criterion. From this information, it is possible to derive estimates of the average genetic, shared environment, and nonshared environment influences on deviant group membership.

Heritability of the Group Deficit in a Standardized Reading Measure

DF analyses have been used to assess average genetic influence on the group deficit (approximately below the local tenth percentile) for a composite measure of word recognition, reading comprehension, and spelling from the Peabody Individual Achievement Test (Dunn & Markwardt, 1970). The most recent analysis by DeFries and Alarcon (1996) estimated that the heritability of the group deficit in this composite measure was 0.56. This means that approximately half of the average group deficit in our sample is due to genetic factors.

Heritability of Group Deficits in Phonological Decoding, Orthographic Coding, and Phoneme Awareness

DF analyses have also been conducted for group deficits in phoneme awareness, phonological decoding, and orthographic coding (Gayan & Olson, 2001; Olson, Forsberg, & Wise, 1994; Olson, Wise, Conners, Rack, & Fulker, 1989). Phoneme awareness is measured by language tasks that require the isolation and manipulation of phonemes within spoken words or nonwords. Performance in these tasks is highly correlated with reading skill, particularly with the component reading skill of phonological decoding. We measure phonological decoding through both the oral and silent reading of nonwords. Our orthographic coding measures assess subjects' sensitivity to the precise spelling patterns for words in the comparison of a word with a homophonic nonword (e.g., rain rane) and in the choice between homophones (bear bare) to fit the meaning of a spoken sentence. All of these tasks are significantly correlated with each other and with measures of printed word recognition.

Olson *et al.* (1989) suggested that the group deficit in orthographic coding might have a lower heritability than for phonological decoding. We argued that accurate recognition of the precise spelling for a word would depend more on the environmental effects of exposure to that word in print, and we inferred that print exposure would be more likely mediated by environmental factors than by genes. However, a trend toward lower heritability in the small sample of twins analyzed by Olson *et al.* (1989) was not replicated in larger samples (Gayan & Olson,

2001). In these later analyses, group accuracy deficits in the orthographic tasks have heritability levels that range between $h^2g = 0.6–0.7$, similar to those for different measures of phonological decoding.

Bivariate DF Analyses for Shared Genetic Influence across Skills

Bivariate DF analyses were employed by Olson *et al.* (1994) and Gayan and Olson (2001) to assess the degree of common genetic influence on the shared variance among deficits in the above measures: The question was, to what degree are the group deficits in the measures due to the same or different genetic influences? The results indicated that there were significant common genetic effects on group deficits in all the measures, but there was also evidence of some independent genetic effects on each measure. Shared and independent genetic effects on individual differences in each of the measures have recently been confirmed in both normal and reading disabled groups using appropriate factor models (Gayan & Olson, 2003).

The partial genetic independence of phonological and orthographic skills is of particular interest to researchers who study individual differences or subtypes within the reading disabled population, and who wonder about the role of genes and environment in these differences (cf. Castles & Coltheart, 1993; Castles, Datta, Gayan, & Olson, 1999; Olson, Kliegl, Davidson, & Foltz, 1985).

Comorbidity for Dyslexia and Attention Deficit Hyperactivity Disorder (ADHD)

The above bivariate genetic analyses have also been performed for reading deficits and ADHD. ADHD has commonly been diagnosed in the United States by asking teachers and/or parents to rate their children on a list of symptoms for attention deficits and hyperactivity. If a sufficiently high number of symptoms are checked, the children are categorized as ADHD. Researchers in the United States have noted a high rate of ADHD among children with dyslexia, ranging from about 20% to 40% across studies. We now have evidence from the Colorado twin study that part of this comorbidity is due to shared genetic influences, particularly for dyslexia and the attention-deficit symptoms (Willcutt *et al.*, 2002).

Does the Degree of Genetic Influence Vary Systematically across Individuals?

The DF analyses for group deficits described above do not provide evidence on the genetic influence for any individual disabled reader within the group. We could begin to make more accurate predictions about the degree of genetic influence on individual deficits if we found evidence that the level of heritability varies, depending on individual variables such as age, gender, or IQ. Fortunately, DF analyses can be extended to assess the size and statistical significance of such effects on the average level of heritability. Gender does not appear to be related to the degree of genetic influence on reading deficits (DeFries, Gillis, & Wadsworth, 1993). Nonsignificant trends in relation to age reported by Wadsworth, Gillis, DeFries, and Fulker (1989) have been further explored by DeFries *et al.* (1997). The latter study found trends toward decreasing heritability with age for word recognition and increasing heritability with age for spelling. The opposing direction of these trends resulted in a statistically significant interaction.

In addition, there is now significant evidence for the importance of IQ in the genetic etiology of deficits in word recognition: Disabled readers with relatively high IQ scores tend to have a stronger genetic etiology than those with relatively low IQ (Olson, Datta, Gayan, & DeFries, 1999; Wadsworth, Olson, Pennington, & DeFries, 2000). The concurrence of low IQ and low reading may be more likely due to some shared family environment that constrains both reading and general cognitive development. With high IQ, the reading environment may tend to be better, and reading failure more likely due to genetic constraints. We are currently conducting additional analyses to better understand this apparent relation between IQ and genetic influence on reading deficits.

MOLECULAR GENETIC APPROACHES TO LOCATING THE GENES FOR DYSLEXIA

Behavioral comparisons of MZ and DZ twins have yielded valuable information about the balance of genetic and environmental influence on group deficits in reading and related language skills. Further analyses of genetic influence on reading disabilities in relation to IQ and other individual characteristics will provide more specific information about this balance across individuals. However, the further specification and understanding of genetic mechanisms at the individual level will ultimately depend on the identification of specific genes that are associated with reading disability.

A first step in locating genes that influence dyslexia is to find regions on chromosomes that are likely to carry those genes. This is done through a method called linkage analysis. DNA markers (not necessarily genes) that vary across individuals are identified in a region of interest. Siblings that share or do not share dyslexia are then compared on whether they share or do not share the specific marker. If sharing the marker means it is significantly more likely that they also share dyslexia, there is linkage evidence for dyslexia in that region. However, the identified region on a chromosome may be very broad and contain many genes. Linkage only shows that having inherited the same large region of DNA on a chromosome from their parents, the siblings are more likely to share dyslexia.

Recent linkage analyses suggest that a gene or genes on the short arm of chromosome 6, close to the HLA region, may account for a significant proportion of reading disabilities. Preliminary evidence for this linkage was first obtained by Smith, Pennington, and Kimberling (1991) using data from 19 extended families with a history of reading problems. Cardon *et al.* (1994) applied more powerful linkage analyses to these family data and added a sample of 46 DZ twin families from the Colorado Reading Project. Taken together, the extended family and twin data provided highly significant evidence for a genetic linkage to reading disability near the HLA region of chromosome 6.

Reading disability in the above linkage studies was ascertained through a composite measure of word recognition, reading comprehension, and spelling from the Peabody Individual Achievement Test (Dunn & Markwardt, 1970). Gayan *et al.* (1999) used the same analytic methods with an expanded DZ twin and sibling sample to look for linkage to deficits in the specific skills of phoneme awareness, phonological decoding, orthographic coding, and fluent word recognition. These results suggest that deficits in phoneme awareness, phonological decoding, and orthographic coding show strong evidence for linkage to genetic markers in the HLA region of chromosome 6. The evidence was less strong for two measures of word recognition.

A similar pattern of linkage results for phoneme awareness and word recognition has been reported by Grigorenko *et al.* (1997). They used data from extended families with a history of reading disabilities and found strong linkage for deficits in phoneme awareness to the same HLA region of chromosome 6. Linkage appeared to be weaker in this region for deficits in word recognition, which were more strongly linked to a region on chromosome 15. The possibility of differential genetic linkages for different component skills in reading and language is intriguing, but a much larger subject sample is needed to test the statistical significance of these differences. This additional data is now being collected by our Center, with additional genetic markers including other regions of the genome.

Most recently, DNA from the Colorado sample and an independent sample collected in England were subjected to a whole-genome scan to search for linkage in other regions. Fisher *et al.* (2002) found evidence in both samples for linkage to the HLA region on chromosome 6, and even stronger evidence for linkage in both samples to a region on chromosome 18. A strong genetic linkage for deficits in phoneme awareness and orthographic coding has now been confirmed in two independent laboratories for the same regions of chromosomes 6 and 18.

A different approach to finding genes related to dyslexia is to see if genes already identified, which are known to be expressed in the brain, have variations (called alleles) that are associated with dyslexia. One example of this approach was used by Smith *et al.* (2001). They looked at different alleles of a gene called MOG in the same region of chromosome 6 that had been supported in previous linkage studies for dyslexia. Unfortunately, the variations in this gene were not significantly associated with dyslexia. Franks *et al.* (2002) looked for association of dyslexia with two genes in a region of chromosome 2 that were in a region that had been previously identified in linkage studies. They also failed to find allelic variations in these genes that were associated with dyslexia. Another related approach is to find variations in DNA that are close enough to a responsible gene that they are nearly always inherited with the gene. This procedure was used by Kaplan *et al.*

(2002). They found preliminary evidence for association with a marker on chromosome 6 in the same region identified by linkage studies. If this result can be confirmed in other independent samples, it could serve as an identifiable marker that is associated with some proportion of the genetic risk for dyslexia.

Once genes that have alleles associated with dyslexia are identified, the real hard work begins. The next steps are to clone the gene(s), identify the coded protein(s), and determine the influence of the protein(s) on the developing nervous system and related behavior, in interaction with the environment. Much more research will be needed to reach the last two goals, but recent advances in methods for locating genes and related markers may soon allow us to identify individuals who have a gene or genes that place them at risk for reading disability. Early information about a genetic risk could be used to provide additional support in a child's early language and reading environment. Providing support prior to school entry could help avoid the stigmatizing effect of school failure in reading.

AVOIDING GENETIC DETERMINISM

A strong genetic influence on some aspect of human behavior or physiology does *not* mean that it cannot be changed by the environment. For example, Diabetes is a disease with a strong genetic influence, but its course can be substantially changed through environmental manipulations of insulin and diet. Similarly, there is much evidence that dyslexics' reading can be substantially improved through intensive individualized instruction that focuses both on phonological skills and accurate reading in context (e.g., Torgesen *et al.*, 2001). However, the cost of this individualized instruction with a human tutor is very high and unavailable to many children with dyslexia. Could computer-based remedial instruction be of help?

Since 1986, we have been exploring the use of talking computers in the schools to support second to fifth grade children with reading disabilities in their word decoding while reading stories. The programs allow children with reading disabilities to read interesting stories on the computer that are more appropriate for their age level, and independently obtain spoken decoding support by targeting difficult words with a mouse. More recently, we have incorporated additional programs designed to improve disabled readers' phoneme awareness and phonological decoding (Wise & Olson, 1995; Wise *et al.*, 1999, 2000). The main point to be made here is that these and other programs can substantially improve disabled readers' phoneme awareness and phonological decoding, skills that are critical for reading development and that have a very strong genetic influence on their group deficits. It is clear that the improvement of these and other reading-related skills in children with reading disabilities often requires extraordinary environmental support. Computer programs incorporating both synthetic and digitized speech can efficiently provide much of this support.

NEW DIRECTIONS IN THE CLDRC TWIN STUDIES

For the 2000–5 funding period, two new components have been added to the Center. Beginning in 2001, a new component directed by Janice Keenan at the University of Denver is focusing on deficits in reading and language comprehension. Keenan is studying the same school-age twins who were tested in Boulder on word-level reading and language skills. Preliminary results from a small twin sample suggest that the heritability for individual differences in listening comprehension may be substantial. The correlation for 30 pairs of identical twins was 0.69, while that for 21 pairs of fraternal twins was 0.01, and these correlations were significantly different by Fisher's z (Keenan, Betjemann, & Fazendeiro, 2002). (We expect that the correlation for fraternal twins will be higher as the sample grows, since they share half their segregating genes on average, as well as their shared family environment.) With a larger sample of twins, we will be able to see how genetic influences on individual differences in reading and listening comprehension are shared with word-level reading skills, phonological awareness, and higher level language skills. Our working hypothesis is that while there will be significant shared genetic influence across levels of reading and language skills, there will also be significant, independent genetic influences for group deficits and individual differences in higher and lower level reading and language skills.

A second, new component of the NIH Center is focused on the etiology of individual differences among children before they begin formal schooling. Because individual differences in prereading skills and environment may have important influences on later reading development in school, we initiated a large longitudinal study of identical and fraternal twins beginning at age 4, with follow-up assessments at the end of kindergarten, first grade, and second grade. A major focus of the study is the assessment of preschool twins' learning rate from training for the awareness of abstract phonemes in speech. The study includes international collaborations with parallel studies of identical and fraternal preschool twins directed by Brian Byrne in Australia, Stefan Samuelsson in Norway, and Richard Olson in the United States. Preliminary results from a combined sample of US and Australian preschool twins revealed significant genetic influence on both static and dynamic measures of phoneme awareness, and on measures of learning and memory. In contrast, individual differences in the preschoolers' print knowledge, including letter identification and concepts about print, were largely due to differences in shared family environment (Byrne *et al.*, 2002). Several measures of vocabulary and grammatical knowledge were also predominantly influenced by shared family environment. It remains to be seen how these individual differences in preschoolers relate to genetic and environmental influences on their subsequent reading development.

The varied and convergent research perspectives on reading disability in the CLDRC and the other NICHD sponsored centers have resulted in some major advances over the past decade. As this work continues, with collaboration of investigators in the United Kingdom and other countries, we look forward to learning much more about the etiology and optimal treatment of different reading disabilities.

ACKNOWLEDGMENTS

This research is being conducted with the support of National Institutes of Health grants 2 P50 HD27802 and 1 R01 HD38526. We are grateful to the many twins and their families who have participated in the research.

REFERENCES

Byrne, B., Delaland, C., Fielding-Barnsley, R., Quain, P., Samuelsson, S., Hoien, T. *et al.* (2002). Longitudinal twin study of early reading development in three countries: Preliminary results. *Annals of Dyslexia, 52*, 49–74.

Cardon, L. R., Smith, S., Fulker, D., Kimberling, W., Pennington, B., & DeFries, J. (1994). Quantitative trait locus for reading disability on chromosome 6. *Science, 266*, 276–279.

Castles, A., & Coltheart, M. (1993). Varieties of developmental dyslexia. *Cognition, 47*, 149–180.

Castles, A., Datta, H., Gayan, J., & Olson, R. K. (1999). Varieties of developmental reading disorder: Genetic and environmental influences. *Journal of Experimental Child Psychology, 72*, 73–94.

DeFries, J. C., & Alarcon, M. (1996). Genetics of specific reading disability. *Mental Retardation and Developmental Disabilities Research Reviews, 2*, 39–47.

DeFries, J. C., Alarcon, M., & Olson, R. K. (1997). Genetics and dyslexia: Developmental differences in the etiologies of reading and spelling deficits. In C. Hulme & M. Snowling (Eds.), *Dyslexia: Biology, cognition, and intervention* (pp. 20–37). London: Whurr.

Defries, J. C., Filipek, P. A., Fulker, D. W., Olson, R. K., Pennington, B. F., Smith, S. D. *et al.* (1997). Colorado Learning Disabilities Research Center. *Learning Disability Quarterly, 7–8*, 19.

DeFries, J. C., & Fulker, D. W. (1985). Multiple regression analysis of twin data. *Behavior Genetics, 15*, 467–473.

DeFries, J. C., Gillis, J. J., & Wadsworth, S. J. (1993). Genes and genders: A twin study of reading disability. In A. M. Galaburda (Ed.), *Dyslexia and development: Neurobiological aspects of extra-ordinary brains* (pp. 187–204). Cambridge, MA: Harvard University Press.

Dunn, L. M., & Markwardt, F. C. (1970). *Examiner's manual: Peabody Individual Achievement Test*, Circle Pines, MN: American Guidance Service.

Fisher, S. E., Francks, C., Marlow, A. J., MacPhie, L., Williams, D. F., Cardon *et al.* (2002). Genome-wide scans in independent samples reveal strong convergent evidence for a chromosome 18 quantitative-trait locus influencing developmental dyslexia. *Nature Genetics, 30*, 86–91.

Franks, C., Fisher, S. E., Olson, R. K., Pennington, B. F., Smith, S. D., DeFries, J. C. *et al.* (2002). Fine mapping of the chromosome 2p12-16 dyslexia susceptibility locus: Quantitative association analysis and positional candidate genes SEMA4F and OTX1. *Psychiatric Genetics, 12*, 35–41.

Gayan, J., & Olson, R. K. (2001). Genetic and environmental influences on orthographic and phonological skills in children with reading disabilities. *Developmental Neuropsychology, 20*(2), 487–511.

Gayan, J., & Olson, R. K. (2003). Genetic and environmental influences on individual differences in printed word recognition. *Journal of Experimental Child Psychology, 84*, 97–123.

Gayan, J., Smith, S. D., Cherny, S. S., Cardon, L. R., Fulker, D. W., Kimberling, W. J. *et al.* (1999). Large quantitative trait locus for specific language and reading deficits in chromosome 6p. *American Journal of Human Genetics, 64*, 157–164.

Grigorenko, E. L., Wood, F. B., Meyer, M. S., Hart, L. A., Speed, W. C., Shuster, B. S. *et al.* (1997). Susceptibility loci for distinct components of developmental dyslexia on chromosomes 6 and 15. *American Journal of Human Genetics, 60*, 27–39.

Kaplan, D. E., Gayan, J., Ahn, J., Won, T. W., Pauls, D., Olson, R. *et al.* (2002). Evidence for linkage and association with reading disability on 6P21.3-22. *American Journal of Human Genetics, 70*, 1287–1298.

Keenan, J. M., Betjemann, R. S., & Fazendeiro, T. (2002, June 29). *Reading disability and inference deficits in listening comprehension.* Poster presented a the meeting of the Society for the Scientific Study of Reading, Chicago.

Olson, R. K., Datta, H., Gayan, J., & DeFries, J. C. (1999). A behavioral-genetic analysis of reading disabilities and component processes. In R. M. Klein, & P. A. McMullen (Eds.), *Converging methods for understanding reading and dyslexia* (pp. 133–153). Cambridge MA: MIT Press.

Olson, R. K., Forsberg, H., & Wise, B. (1994). Genes, environment, and the development of orthographic skills. In V. W. Berninger (Ed.), *The varieties of orthographic knowledge I: Theoretical and developmental issues* (pp. 27–71). Dordrecht, The Netherlands: Kluwer Academic.

Olson, R. K., Kliegl, R., Davidson, B. J., & Foltz, G. (1985). Individual and developmental differences in reading disability. In G. E. MacKinnon, & T. G. Waller (Eds.), *Reading research: Advances in theory and practice* (Vol. 4, pp. 1–64). New York: Academic Press.

Olson, R. K., Wise, B., Conners, F., Rack, J., & Fulker, D. (1989). Specific deficits in component reading and language skills: Genetic and environmental influences. *Journal of Learning Disabilities, 22*(6), 339–348.

Smith, S. D., Pennington, B. F., & Kimberling, W. J. (1991). Screening for multiple genes influencing dyslexia. *Reading and Writing: An Interdisciplinary Journal, 3*, 285–298.

Smith, S. D., Kelley, P. M., Askew, J. W., Hoover, D. M., Deffenbacher, K. E., Gayan *et al.* (2001). Reading disability and chromosome 6p21.3: Evaluation of MOG as a candidate gene. *Journal of Learning Disabilities, 34*, 512–519.

Torgesen, J. K., Alexander, A. W., Wagner, R. K., Voeller, K., Conway, T., & Rose, E. (2001). Intensive remedial instruction for children with severe reading disabilities: Immediate and long-term outcomes from two instructional approaches. *Journal of Learning Disabilities, 34*(1), 33–58.

Wadsworth, S. J., Olson, R. K., Pennington, B. F., & DeFries, J. C. (2000). Differential genetic etiology of reading disability as a function of IQ. *Journal of Learning Disabilities, 33*, 192–199.

Wadsworth, S. J., Gillis, J. J., DeFries, J. C., & Fulker, D. W. (1989). Differential genetic aetiology of reading disability as a function of age. *Irish Journal of Psychology, 10*, 509–520.

Wilcutt, E. G., Pennington, B. F., Smith, S. D., Cardon, L. R., Gayan, J., Knopic, V. S. *et al.* (2002). Quantitative trait locus for reading disability on chromosome 6p is pleiotropic for attention-deficit/hyperactivity disorder. *American Journal of Medical Genetics (Neuropsychiatric Genetics), 114*, 260–268.

Wise, B. W., & Olson, R. K. (1995). Computer-based phonological awareness and reading instruction. *Annals of Dyslexia, 45*, 99–122.

Wise, B. W., Ring, J., & Olson, R. K. (1999). Training phonological awareness with and without attention to articulation. *Journal of Experimental Child Psychology, 72*, 271–304.

Wise, B. W., Ring, J., & Olson, R. K. (2000). Individual differences in gains from computer-assisted remedial reading with more emphasis on phonological analysis or accurate reading in context. *Journal of Experimental Child Psychology, 77*, 197–235.

3

The Neurobiology of Dyslexia

Alan A. Beaton

It is now almost universally accepted (but see Ehri, 1992) that dyslexia is a constitutional condition, almost certainly genetic, rooted in the biology of the central nervous system. The question is, how do the relevant genes have the effect that they do? To show that a particular gene locus is implicated in, for example, phonological awareness does not explain how the gene (or more probably genes) code for biological mechanisms which underlie this ability. One step towards solving this problem might be to relate relevant gene loci to particular brain structures and mechanisms. Presumably, genetic factors determine in part the relative size or efficiency of regions of the brain involved in language and other processes relevant to reading and its component skills.

Figure 1 shows a diagram of the eyes and brain. Defects at every stage from the retina (Grosser & Spafford, 1989, 1992; Stordy, 2000) through the midbrain to the cerebral cortex and the cerebellum have been said to be involved in dyslexia. Figure 2 shows the main features of the human brain which are referred to in this chapter.

The brain consists of two apparently symmetrical halves or hemispheres. It has an uneven surface made up of ridges, known as convolutions, or gyri, and troughs, known as sulci. There is some individual variability in the pattern of convolutions, but there is also sufficient similarity between different brains for individual gyri and sulci to be identified and named. To the naked eye, there is no

---

Alan A. Beaton, Senior Lecturer, Department of Psychology, University of Wales, Swansea SA2 8PP, UK.

The Study of Dyslexia, edited by Turner and Rack.
Kluwer Academic Publishers, New York, 2004.

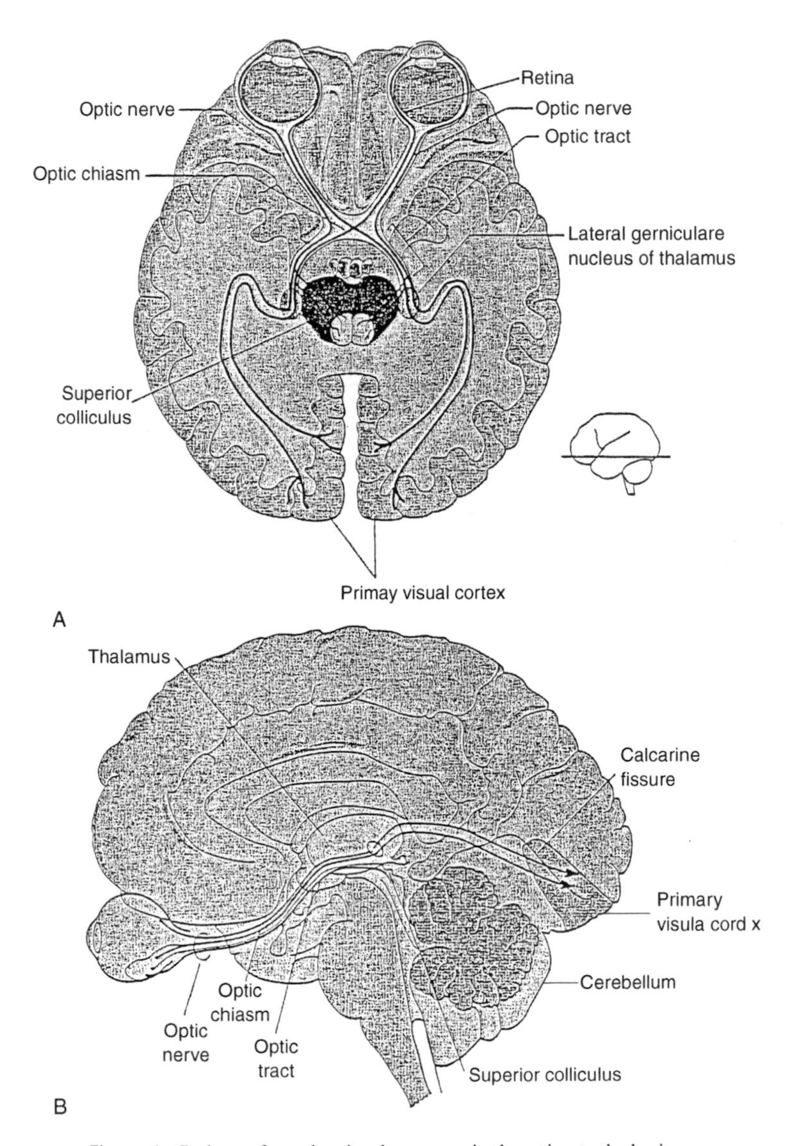

Figure 1. Pathway from the visual receptors in the retina to the brain.

conspicuous difference between the brains of dyslexic and other people although it has been claimed that one kind of gyral pattern is relatively more frequent in dyslexics than controls (Hynd & Hiemenz, 1997). This pattern is also found in non-dyslexics, but the functional significance of small variations in gyral patterning is unknown.

The first published report of a postmortem examination of the brain of a purportedly dyslexic individual concerned a boy named Billy who died suddenly of a brain haemorrhage at approximately 12 years of age. The author of the report

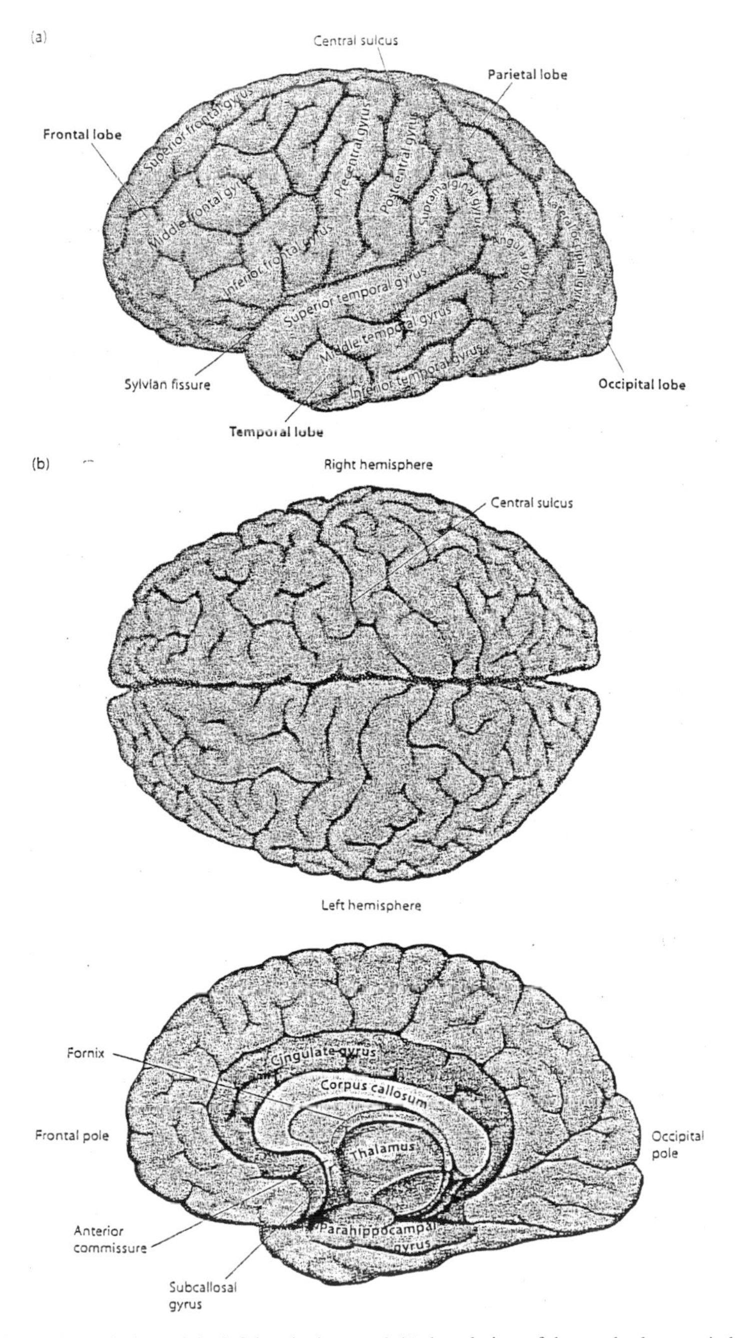

Figure 2. (a) Lateral view of the left hemisphere and (b) dorsal view of the cerebral cortex in humans. The major features of the cortex include the four cortical lobes and various key gyri. Gyri (singular is *gyrus*) are separated by sulci (singular is *sulcus*) and result from the folding of the cerebral cortex that occurs during development of the nervous system, to achieve an economy of size.

(Drake, 1968) wrote: "in the cerebral hemispheres, anomalies were noted in the convolution pattern of the parietal lobes bilaterally. The cortical pattern was disrupted by penetrating deep gyri that appeared disconnected. Related areas of the corpus callosum appeared thin" (p. 496). Billy was said to have marked difficulty with reading and writing, and some difficulty with arithmetic, although by age 12 years 2 months his performance on standard tests suggests that he performed at a more or less satisfactory level in reading and spelling. Other aspects of the report suggest that he would today be classified as showing attention-deficit hyperactivity disorder (ADHD). Billy's medical history included "dizzy spells" and "blackouts" occurring from age 6 as well as recurring left frontal headaches during the 2 years prior to his death. In short, the extent to which Billy could be regarded as a "representative" dyslexic person is unclear.

It has been known for over 100 years that the majority of people speak with only one side of their brain. A stroke affecting the left hand side of the brain (left cerebral hemisphere) may produce devastating effects on language, whereas damage affecting the right hand side (right cerebral hemisphere) may leave language functions relatively intact. The qualification "relatively" is important. It seems that some aspects of language, such as prosody or the appreciation of metaphor or humour, for example, can be impaired by right-sided brain damage.

Damage due to injury, stroke, or other disease (referred to simply as lesions) to two so-called "classical" brain areas produce different effects. Lesions involving the inferior frontal gyrus, or Broca's area, tend to make speech effortful and dysfluent; the patient has difficulty in finding the right word and omits "little" words (prepositions, conjunctions) so that speech takes on the character of a telegram in which only the main words are retained. This is known as Broca's aphasia after the 19th century French neuroanatomist and physician who first linked speech difficulties to damage in the inferior, or third, frontal gyrus of the left hemisphere.

In contrast to Broca's aphasia, damage to the posterior region of the superior temporal gyrus on the left side, Wernicke's area, tends to produce fluent language which may be devoid of meaning, or even gibberish. Patients with this form of aphasia, named after a German neurologist of the late 19th century, have difficulty in understanding what is said to them. Neither of these descriptions should be regarded as providing hard and fast rules—Broca's aphasics may have difficulty in understanding certain kinds of grammatical constructions and Wernicke's aphasics may have word-finding difficulties. Nonetheless, the two classifications are useful approximations and are widely used among neurologists and speech professionals. Nor should it be thought that Broca's and Wernicke's areas are the only two regions of the brain involved in language. Deficits in spoken (or written) language can be produced by lesions in other cortical and sub-cortical areas. Damage to the thalamus, for example, has been linked to particular aphasic symptoms (Crosson, 1985) while damage to the cerebellum is associated with impairments of articulation.

If the fundamental problem of dyslexia is, as has been claimed, in the phonological domain (rather than visual as on the face of it might naively be imagined), then it makes sense to look for differences between dyslexics and controls in language-related areas of the brain.

NEUROIMAGING AND NEUROMAGNETIC STUDIES OF DYSLEXIA

Structural magnetic resonance imagery (MRI) enables the investigator to look at the physical characteristics of the living brain. For many purposes, however, one wants to know what the brain is doing at any moment in time. When the brain is active, blood flow increases to those parts of the brain where the activity occurs. Using a positron emission tomography (PET) scanner, regional differences in blood flow between any two experimental conditions can be displayed in colour on a computerised image of the brain.

An alternative technique for measuring brain activity is functional MRI (fMRI). The assumption is that changes in blood flow are closely correlated with changes in fMRI activity although the exact nature of the relationship between fMRI responses and the underlying neural activity is unclear at present (but see Logothetis, Pauls, Augath, Trinath, & Oeltermann, 2001).

During the past decade, PET and fMRI techniques have been applied in a number of laboratories to the study of reading disability. These normally take the form of comparing patterns of regional blood flow between dyslexic and non-dyslexic groups. There are, however, a number of important methodological issues which make imaging data difficult to evaluate (Bub, 2000). In particular, changes in blood flow are shown as an increase or decrease in relation to some control condition. If the same region of the brain is activated to the same extent in both a control condition and an experimental condition, this will not show up as a regional difference when the two conditions are compared. A region of activation or de-activation represents a difference between conditions, not some absolute level of activity, so the choice of experimental conditions to compare is crucial. Note, incidentally, that (de)activation of a particular brain area may represent either excitation or inhibition—it is not possible to tell which.

Furthermore, differences between dyslexics and controls in patterns of regional cerebral blood flow may relate to any of a number of possible differences between the two groups. These may involve the strategies employed in carrying out the experimental tasks, the level of difficulty experienced, the degree of anxiety felt in response to the experimental procedure, or to differences in IQ or other subject variables (Flowers, Wood, & Naylor, 1991). Dyslexia might cause the differences, or they may cause the dyslexia, or both might be due to some third common factor. (This applies to differences in brain structure as well as function—see Castro-Caldas *et al.*, 1999).

Despite the interpretative difficulties, recent research with adult dyslexics clearly points to some abnormality of function in certain regions of the left hemisphere. Rumsey *et al.* (1992) reported that in comparison with 14 control subjects and resting levels, 14 dyslexic men showed reduced activation at the left temporo-parietal region in response to a simple task in which they had to press a button if two words rhymed. However, the dyslexics showed increased activation at the right temporal region. In another PET study, increased activation was seen in 10 dyslexic adults when they had to identify an auditorily-presented target presented within a stream of distractors. This time, the increase was not confined to the right hemisphere, but occurred at both sides of the brain close to the midline (Hagman *et al.*, 1992).

In a study by Rumsey *et al.* (1997), the tasks included reading aloud pseudowords (said to involve phonological processing) and irregular/inconsistent words (e.g., save vs have, reflecting orthographic processing). Dyslexics showed more widespread activation and de-activation than controls, which was attributed to greater subjective difficulty for the dyslexics. The latter also showed reduced activation in posterior temporal/inferior parietal regions relative to controls, especially in the "phonological" condition and especially on the left side.

It was subsequently reported that regional blood flow volume in the left angular gyrus was significantly correlated with reading skill in control subjects, but in dyslexics there was an inverse correlation (Rumsey *et al.*, 1999). This was seen as confirming the important role of the angular gyrus in normal readers and its probable impairment in developmental (as distinct from acquired) dyslexia. The left angular gyrus has for long been thought on the basis of neuropsychological studies to be involved in reading (Dejerine, 1891; Henderson, 1986).

An analysis of fMRI data obtained from 29 adult dyslexic readers and 32 normal control readers was conducted by Pugh *et al.* (2000). These authors argued that "a basic weakness in phonological representations … limits dyslexic readers' ability to build efficient structures within the angular gyrus that link orthographic codes computed in the extra-striate areas of the occipital lobe to phonological codes represented in the superior temporal gyrus" (p. 54–55).

In a much quoted study, Paulesu *et al.* (1996) observed that, in comparison with controls, a small group of adult (compensated) dyslexic men with residual phonological processing difficulties showed reduced PET activation of the left insular region. It was suggested that the insula acts as a bridge between anterior and posterior language processing areas, and that there is "reduced connectivity" in dyslexia. Contrariwise, Rumsey *et al.* (1997b) found increased activation in the insula region bilaterally in dyslexics compared with controls. The difference in the results of the two studies might relate to better levels of performance on word and non-word reading tasks by the dyslexics in the Paulesu *et al.* study than in the Rumsey study, or to some unidentified difference in the composition of the two groups of dyslexics. Alternatively, the findings might relate to methodological differences. Using fMRI in a study in which participants judged whether visually presented stimuli of various kinds were the same or different, a failure to confirm the PET finding of reduced activation of the insula in adult dyslexics was also reported by Shaywitz *et al.* (1998). Incidentally, these participants formed the basis of the report by Pugh *et al.* (2000).

Brunswick, McCroy, Price, Frith, and Frith (1999) reported that, while reading words and pseudo-words aloud, six adult (compensated) dyslexic students showed significantly reduced PET activation (relative to a rest condition) in the left posterior inferior temporal region (and elsewhere) in comparison with control participants. Since this region has been implicated in phonological processing tasks (Price, Moore, & Frackowiak, 1996; but see also Petersen, Fox, Snyder, & Raichle, 1990), Brunswick *et al.* saw their findings as "consistent with dyslexia involving a core deficit in accessing phonological word forms" (p. 1913).

For a number of reasons, functional activation studies of dyslexics almost invariably involve adults rather than children. We know little about the functional activation patterns of children which arguably is the group for whom we most need it. In a rare fMRI study carried out with children, German-speaking dyslexics showed under-activation in left inferior temporal regions, as has been found with adults, but frontal regions, Broca's area in particular, were also under-activated in comparison with control readers (Georgiewa *et al.*, 1999).

While PET and fMRI reveal different brain areas that are simultaneously activated during cognitive activity, they lack good temporal resolution. Magnetoencephalography (MEG) has relatively good spatial resolution and excellent temporal resolution. Using this technique, Salmelin, Service, Kiesilä, Uutela, and Salonen (1996) compared the time course of cortical activation in six adult dyslexic and eight control subjects as they passively viewed single Finnish words and non-words (i.e., no response was required). Whereas controls showed a sharp activation in the left inferior temporo-occipital region at about 180 ms after stimulus onset, dyslexics either failed to show any activation or showed a considerably later response. Within 200–400 ms, the left temporal lobe, including Wernicke's area, was said to be strongly activated in controls but not dyslexics. This was seen as consistent with impaired phonological processing in this region of the brain of dyslexic subjects, as suggested by the PET study of Rumsey *et al.* (1992). A left inferior frontal region was also activated within 400 ms in four of the six dyslexics, but in none of the eight controls. This is reminiscent of the finding of Brunswick *et al.* (1999) who reported over-activation (while reading words and non-words) in the left pre-motor area in dyslexics compared with controls.

In summary, neuroimaging and neuromagnetic studies strongly suggest that there is some processing abnormality in posterior temporal-parietal areas, and perhaps elsewhere, in the left hemisphere of dyslexics (Beaton, 2004). Under-activation of these posterior regions together, perhaps, with over-activation of certain frontal regions suggests that they form a widely distributed system which includes those areas, such as the angular gyrus, classically identified as involved in acquired disorders of reading and writing. Because there is still considerable uncertainty as to the precise localisation of specific components of reading in normal readers (see Howard *et al.*, 1992; Indefrey *et al.*, 1997; Poldrack *et al.*, 1999; Pugh *et al.*, 1996; Rumsey *et al.*, 1997a), the findings do not at present permit one to arrive with confidence at any more detailed conclusions with regard to dyslexia. Nonetheless, the findings generally support the long-held view based purely on behavioural studies that there is some functional impairment of language-related areas of the lefthemisphere, specifically of posterior regions (Rumsey *et al.*, 1994), and possibly of mid-posterior cortex bilaterally (Rumsey *et al.*, 1997b).

LATERALITY AND DYSLEXIA

It used to be thought that all right-handers speak with the left hemisphere, while left-handers speak with the right hemisphere. This is often referred to in the literature as Broca's rule though not in fact formulated by Broca himself (Harris, 1991). In any event, we now know that this is not true; most left-handers speak with the left hemisphere. However, it is the case that relatively few right-handers have language lateralized to the right hemisphere, while a greater proportion (though still a minority) of left-handers do. The figures often quoted are that fewer than 4% of right-handers have right-sided speech whereas in left-handers, approximately 15% have speech controlled from the right hemisphere (with another 15% having speech distributed across both hemispheres). For reasons that will not be discussed here, these figures are unlikely to be reliable. What is true is that right hemisphere speech in both left- and right-handers is more common than is usually supposed (see Annett, 1975; Annett & Alexander, 1996). Using the new technique of functional transcranial Doppler (fTCD) ultrasonography, it has been estimated that approximately one in 13 neurologically normal volunteers has speech represented in the right hemisphere (Knecht *et al.*, 2000a), and that the proportion of right-hemisphere speakers varies systematically as a function of strength of handedness. Approximately 25% of strong left-handers appear to have right-hemisphere speech while for strong right-handers, the figure is about 10% (Knecht *et al.*, 2000b).

During the 19th and early part of the 20th centuries, the idea was prevalent that one half of the brain, usually the left, is dominant for all psychological functions, not just speech. A dominant cerebral hemisphere seemed to follow naturally from the idea of a dominant hand. However, the notion of cerebral dominance was taken even further in that one half of the brain was said to inhibit or dominate the other. Lack of clear-cut handedness was regarded as showing a failure to establish dominance at the cerebral level.

Samuel Orton believed that the memory traces of letters presented to the brain were stored in mirror-image fashion in the left and right hemispheres. Normally, the "engrams" in the so-called minor (right) hemisphere were "suppressed" by the dominant hemisphere, but in cases of incomplete or weak cerebral dominance there was incomplete suppression. The result was confusion as to the correct orientation of letters. This "strephosymbolia" accounted for what Orton regarded as the relatively high number of letter reversals (e.g., b for d) shown by the poor readers referred to him (e.g., Orton, 1928). Orton's theory has logical as well as empirical shortcomings (Beaton, 1985, 2004) and is no longer taken seriously today. Like Freud, perhaps, his contribution to the field is valued not so much for the detail of his theories as for his recognition of the problems faced by dyslexic individuals.

Orton drew attention to what he believed was a raised frequency of left-handedness among children with dyslexia (and among those who stuttered). He also believed that children with severe reading difficulties showed crossed hand–eye dominance more often than expected. This refers to a dominant right hand in combination with a dominant left eye or a dominant left hand in combination with a dominant right eye. In actual fact, crossed-laterality is exceedingly common in the general population and its occurrence should no more be considered diagnostic of dyslexia than the presence of left-handedness (which is not to say that an association between crossed-laterality and dyslexia will not be found if one looks at a sufficiently large number of cases). There is certainly a genetic connection between language lateralisation and handedness although the exact nature of the relationship is controversial (see Beaton, 2003), and handedness is probably related in some way to reading and dyslexia (Eglinton & Annett, 1994), if not in precisely the way envisaged by Orton.

Handedness itself cannot be a cause of poor reading (any more than it can be a cause of language lateralisation, though presumably the two phenomena have mechanisms in common), but could there be a relationship between handedness and those language areas of the brain implicated in aphasia? Differences between left- and right-handers have been observed in part of Broca's area (Foundas *et al.*, 1998). However, it is in relation to a region encroaching on Wernicke's area that most interest has centred. This region is known as the planum temporale (PT). It forms a roughly triangular area on the upper or superior surface of the temporal lobe and is exposed by a horizontal knife-cut through the Sylvian fissure. In 1968, a celebrated report by Geschwind and Levitsky appeared showing that the length of the PT on the left was larger than that on the right in 65 of 100 specimens studied postmortem. This appeared to confirm, in a large sample of brains, the similar observations made earlier in the century.

The borders of the PT have been defined somewhat differently by different researchers and the size of the PT has been variously measured in terms of length, area, and volume both postmortem and using *in vivo* neuroimaging techniques. The exact delineation and methods of measurement of the PT are technical matters involving some controversy (see Barta *et al.*, 1995; Shapleske, Rossell, Woodruff, & David, 1999; Zetzche *et al.*, 2001) but most authors seem to accept that there is an asymmetry in the size of the planum on the left and right sides of the brain. Given the location of the planum and reports that the asymmetry apparently varies with handedness (for review, see Beaton, 1997), a common assumption is that this asymmetry has something to do with language lateralisation. There are in fact some reasons to doubt this (Beaton, 1997, 2002) even though a correlation has been reported for 5–12-year-old children between asymmetry of the planum and scores on a test thought to require an appreciation of the phonemic structure of words (phonemic awareness) by Leonard *et al.* (1996).

The putative relevance of the PT for dyslexia is that it has been reported that PT asymmetry is reduced or reversed among dyslexic persons (for reviews, see Beaton, 1997; Hynd & Semrud-Clikeman, 1989). On the basis of postmortem studies of a small number of brains, Galaburda and his collaborators suggested that among dyslexics, the PT is symmetrical (Galaburda, Sherman, Rosen, Aboitiz, & Geschwind, 1985; Humphreys, Kaufmann, & Galaburda, 1990). Subsequently, reports of a number of neuroimaging studies appeared claiming to show that right-sided asymmetry or symmetry was more frequent among dyslexics than controls (Duara *et al.*, 1991; Rumsey *et al.*, 1986), although this has not always been found (Leonard *et al.*, 1993; Schultz *et al.*, 1994). One study in particular, that of Larsen, Höien, Lundberg, and Ödegaard (1990), has been much cited as it appeared to distinguish between dyslexics with phonological problems and one participant who had "purely orthographic dysfunction." The latter, in contrast to the majority of the former, was said to show asymmetry of the planum. However, the definition of "orthographic" problems was unsound and the data were evaluated purely by visual inspection rather than precise measurement.

If it is the case that dyslexia is characterized by reversed asymmetry or symmetry of the PT, how might this come about, and what are the implications? One theory concerns a link between genetic mechanisms under-lying speech lateralisation and handedness.

We are accustomed to thinking of left- and right-handedness in terms of a simple dichotomy. In fact, between those people who do everything with either one hand or the other, extremely right- or left-handed individuals, there are others who exhibit different preferences for different activities. The different preferences span the entire range between the two extremes. In short, it is helpful to consider hand preference as constituting a continuum rather than a dichotomy (Annett, 2002). This is especially so if one considers hand skill rather than preference (but see McManus, 1991; McManus, Murray, Doyle, & Baron-Cohen, 1992).

There are a number of competing theories of handedness (for review, see Beaton, 2003). Marian Annett has for many years championed what she terms the Right Shift Theory of cerebral lateralisation and handedness (Annett, 2002). She argues that, as for most mammals, human handedness is largely a matter of chance. If chance alone were the sole factor, then half the population would be left-handed and the other half right-handed (to an approximate extent). Similarly, half of us would speak with the left hemisphere and half with the right hemisphere. However, chance is not the only factor. In Annett's view, there is an additional genetic factor at work in the majority of the population (she calculates about 81.6%) which "forces" the left hemisphere to be the speech hemisphere by "impairing" the right hemisphere in some way. This in turn acts coincidentally in some way to make it more probable that the individual will be right-handed. This mechanism operates in addition to chance factors. Thus, a frequency distribution (as in a bell curve) of the relative skill of the two hands is symmetrical with its mean (average) value shifted somewhat to the right of the point at which the two hands are equal in skill.

Some people (approximately 18.4% of the population) lack the genetic factor which brings about left-sided speech in the majority (call them right shift minus, RS−). These people have speech distributed in the left or right hemisphere randomly (i.e., according to chance), and their handedness is also distributed randomly and independently of their speech lateralisation. The proportion of RS− individuals who are left-handed (approximately 50%, some "natural" left-handers become right-handed due to social or environmental pressure) is higher than the proportion of left-handers among those who have the right-shift factor (RS+). The latter proportion is very small, consisting of those in whom the right-shift factor was not sufficiently strong to counter chance factors which otherwise predisposed them to becoming left-handed. (Note that we are talking here of relative proportions and not of absolute numbers.)

With regard to the PT, Annett (1992) drew attention to interesting parallels between the distribution of handedness and the distribution of planum asymmetry. Without going into too much detail, it is possible to think of some proportion of the population as having planum asymmetry distributed by chance (half having a larger planum on the left and half on the right) and the remainder having a larger planum on the left. In short, Annett's theory might apply to the distribution of PT in a way which directly parallels the distribution of handedness (without implying that the one causes the other).

According to Annett's theory, those people who lack the right-shift factor also lack some subtle advantage in speech processing enjoyed by the majority (but have compensating advantages in other spheres). It is these people, on Annett's view, who are most at risk of a phonological form of developmental dyslexia (Annett, 1996, 2002; Annett, Eglinton, & Smythe, 1996).

If Annett is correct that (some proportion of) RS− individuals are at risk of phonological dyslexia, and if RS− individuals have hemispheric speech lateralisation, and perhaps therefore planum asymmetry, distributed at random, then one would expect to find that these individuals have rightward asymmetry of the planum in approximately 50% of cases (unlike controls who would be expected to show leftward asymmetry in the large majority of cases). Although no one has identified the relevant right-shift gene (rs+ and rs− alleles), the data on dyslexia and the planum are consistent with this expectation if it is considered that "symmetry" of the planum reported in some papers is in fact a statistical artifact of averaging over roughly equal numbers of rightward and leftward cases (leading to a mean asymmetry of zero).

Annett's theory is highly controversial. There are many who disagree with her, both in regard to her genetic theory of cerebral lateralisation and handedness (McManus, 1985; McManus, Shergill, & Bryden, 1993) and more particularly with regard to her extension of the theory to encompass cognitive abilities such as reading (Beaton, 1995). An equally controversial theory, as it applies to planum asymmetry and to dyslexia, is based not on genetic theory but on hormones (though the two approaches are not incompatible).

THE HORMONAL HYPOTHESIS

In a series of influential papers Geschwind, Galaburda, and their colleagues proposed that excessively high levels of pre-natal foetal testosterone, a male sex hormone circulating during early maturation of the brain but present also in female foetuses in much smaller amounts, are associated with increased rates of left-handedness, and dyslexia. This was intended to explain not only an association between dyslexia and non-right handedness, but also the higher incidence of dyslexia and other learning difficulties in males than females. Furthermore, by acting on the thymus gland, involved in responses of the immune system, high levels of testosterone would account for a reported association between dyslexia, left-handedness and immune disease. According to the most recent version of this hypothesis, the left planum is usually larger than the right because of greater prenatal pruning of cortical connections on the right side. In the presence of "too much" testosterone, this process is arrested leading to symmetry of the plana on the two sides (Geschwind & Galaburda, 1985).

The hypothesis (known as the Geschwind-Behan–Galaburda or GBG hypothesis) generated a huge amount of research which was reviewed comprehensively by Bryden, McManus, and Bulman-Fleming (1994). It was not until 1995, however, that anyone was able to test the theory directly. Grimshaw, Bryden, and Finegan (1995) measured testosterone levels in the amniotic fluid of mothers undergoing amniocentesis at 16 gestational weeks. The levels of testosterone are assumed to reflect largely though not exclusively foetal rather than maternal levels of the hormone. The results were contrary to the predictions of the GBG hypothesis. When the children of the pregnancies which went to full term were tested at the age of 10 years, high levels of foetal testosterone were found to be associated with stronger right-handedness and stronger left lateralisation of language (measured by a non-invasive technique known as dichotic listening) than low levels of testosterone.

Although there is some research linking testosterone from the mother's umbilical artery to infant patterns of handedness (Tan & Tan, 1999, 2001) and testosterone levels in adult saliva to handedness (Moffat & Hampson, 1996), I know of no findings directly linking testosterone to dyslexia. Interestingly, however, testosterone has been linked (Moffat, Hampson, Wickett, Vernon, & Lee, 1997) to the size of another brain structure which some have seen as having a role to play in dyslexic symptomatology. This structure is the corpus callosum.

DYSLEXIA AND INTERHEMISPHERIC TRANSFER: THE ROLE OF THE CORPUS CALLOSUM

The adult corpus callosum is a band of approximately 175 million fibres connecting the two sides of the brain. Its role as demonstrated by the classic split-brain studies is predominantly to convey information from one cerebral hemisphere to the other. If the callosum is surgically sectioned for the relief of otherwise intractable epilepsy, then information directed to only one side of the brain, which is possible by virtue of the anatomic arrangement of the major sensory pathways, remains confined, in the main, to the hemisphere of input (for review, see Gazzaniga, 2000).

The left and right sides of visual space are represented in the occipital lobes of the right and left hemispheres of the brain, respectively. Asking someone to compare two stimuli flashed briefly to the left and right of a central point at which he or she is staring (fixating) presumably requires some transfer of information between the hemispheres for the task to be performed. In comparison with controls, dyslexics have sometimes been found to show deficits in cross-field matching thereby suggesting an impairment in inter-hemispheric transfer of visual information. Analogous findings for the auditory and tactile modalities have been interpreted in the same way, although in these modalities, sensory input from one side is not restricted to the opposite hemisphere as it is in vision. Although there are problems in interpreting the relevant data (see reviews by Beaton, 1985; Young & Ellis, 1981) the idea of an inter-hemispheric transfer deficit is supported by some electrophysiological research (Leisman, 2002).

Another paradigm which has been adopted to investigate inter-hemispheric transfer involves bimanual co-ordination tasks. The initial impetus for this research was a study by Preilowski (1972) showing that patients who have undergone surgical division of the anterior corpus callosum (commissurotomized or so-called split-brain patients) were impaired in certain aspects of bimanual co-ordination. However, it should not be thought that all bimanual tasks are impaired after commissurotomy. Well-practised tasks may be carried out as efficiently as before the operation though novel tasks are difficult to learn (Franz, Waldie, & Smith, 2000).

Preilowski's (1972) task involved using two knobs, controlled by left and right hands, to operate a single cursor. In some circumstances, each hand needed to turn the knobs in the same direction, either clockwise or anticlockwise, to succeed. In other circumstances, the two hands had to turn the knobs in opposite (mirror) directions, one turning clockwise, the other turning counter-clockwise. Using this task, Gladstone, Best, and Davidson (1989) found dyslexic boys to be slower and less accurate in the latter condition than were age-matched control subjects. Unlike the controls, the dyslexics were significantly slower in the mirror-movements condition than when the knobs had to be turned in the same direction. Gladstone *et al.* considered that co-operation between the hemispheres is compromised in dyslexia but that neural control of hand movements was also abnormal. Somewhat similar deficits in bimanual co-ordination among dyslexic adults have also been seen as consistent with an impairment of callosal function (Moore, Brown, Markee, Theberge, & Zvi, 1995).

Wolff and his colleagues have used unimanual and bimanual tapping tasks to investigate inter-hemispheric co-operation in young disabled readers. Participants tapped in time to a metronome with one or both hands either synchronously or asynchronously. In general, deficits were observed only at fast rates of tapping and when integration of timed responses between the two sides of the body were required. The motor tasks which most clearly discriminate between adult dyslexics and control subjects involve the integration of asymmetric or asynchronously timed movements between the right and left hands (Rousselle & Wolff,

1991). As each hand is controlled predominantly, but not exclusively, from the opposite hemisphere, this presumably depends upon efficient transmission of motor commands between the hemispheres.

FINGER LOCALISATION AND DYSLEXIA

In addition to those results obtained from bimanual co-ordination tasks, which suggest that inter-hemispheric transfer is less efficient in reading disabled individuals, the findings from another task have been interpreted in the same way. This is the cross-hand finger localisation task. An experimenter, unseen by the subject, touches one or more of the fingers of one hand and participants respond by indicating in some way, by using the thumb, for example, which fingers have been touched. In the within-hand condition, the same hand is used to respond as was stimulated by the experimenter. In the cross-hand condition, the opposite hand from that stimulated is used for responding. The assumption is that touching one hand stimulates primarily or exclusively the somatosensory cortex of the contralateral (opposite) hemisphere. Patients with damage to certain parts of the callosum do not perform well on this task (e.g., Geffen, Nilsson, & Quinn, 1985) and commissurotomised patients fail in the cross-hand condition even though within-hand localisation is unimpaired (Volpe, Sidtis, Holtzman, Wilson, & Gazzaniga, 1982). This implies that an intact corpus callosum is essential for successful performance in the cross-hand condition.

An impairment specifically in the cross-hand finger localisation condition is taken as showing a loss of information during inter-hemispheric transfer due either to a specific deficit or to a lack of callosal development. It is thought that children do not have a fully developed corpus callosum until they are approximately 12 years old—and it may take much longer to fully mature (Giedd *et al.*, 1999; Pujol, Vendrell, Junqué, Marti-Vilalta, & Capdevila, 1993). Consistent with this idea, Quinn and Geffen (1986) showed that cross-hand localisation was less accurate than within-hand hand localisation but that the magnitude of the cross-hand deficit decreased with age in normal children aged 5–11 years. However, even adults may show a deficit in the cross-hand compared with the within-hand condition, if the task is made difficult enough by stimulating three or four fingers in sequence (Beaton & Yearley, 2000, unpublished).

The issue with regard to dyslexia, then, is whether dyslexics show a greater cross-hand relative to within-hand deficit than do control participants. This has been reported to be the case for dyslexic adolescents by Gross-Glenn and Rothenberg (1984), and for a sub-group of Italian dyslexic children by Fabbro *et al.* (2001).

Amanda Puddifer and I used the finger localisation task with a group of reading disabled children who were individually matched to two control groups based on chronological and reading age (Beaton & Puddifer, unpublished). Participants indicated where their fingers had been touched by repeating the sequence using the thumb of the same or the opposite hand. All three groups showed reduced levels of accuracy in the cross-hand condition compared to the within-hand condition, but the percentage transfer was significantly less for the reading age-matched and reading disabled groups than for the chronological age controls.

Across all participants combined there were statistically significant relationships between success in the cross-hand condition (relative to the with-hand condition) and performance on both non-word reading and phoneme deletion tasks. These findings support the view that the degree to which tactile information is transferred across the corpus callosum is related to phonological ability. Intermanual transfer of finger localisation information has been reported to be more efficient among those with higher rhyming skills even in college-age dyslexics (Moore, Brown, Markee, Theberge, & Zvi, 1996).

It is possible that some explanation other than defective inter-hemispheric transfer may account for at least some of the reduced level of performance seen in the cross-hand condition by dyslexic compared with control readers (Pipe, 1991; Quinn & Geffen, 1986). When a single hand is used for both stimulation and response, simple tactile discrimination of where on the fingers one has been stimulated is sufficient for accurate performance. However, because the fingers of the left and right hands occupy mirror-image positions, an awareness of the relative position of the fingers of the two hands is also required if a correct response is to be made by the opposite hand.

Pipe (1991) carried out a study with 6-, 8-, and 10-year-olds. In one condition, children responded using either the same or their opposite hand to that which had been stimulated by successively opposing the thumb with the fingers which had been stimulated. In the cross-hand condition, the fingers are, of course, in mirror-image positions. In another condition of the experiment, children were asked to respond using their own hand (within-hand condition) or a model hand on which the fingers were in the mirror-image position of the stimulated hand. This latter condition is analogous to the conventional cross-hand condition but does not require inter-hemispheric transfer. Pipe discovered that whether children used a model to respond with their stimulated hand or used the thumb of their opposite hand to indicate the sequence of stimulation did not matter. Accuracy of performance was equivalent in these conditions and significantly worse than in a within-hand condition in which children indicated with thumb and finger the sequence of finger stimulation on that same hand. These findings suggest that the cross-hand deficit that is usually observed may have to do not so much with loss of information during inter-hemispheric transfer as with maintaining a mental representation of the relative position of the fingers on each hand. Dyslexics (like younger control readers) may be less able or ready to construct relevant mental representations which can be used to carry out the task.

In a recent study conducted by Rebecca Edwards in my laboratory, dyslexic adult students and control students were administered the within-hand and cross-hand finger-localisation tasks (Beaton, Puddifer, & Edwards, in preparation) in both the conventional mode and using photographs of the left and right hand on which to respond. That is, we extended Pipe's set-up to include a condition in which a photograph (hereafter a model) was used to test the same hand (as well as cross-hand) condition of stimulation. What we found overall was that dyslexic students made significantly more errors than controls, but both groups showed the usual cross-hand deficit. In the cross-hand (but not within-hand) condition, both groups made more errors using the conventional response mode than using the models. While this finding is consistent with the idea that some information is lost during inter-hemispheric transfer in both dyslexics and controls, we obtained no evidence from this group of adult students to suggest that this was any greater for dyslexics than controls. On the other hand, dyslexics generally benefited more from the models than did the controls.

We also found that for both groups, scores on a phonological test (Spoonerisms) correlated significantly with accuracy on cross-hand finger localisation scores, confirming our own earlier findings and those of Moore *et al.* (1996) with children (see also Fletcher, Taylor, Morris, & Satz, 1982; Lindgren, 1978). A possible link between callosal transfer and cognitive processes relevant to dyslexia is further suggested by the finding that at least some individuals in whom the corpus callosum is congenitally absent have shown deficits in production of words beginning with a given letter (Dennis, 1981; Jeeves & Temple, 1987) and in certain other aspects of phonological processing (Temple, Jeeves, & Villaroya, 1989, 1990; Temple & Ilsley, 1993). Whether these impairments are to be found in all acallosal individuals and whether they are due merely to the absence of the corpus callosum or to associated cerebral pathology cannot be decided on the basis of the data presently available.

MORPHOLOGY OF THE CORPUS CALLOSUM IN DYSLEXIA

As a result of the behavioural findings, some investigators have compared structural aspects of the corpus callosum between dyslexics and controls. In particular, the relative size of the corpus callosum or of its sub-divisions has been compared. These studies have all been carried out in the intact brain; I know of no post-mortem investigations of the morphology (size and shape) of the callosum in dyslexia. The usual approach is to compare the size of the callosum between groups (controlling for overall mid-sagittal brain size) and to express the size of individual callosal segments as a proportion of total callosal size. The posterior section of the callosum is the splenium; the centre section is the trunk (including the isthmus); the curved anterior section is known as the genu.

One study reported finding a proportionally larger splenium in dyslexic adults (Duara *et al.*, 1991), one found a smaller genu in dyslexic children (Hynd *et al.*, 1995), and one found no difference between adolescent dyslexics and controls in the proportional size of the splenium or in total callosal area (Larsen, Höien, Lundberg, & Ödegaard, 1992). A recent study found no difference between dyslexic children and controls in total callosal area or any of its segments

(Von Plessen *et al.*, 2002). An enlarged callosum except for the genu and splenium was reported by Robichon and Habib (1998) for 16 dyslexic men in comparison with 12 controls. In the largest study to date, 75 reading-disabled individuals (all from separate pairs of twins) were compared with 22 unrelated control participants. Neither total callosal area nor any sub-division differed between the two groups (Pennington *et al.*, 1999).

In summary, at least seven studies have measured *in vivo* the size of the corpus callosum in dyslexic participants. Three reported proportionally larger callosal segments in dyslexics than controls, one reported a smaller genu in dyslexics, and three reported no size difference between dyslexics and controls.

Individual variation in callosal size is striking (Giedd *et al.*, 1996, 1999; Parashos, Wilkinson, & Coffey, 1995) and large samples are required to adequately assess morphological characteristics of the callosum in relation to other variables such as handedness, gender, or presence of dyslexia (for reviews, see Beaton, 1997; Bishop & Wahlsten, 1997; Driesen & Raz, 1995). Given that Pennington *et al.* (1999) investigated the largest number of poor readers studied to date, their negative results might indicate that the positive findings reported in other studies represent merely chance findings. It is possible, of course, that individuals diagnosed as dyslexic according to differing criteria or with different accompanying difficulties may differ in their callosal morphology. However, in the study by Pennington *et al.* (1999), the findings were not affected by excluding those individuals who also had a diagnosis of ADHD.

A gross difference in size of the callosum may not be the relevant factor. It may be a much more subtle regional distribution in the relative number or size of fibres throughout the callosum that is important and this may not be picked up by simply dividing the callosum into a number of equal-sized segments. Although in the study by Von Plessen *et al.* (2002) there were no differences between dyslexic children and controls in overall callosal area or in the cross-sectional area of regional segments of the callosum, the posterior section of the trunk (isthmus) was on average shorter in 20 dyslexic boys than in controls. This suggests some difference in the shape of the callosum in the two groups. This might be associated with a difference in fibre distribution and neural transmission characteristics. At the time of writing, there are no data regarding the relationship of callosal fibre size or density to efficiency of information transfer and hence, presumably, to level of performance on various tasks. However, in the study by Robichon and Habib (1998), the number of errors made on a sound categorization task correlated with the area of two callosal regions as well as with two indices of callosal shape, "circularity" and "slenderness." Other correlations between performance measures and the size of callosal segments were also reported but for statistical reasons some caution is necessary in accepting the findings.

To summarise the above findings, although some studies indicate that dyslexics differ from controls in terms of callosal structure, there is little or no consistent evidence at present that gross morphological characteristics are associated with group differences in cognitive or motor performance. Having said this, it may not be too fanciful to suggest that difficulty in co-ordinating the activities of the two sides of the brain may help to explain some of the clumsiness said by some authors to be common among dyslexic children.

FURTHER MOTOR DEFICITS IN DYSLEXIA

It is not only bimanual co-ordination tasks on which dyslexics may be impaired. The handwriting of dyslexic children is notoriously poor and on unimanual tasks, such as peg-moving or finger-tapping, disabled readers have sometimes been reported to perform worse than control participants of the same age (e.g., Badian & Wolff, 1977; Fawcett & Nicholson, 1995; Gardner & Broman, 1979; Wolff, Cohen, & Drake, 1984). It is not only dyslexic children who show sensori-motor impairments. Motor deficits reported by Fawcett and Nicolson (1995) were seen in dyslexics up to 17 years of age and adult dyslexics are slower in pointing with one hand to an unpredictable visual target (Velay, Daffaure, Giraud, & Habib, 2002). Findings such as these imply a persisting difficulty with certain fine motor skills.

Gross motor functions can also be affected. According to Miles (1993), dyslexics often have a history of difficulties in everyday tasks such as learning to ride a bicycle. Orton (1925) observed that "some of them give a history of delay in learning to talk and walk and of a lack of nicety of balance and consequent frequent falls and of indecision in the choice of the right or left hand in using the knife, fork and spoon, all of which speak for a definite delay in decisive dominant control of the motor mechanisms" (p. 595). He also noted that some of the children referred to him showed "evidence of mild apraxia" although, as Denckla (1985) remarked, there is a common tendency among some clinicians to use the word dyspraxia (or apraxia) rather liberally.

Orton's reference to a "lack of nicety of balance" is interesting in the light of research by Fawcett and Nicolson (1992) showing that dyslexic children find it more difficult than controls to balance on a beam with one foot while performing a secondary task (such as mental arithmetic). No difference between groups was observed when balancing was carried out without a concurrent task (see also Nicolson & Fawcett, 1990, 1994). However, the former result has proved difficult to replicate (Yap & van der Leij, 1994) and it may be that the deficit is more likely to show up in children who have attentional problems as well as reading difficulties (Wimmer, Mayringer, & Raberger, 1999).

Reporting the results of a series of experiments (Fawcett & Nicolson, 1992, 1995; Nicolson & Fawcett, 1990, 1994) carried out with children and adolescents drawn from local dyslexia associations, Fawcett and Nicolson (1995) claimed that dyslexic children (predominantly male) were significantly worse than reading age controls at threading beads and on a range of reaction time, balancing, and other motor tasks. Fawcett and Nicolson (1995) pointed out that the hypothesis that dyslexia involves primarily a phonological deficit cannot handle these findings. They therefore proposed that dyslexics are impaired in mastering skills which eventually come to be carried out automatically rather than under conscious control. Subsequently this "automatisation deficit" hypothesis became incorporated in their "cerebellar deficit" hypothesis (Nicolson, Fawcett, & Dean, 1995) although they were not the first authors to suggest involvement of the cerebellum in dyslexia (see Frank & Levinson, 1973; Levinson, 1988). Indeed, it has been argued in recent years that the cerebellum is involved not only in motor functions, as traditionally conceived (e.g., Holmes, 1917, 1939; Ito, 1993), but also in certain cognitive functions (Allen, Buxton, Wong, & Courchesne, 1997; De Schutter & Maex, 1996; Fabbro, Moretti, & Bava, 2000; Leiner, Leiner, & Dow, 1989, 1993; Schmahmann & Sherman, 1998) including reading (Brunswick *et al.*, 1999; Fulbright *et al.*, 1999), although this is a contentious issue (Glickstein, 1993; Ivry, 1996; Thach, 1997).

THE CEREBELLAR DEFICIT HYPOTHESIS

Nicolson *et al.* (1995) argued that "A search for the underlying cause of deficits in balance, in motor skill and in automatisation would generally point strongly to the cerebellum" (p. 43). In this paper, they report research based on previous findings with patients who had damage to the cerebellum. Ivry and Keele (1989) had found that in comparison with Parkinson's disease patients, cerebellar patients were impaired on a task requiring them to judge which of two auditorily presented intervals of time was the longer. Nicolson *et al.* gave the same task to dyslexic children and found them to be impaired relative to age and IQ matched control children. Like cerebellar patients, the dyslexics were not impaired on a comparable loudness-estimation task.

In a replication of their earlier work, Fawcett and Nicolson (1999) tested a new sample of dyslexic children on a battery of tests which have been associated with cerebellar damage. The dyslexics were drawn from a school for dyslexics and two schools with large dyslexic units. The "cerebellar" tests involved assessment of postural ability (degree of sway or movement in response to a gentle push in the back), arm shake (degree of arm movement when the wrist was passively shaken by the experimenter, a measure of muscle tone), and speed of toe tapping. Other tests administered were of accuracy in phonemic segmentation, repetition of nonsense words, and speed of picture naming. The dyslexics were said to be particularly impaired on the "cerebellar" tests and it was concluded therefore that the cerebellum is a key structure involved in dyslexia. A summary of the cerebellar hypothesis was presented by Nicolson, Fawcett, and Dean (2001).

In a PET imaging study, Nicolson *et al.* (1999) measured brain activation levels while six adult dyslexic subjects and six control subjects performed motor tasks. The tasks involved carrying out a sequence of finger movements with the right hand with the eyes closed. In one condition, the sequence was highly overlearned; in another condition, participants learned a new sequence.

Compared with a resting condition, controls showed greater activation than dyslexics in the right cerebellum during performance of both the overlearned task and during acquisition of the novel sequence. This was taken as supporting the cerebellar deficit hypothesis.

Rae *et al.* (2002) used MRI with 11 adult male dyslexics and 9 controls. Among controls there was a significant left–right asymmetry in the proportion of grey matter (reflecting cell bodies) to total cerebellar volume. Among the dyslexics, there was no such asymmetry. The authors saw their finding of no cerebellar asymmetry in dyslexics as consistent with previous reports of symmetrical temporal plana in dyslexia. In a previous paper, this group claimed to find evidence of a biochemical asymmetry in the cerebellum of dyslexic men (Rae *et al.*, 1998). Although they interpreted their magnetic resonance spectroscopy findings in terms of "an altered pattern of cell density in the cerebellum of a dyslexic individual" (p. 1852), they did not directly measure cell size. Recently, Finch, Nicolson, and Fawcett (2002) have reported that there are relatively fewer small cells and more large cells in the cerebellum of "dyslexic" brains (and in a part of the auditory pathway) as compared with control brains. They did not find any significant cerebellar asymmetry.

The brains examined by Finch *et al.* (2002) were the four male brains reported on by Galaburda *et al.* (1985) in relation to asymmetry of the PT (and to certain sub-cortical features—see below). This places a heavy interpretative load on the brains of only a few individuals, yet there are reasons to doubt whether these specimens can be considered representative of dyslexic brains in general (Beaton, 1997). In addition, certain other aspects of the Finch *et al.* (2002) study are less than ideal. For further discussion of this paper, and of the cerebellar deficit hypothesis in general, the reader is referred to Beaton (2002), Bishop (2002), and to the commentaries which follow the paper by Nicolson *et al.* (2001).

Given that dyslexics exhibit certain difficulties in the sphere of motor activity, the question has been posed as to whether remediation of motor deficits will have any impact on reading. McPhillips, Hepper, and Mulhern (2000) recently examined the effects of movement training with 3 groups each of 20 dyslexic children. One group was trained by asking them to repeat movements associated with what is known as the asymmetrical tonic neck reflex (ATNR). This reflex is elicited in neonates by a sideways turning of the head when the infant is lying on its back. It consists of extension of the arm and leg on the side to which the head turns and flexion of the upper and lower limb on the opposite side of the body. Normally the ATNR is not seen after the age of about 6 months and its persistence beyond this time is associated with central nervous system abnormality. In the children of this study, a persistent ATNR was assessed by asking them to stand upright with their feet together while an experimenter slowly turned the head one way or the other. Positive indicators of this reflex were said to include extending arms in the same direction as the head turn, dropping the arms, or swaying and loss of balance.

The study by McPhillips *et al.* was intended to see whether the ATNR could be inhibited by a training regime involving rehearsal and voluntary repetition of the movements involved in the TNR and other primary reflexes. The rationale was that, during infancy, the occurrence of the reflex might contribute in some way to its eventual inhibition. The experimental group was provided with a regime of specific movements to carry out each day for about 10 min for a period of 12 months. A placebo control group was also given a sequence of movements to carry out but these were not based on the primary reflexes. A control group of children were given no specific instructions. The results seem to indicate that the experimental group made significantly greater gains in reading than either of the other two groups which did not differ between themselves. It will be of great interest to see whether the results prove replicable by other groups of workers (see Reynolds *et al.*, 2003), especially since ideas as to why movement "therapy" should enhance reading ability are somewhat vague. The conclusion of McPhillips *et al.* was simply that "persistent primary reflexes may have a critical role in early neurological maturation which, in turn, has repercussions for later reading development" (p. 540).

BEYOND THE CORTEX

Anatomical investigations have suggested some form of abnormality in the cortex of dyslexic brains (Galaburda *et al.*, 1985), yet other areas of the brain may play a part. The thalamus is almost certainly implicated in some aspect of language functions (Crosson, 1985) and has been shown to be under-activated in dyslexic adults in comparison with controls in some PET and fMRI studies (see Roush, 1995). Indeed, particular parts of the thalamus, the lateral and medial geniculate nuclei, have been said to be abnormal in dyslexia. To appreciate the significance of the findings, it is first necessary to consider the organization of the visual system.

THE M AND P VISUAL SYSTEMS

Almost all we know about the neurophysiology of vision comes from animal research but converging evidence from different sources, principally psychophysical experiments, strongly implies that the human visual system is organised in a similar manner to that of the rhesus macaque monkey. The visual system of this species is divided into two main streams: the magnocellular or M pathway and the parvocellular or P pathway. Cells of the M sub-division are larger than those of the P sub-division and the two classes of cell show different neurophysiological responses (for review in relation to reading see Breitmeyer, 1993). For example, M system cells respond to the onset and/or offset of a stimulus whereas P system cells respond with a more sustained response. M cells respond best to stimuli of low luminance and low spatial frequency (coarse detail) but high temporal frequency (rapidly changing). They are relatively insensitive to colour. Cells of the P system respond best to static stimuli of high spatial frequency (fine detail) at high luminance levels and are sensitive to colour. Given the distinctive neurophysiological properties of the two systems, it is generally considered that the M system is specialised for the perception of movement whereas the P system is specialised for the analysis of form and colour. The cells of these two subdivisions are physically segregated in different layers of the lateral geniculate nucleus (LGN) and to some extent, though perhaps less than was at one time thought (see Yabuta, Sawatari, & Callaway, 2001), at the primary visual cortex (area V1) and perhaps even beyond.

A considerable amount of experimental work has accumulated to suggest that there is an impairment of the M system in dyslexia (for reviews, see Beaton, 2004; Lovegrove, 1991, 1996; Lovegrove, Martin, & Slaghuis, 1986; Stein, 2000; Stein & Walsh, 1997). Much of this work originated from William Lovegrove's laboratory in Australia and involved presentation of stationary stimuli such as gratings (patterns of alternating black and white bars). Differences between reading-disabled and control children were found in their ability to detect subtle differences in contrast between the luminance of the alternating light and dark bars (Lovegrove, Bowling, Badcock, & Blackwood, 1980; Lovegrove *et al.*, 1982; Spafford, Grosser, Donatelle, Squillace, & Dana, 1995; but see Gross-Glenn *et al.*, 1995). This is known as contrast sensitivity and it varies with the width of the light and dark bars (in technical terms, with spatial frequency). Differences between reading impaired and control children in contrast sensitivity functions have been reported to vary with overall luminance level (Martin & Lovegrove, 1984) and spatial frequency (Lovegrove *et al.*, 1982) in a manner suggestive (to some) of a defect in the M pathway of dyslexics. This view is supported by the findings of one study showing that the difference in sensitivity to temporal frequency between dyslexics and control subjects increased with the frequency of flicker of a sinusoidal grating (Felmingham & Jakobson, 1995).

Recent experiments have involved presentation of moving stimuli (usually random dots), sometimes in combination with neuroimaging techniques. Participants are required to make judgements about the relative speed or direction of movement of different sets of dots. It turns out that at least some dyslexic children (Cornelissen, Richardson, Mason, Fowler, & Stein, 1995) and adults (Eden *et al.*, 1996; Talcott *et al.*, 1998) are impaired in various aspects of movement processing.

The M system carries information from the primary receiving area of the visual cortex (V1) to other cortical areas and in particular to an area in the middle temporal lobe known as MT/V5. Neuroimaging studies with humans indicate that this area is active in response to certain aspects of movement (Tootell *et al.*, 1995; Watson *et al.*, 1993; Zeki *et al.*, 1991). It is widely thought of as being a "center" for motion detection and perception. However, some degree of motion processing probably also takes place at cortical areas which provide input to MT/V5.

Eden *et al.* (1996) reported that six dyslexic adults were significantly worse than eight control subjects in discriminating between the relative velocities of two sets of moving dots. Using fMRI it was also found that area MT/V5 was not activated in the dyslexic subjects though it was in the controls. In contrast to these results, using the new technique of MEG, Vanni, Uusitalo, Kiesila, and Hari (1997) found that similar responses were evoked in area V5 of Finnish adult dyslexic and control subjects by a low-brightness stimulus which slightly shifted position back and forth every 45 ms. Lack of activation in area MT/V5 of the dyslexics studied by Eden *et al.* (1996) cannot have been due to an overall problem with their visual system since they showed normal responses to stationary stimulus patterns in other visual areas. It is therefore possible that the different methodologies (MEG vs fMRI) account for the discrepancy in findings reported by Eden *et al.* (1996) and Vanni *et al.* (1997), or that faster movement of the stimulus in the study by Vanni *et al.* than in the one by Eden *et al.* led to equal degrees of activation of MT/V5 in dyslexics and controls.

Exactly what leads to poor motion processing in dyslexia is not known—there are a number of conceivable mechanisms (Walther-Müller, 1995); some studies have provided evidence linking movement-related (putatively magnocellular) functions to reading performance. Cornelissen, Hansen, Hutton, Evangelinou, and Stein (1998) presented a motion detection task to a group of 58 unselected children aged from 9 to 11 years. Performance on this task correlated with the proportion of certain kinds of errors made in reading a set of regular words. It was argued that this kind of error reflected faulty coding of the position of letters within words and that this information is carried by the M system. Impaired magnocellular function might lead to a degraded encoding of letter position during reading such that "positional uncertainty of this kind could cause letters or parts of letters to be lost or duplicated, or even incorrectly bound together, leading to a scrambled or nonsense version of what is actually printed on the page" (p. 473).

In another study of motion detection thresholds, adult dyslexics and control participants were given tests of non-word reading (Talcott *et al.*, 1998). Within each of the participant groups, poorer performance on putative tests of M pathway function was associated with a greater number of non-word reading errors. Furthermore, velocity discrimination thresholds have been reported to correlate with the reading rate for words on a standard test of reading (Demb, Boynton, Best, & Heeger, 1998a).

THE CONSEQUENCES OF A MAGNO SYSTEM DEFICIT

While correlations between reading and performance on movement perception tasks are interesting, they do not tell us how a magnocellular deficit might lead to difficulties in learning to read and to spell. Breitmeyer and Ganz (1976) suggested that in the brief interval between successive fixations of a visual scene, when the eyes move and vision is impaired (although we are not usually aware of this), the M pathway suppresses or masks activity in the P pathway. This would ensure that the images received during each fixation would not interfere with each other. Any anomalous operation of the suppression mechanism in relation to the

M system would thus be expected to have consequences for reading. However, recent research has provided evidence for the opposite effect, namely that during saccades the magnocellular system is selectively suppressed (Burr, Morrone, & Ross, 1994; see also review by Ross, Burr, & Morrone, 1996) which rather undermines Breitmeyer's theory.

Talcott *et al.* (1998) suggested that impaired M pathway functions do not interfere directly with reading, but rather indirectly, perhaps via effects on persistence of retinal images, visual stability, or eye movements. Stein (2000) recently expressed the argument as follows:

> One problem that constantly bedevils the hypothesis that dyslexics have impaired magnocellular function is that people find it very difficult to understand how a system devoted to detecting visual motion could possibly be relevant to reading. After all, we don't usually have to track moving targets when reading; the page is usually kept stationary. In fact, the retinal images of print are not stationary, and many dyslexic children complain that letters seem to move around when they are trying to read, i.e. their visual world is highly unstable. This is because during reading visual images are actually very far from being stationary on the retina, and dyslexics fail to compensate for this ... We believe that their unstable visual perceptions are the result of the insensitivity of their visual magnocellular systems (p. 111).

In an earlier review, Stein and Walsh (1997) argued that "Slight impairments of mLGN (magnocellular laminae of the LGN) performance or organization might ... multiply up to greater deficits in PPC (posterior parietal cortex) function. The PPC is known to be important for normal eye movement control, visuo-spatial attention and peripheral vision—all important components of reading" (p. 149).

They are not the only authors to implicate eye movements. Borsting *et al.* (1996) suggested an abnormal saccade mechanism could adversely affect reading (p. 1052). It is also possible that attentional factors might be relevant. Certainly, the frequent co-occurrence of dyslexia and attentional problems poses a methodological challenge to those attempting to investigate visual deficits in dyslexia.

The magnocellular deficit hypothesis of dyslexia has generated considerable scepticism and controversy (see Hayduk, Bruck, & Cavanagh, 1996; Hulme, 1988; Lovegrove, 1991; Skottun, 1997, 2000; Walther-Müller, 1995) and there have been a number of failures to support it. Skottun (2000) reviewed nearly two dozen studies involving measurement of contrast sensitivity. He concluded that only 4 out of 22 studies support the magnocellular deficit hypothesis, 11 provide evidence against it, while 7 studies have provided inconclusive results. Even with regard to experiments involving moving stimuli, Skottun notes that random dots can stimulate both M and P pathways and therefore any deficit cannot be unambiguously attributed to the magnocellular pathway. The conclusion of Skottun's review is that there is little compelling evidence for a deficit in dyslexia that can be related exclusively to the magnocellular system. This does not, of course, invalidate those findings said to support the hypothesis even if their interpretation is in doubt. The findings themselves still have to be explained.

Some of the early research reported in the scientific journals was sloppily executed or carried out with inadequate methodological control procedures or with unmatched participant groups. However, not all the research was flawed (see, e.g., Evans, Drasdo, & Richards, 1994) and this is unlikely to be the reason why the idea of a magnocellular deficit has met with so much resistance (as with so much in the field of dyslexia!). The predominant view (see Ramus *et al.*, 2003) of dyslexics' difficulties is that they are caused by phonological problems; the idea that visual problems might be crucial has been largely rejected by many researchers (if not by practitioners).

As always, much hangs on the meaning attributed to a particular word, in this case "visual." All sorts of phenomena can be and have been classified under this heading, including visuo-spatial perception, ocular factors, memory for visually presented events and so on (see Chapter 1 in this volume by Bruce Evans). Many people are reluctant to accept that "visual" problems contribute to dyslexic difficulties. Yet there is no necessary incompatibility between phonological problems on the one hand and M system problems on the other, especially, as we shall see, if the designation "M system" is broadened to include sensory systems other than vision.

DYSLEXIA AND THE THALAMUS

In an experiment concerned with speed discrimination, Demb, Boynton, and Heeger (1998b) found a significant positive correlation between reading rate and level of brain response recorded by fMRI in the middle temporal region of the brain and between reading comprehension and activity at a number of other sites, including the primary visual cortex (V1), the cortical area at which most visual information is first received.

The finding of a group difference in brain activation as early in the visual system as the striate or visual cortex (V1), that is lower than the middle temporal area, is consistent with the idea that the deficit may arise at a pre-cortical (geniculate or retinal) level. In this context, the findings of a postmortem study by Livingstone, Rosen, Drislane, and Galaburda (1991) are of interest. These authors examined the LGN from five purportedly dyslexic individuals and reported that there were proportionally fewer large cells in the magnocellular division than in five control brains. A parallel study of five (living) dyslexic adults showed that there was a reduced electrophysiological response from this area under conditions of low luminance and high temporal frequency. The obvious, if logically tenuous, conclusion was that the reduced visual evoked potential (VEP) and anomalous cellular morphology are related to at least some aspects of dyslexic symptomato-logy. However, the finding of a reduced response specific to dyslexia was not replicated in a group of dyslexic adults and children by Victor, Conte, Burton, and Nass (1993). Nonetheless, it is possible to interpret certain other electrophysiological results (e.g., Duffy, Denekla, Bartels, & Sandini, 1980; Lemkuhle, Garzia, Turner, Hash, & Baro, 1993) as favouring the magnocellular deficit hypothesis.

The brains examined by Livingstone *et al.* (1991) were all said to be "symmetrical" as far as the PT was concerned (Galaburda *et al.*, 1985; Humphreys *et al.*, 1990). The same five brains from dyslexics were examined by Jenner, Rosen, and Galaburda (1999). On the assumption that different layers of the primary visual cortex receive input primarily from magno- and parvocellular layers of the LGN, it was anticipated that cells in the cortical magno layer of dyslexic brains would differ from those of control brains while cells in the parvo layer would not. Jenner *et al.* reported that over all cortical layers combined, there was no hemispheric asymmetry in mean cross-sectional neuronal area of the primary visual cortex (V1) in dyslexic brains, unlike the leftward asymmetry seen in control brains, although this pattern occurred for both layers. On the other hand, controls were said to be more biased towards the left in cell size (as opposed to mean cross-sectional area) than were the dyslexics, both taken over all cortical layers combined and in the magno cortical layer alone but not in the parvo layer. It was concluded that controls had larger sized cells in the left than the right hemisphere whereas the dyslexics had similar sized cells in both hemispheres. Thus, as was claimed for the PT, symmetry rather than asymmetry was thought to characterise this sample of "dyslexic" brains but the reservations made above obviously apply here too.

A MAGNO DEFICIT AND (A)SYMMETRY OF THE PLANUM TEMPORALE

Finally, the relation of a putative magnocellular deficit to neuroanatomical factors merits brief discussion. Given that PT symmetry or reversed asymmetry and a magnocellular deficit have both been hypothesised to underlie dyslexic difficulties in some way, it is of interest to ask whether these two "anomalies" tend to co-occur in the same individual.

This question was taken up by Best and Demb (1999) who investigated PT asymmetry in five dyslexic subjects all of whom showed reduced brain activity in area MT/V5 in response to stimuli designed to elicit strong activation. In this small sample of dyslexic participants, there was no obvious association between anomalous PT asymmetry and putative magnocellular deficits. Similarly, six dyslexics who all showed deficits in motion sensitivity in the study by Eden *et al.* (1996) were drawn from a group among whom the majority were known to have normal leftward PT asymmetry (Rumsey *et al.*, 1997). It thus seems unlikely that there is any relation between magno deficits and PT symmetry although according to Stein (1994) "one can speculate that magnocellular input and lateralisation characteristics are causally connected; in other words that the favoured access of magnocellular input to the left hemisphere is what causes it to become the language hemisphere and the PT to be larger on that side" (p. 247).

The magnocellular division of the visual system is usually thought of as terminating in the motion sensitive area of the temporal lobe. However, fibres descend from the middle temporal area to several sub-cortical sites including the pons (Maunsell & Van Essen, 1983; Merigan & Maunsell, 1993) which in turn projects to the cerebellum. While damage to area MT/V5 in humans has been linked to impairments of motion detection and velocity perception (Zeki, 1991; Zihl, Von Cramon, & Mai, 1983; Zihl, Von Cramon, Mai, & Schmid, 1991), damage to some parts of the cerebellum has also been associated in humans with deficits in discriminating velocity (Ivry & Deiner, 1991) or direction (Nawrot & Rizzo, 1998) of movement. It is tempting to wonder whether some of the findings in regard to movement may actually reflect a cerebellar rather than a cortical deficit.

AUDITORY FUNCTIONS AND DYSLEXIA

The claim that relatively low-level visual deficits of the magnocellular system are found in association with reading difficulties has often been seen as contradicting the view that phonological deficits constitute the core causal problem in dyslexia. It is possible, of course, that visual deficits of one kind or other are not found in all dyslexics but only in some and perhaps only within a particular sub-type. Visual deficits do not appear to be characteristic of so-called "dyseidetic" (Borsting *et al.*, 1996; Ridder, Borsting, Cooper, McNeel, & Huang, 1997) or "surface" (Spinelli *et al.* 1997) dyslexics. Yet there is no fundamental incompatibility between M pathway deficits and phonological impairments and visual and language difficulties are frequently found together in dyslexia (Slaghuis, Lovegrove, & Davidson, 1993; Slaghuis, Twell, & Kingston, 1996; Cestnick & Coltheart, 1999). Conceivably, both are the result of a third underlying factor involving sensory processing in both visual and sensory modalities.

Research has indicated that some dyslexic children (Adlard & Hazan, 1998; Hurford & Sanders, 1990; Masterson, Hazan & Wijayatilake, 1995) and adults (Cornelissen, Hansen, Bradley, & Stein, 1996) have problems in subtle aspects of speech perception or discrimination which might underpin their phonological problems (Mody, Studdert Kennedy, & Brady, 1997). In particular, it seems that some dyslexics are less consistent, or are slower, in classifying certain sounds as belonging to one category than another. As certain parameters of a synthetically produced speech stimulus are gradually changed, there comes a point at which the classification of a stimulus by a listener switches from one category to another. Despite the gradually changing nature of the stimulus, listeners do not perceive sounds intermediate between the two categories, only one category or the other. This is referred to as "categorical perception."

Dyslexics, or at least some proportion of them, show less categorical speech perception than controls (Brandt & Rosen, 1980; Godfrey, Syrdal-Lasky, Millay, & Knox, 1981; Manis *et al.*, 1997; Reed, 1989; Steffens, Eilers,

Gross-Glenn, & Jallad, 1992; Werker & Tees, 1987). Together with indications of impaired phoneme discrimination in dyslexia, this implies that the boundaries between different speech sounds are less distinct in some dyslexics than in normally reading individuals. This might relate to their phonemic awareness and hence to learning to read. Incidentally, some disruption of categorical speech perception has also been reported to occur in patients with bilateral damage to the cerebellum (Ackermann, Gräber, Hertrich, & Daum, 1997).

Impaired speech perception in dyslexia is indicated by other research. Menell, McAnally, and Stein (1999) found that on average 20 adult dyslexic participants were less sensitive to certain changes (amplitude modulation, AM) in acoustic stimuli than 20 control participants. The dyslexic subjects also showed reduced auditory evoked potentials measured at the scalp. The electrophysiological and psychophysical measures correlated positively and significantly. Performance on the psychophysical task also correlated positively and significantly with reading ability. Menell *et al.* concluded that "Because AM in speech is important for its intelligibility, the insensitivity of dyslexic listeners to AM is likely to impair their identification of speech" (p. 802).

An electrophysiological paradigm was used by Schulte-Körne, Deimel, Bartling, and Remschmidt (1998) to investigate whether young poor readers and spellers (mean age 12.5 years) differed from age-matched controls in what is known as mismatch-negativity (MMN). This is a component of the event-related potential (ERP) recorded from the scalp which occurs in response to a change occurring in a sequence of repetitive auditory stimuli. Schulte-Körne *et al.* (1998) presented synthetic speech stimuli consisting of a series of sounds (/da/) among which a different sound (/ba/) was occasionally presented (an "odd-ball" passive discrimination task). The results showed that, in comparison with, 15 control readers, a group of 19 poor spellers/readers had an attenuated MMN (measured as area under the averaged curves) for speech stimuli, but not for non-speech stimuli (pure tones), over the fronto-central regions of the brain.

The MMN is thought to index pre-attentive processing and thus group differences in MMN are considered unlikely to result from differences in attention or motivation. Schulte-Körne *et al.* therefore considered their results to be consistent with other research showing deficits in phoneme perception and with the view there that deficits in pre-attentive speech processing can be considered a cause of dyslexia. Their study, however, was correlational in nature; a longitudinal study with pre-readers would have provided more convincing evidence. There is also the possibility that their results with speech stimuli (indeed all such results) are a reflection of experience or skill in reading rather than a cause of reading. Moreover, using MEG, the finding of an attenuated response to real-speech deviant syllables was not replicated by Heim *et al.* (2000), perhaps due to the different stimuli or electrophysiological responses used in the two studies. Heim *et al.* found that the source of their main waveform was located more anteriorly in 11 dyslexic children than in 9 controls. It was concluded that the neural organization of the auditory cortex differs between dyslexics and controls.

The MMN paradigm was also used in a study of eight adult dyslexics by Kujala *et al.* (2000). There were two conditions. In one condition, a pattern of four tones was presented (pattern condition) and in the other pairs of tones (tone-pair condition) were presented. It was found that at latencies within 400–450 ms, the MMN response distinguished the dyslexic participants from controls only in the pattern condition. Whereas controls showed two consecutive MMN responses in deviant trials, dyslexics did not. Kujala *et al.* interpreted their findings as demonstrating a basic auditory processing defect, as opposed to a linguistic defect, in dyslexic adults.

The medial geniculate body receives fibres from the olivary body, part of the auditory pathway, and projects to the superior temporal gyrus, the primary auditory receiving area. The medial geniculate nucleus (MGN) is the auditory equivalent of the LGN in the visual system. They lie in close proximity but are independent of each other with regard to their functions. The cells of the lateral and medial geniculate nuclei receive input from the optic and auditory nerves respectively. In turn, they transmit information to the visual (striate) cortex and the primary auditory cortex. Thus, the two geniculate nuclei act as relay stations between the relevant sensory organ (the eye and the ear) and the principal areas of the cortex to which they project.

In a postmortem study of the same five dyslexic brains that showed an anomaly of the magnocellular layer of the LGN (and apparent symmetry of the PT), Galaburda, Menard, and Rosen (1994) found that there were more small and fewer large neurones in the left MGN than in the right MGN. In seven control brains, there was no difference in the distribution of small and large cells between the MGN at the two sides. The authors concluded: "It would not be surprising to find that in dyslexics other sensory (and perhaps motor and cognitive) systems also showed differences in large-cell, possibly fast processing, subsystems as well" (p. 8012).

THE TEMPORAL DEFICIT HYPOTHESIS

If skilled motor actions and language share neural mechanisms (Tzeng & Wang, 1984) associated with the control of serial order and/or precision of timing mechanisms, this might help to explain the frequent co-occurrence of phonological and motor difficulties in dyslexia. Certainly, many of the deficits in motor function seen in dyslexia, such as in maintaining asynchronous tapping rates with the two hands, might be regarded as impairments in implementing motor control commands at the right time and in the correct order (Wolff, 1993).

One of the most interesting but controversial of current theories surrounding dyslexia is that it involves a deficit of temporal processing. The theory owes much to the work of Paula Tallal and her collaborators carried out mainly with children diagnosed as having specific language impairment (SLI). Such children often, but not invariably, have difficulty learning to read. Conversely, many dyslexic children have a history of difficulty or delay in achieving normal milestones in relation to speech. It is increasingly being suggested (e.g., Gallagher, Frith, & Snowling, 2000) that the impairments shown by SLI children lie on a continuum with those of dyslexic children (but see Snowling, Bishop, & Stothard, 2000).

Tallal's theory is based largely on what has come to be known as her auditory repetition test. In a series of studies (reviewed in Tallal, Miller, & Fitch, 1993), SLI children were given tests of temporal discrimination and temporal order judgement using pairs of sounds which were either short or long in duration. These stimuli were presented with either relatively short or long intervals of time (the inter-stimulus interval) between them. Language impaired children were found to differ from control children only with short stimuli and short inter-stimulus intervals. The general conclusion which Tallal and her colleagues drew from their early studies was that language impaired (and dyslexic) children have a perceptual deficit which affects the rate at which they can process incoming auditory information. Subsequently, the deficit was said to consist in processing of "speech sounds that incorporate rapidly changing acoustic spectra." This subtle shift of meaning has been the source of a good deal of controversy in the literature (see e.g., Studdert-Kennedy & Mody, 1995).

On one level, the temporal processing deficit hypothesis (reviewed by Farmer & Klein, 1995) seems to fit with the long established view that dyslexics have a particular weakness in the perception or retention of the temporal order of events. It is this that is said to underlie their often-claimed inability to organise their time. However, this is probably too facile a connection to draw. Difficulty in resolving short temporal durations presented at rapid inter-stimulus intervals is a far cry from correctly recalling the order of relatively long duration events separated by relatively long ISIs, as used in early experiments on temporal order judgement, or, more informally, from remembering to do something at the right time of the day, though some have found the notion of a single temporal defect seductive (Farmer & Klein, 1995; Habib, 2000).

The reason why the temporal processing deficit hypothesis has excited so much interest is that, despite its critics, it appears to offer an account of how the well-established difficulties of dyslexics in the phonological domain arise in the first place (see also Goswami *et al.*, 2002). At least some of these pre-date reading, and are held to be causally related to difficulty in acquiring adequate reading skills. The conjecture is that difficulties in processing the extremely short temporal durations involved in speech perception lead to impaired auditory discrimination or categorisation of phonemes of which the speech signal is composed. Thus, a temporal deficit becomes an impairment of phoneme perception which in turn leads to difficulties in learning the relations between components of the spoken and the written word. Such a view neatly explains why aspects of reading disability are heritable despite the fact that it is highly unlikely that we have genes specifically for reading. It is the genes for speech which are inherited. If in some people these make certain aspects of speech perception less efficient than usual, there may well be a knock-on effect in relation to literacy skills. Even in the absence of perceptual difficulties, performance in reading and speech processing rashes may relate to the quality or integrity of a child's phonological representations (Metsala, 1997; Griffiths & Snowling, 2002).

CONCLUDING COMMENTS

This chapter has reviewed research indicating that anomalies in a variety of brain areas are associated with, or might even contribute to, developmental dyslexia. We should not forget, however, that experiential factors might be responsible for certain findings rather than the other way around. Research in the neurosciences is currently revealing how much experience and biology interact to determine functional and even structural properties of the brain.

Different regions of the brain do not function in isolation; what happens in one part of the brain can have effects at other parts far removed. For example, neuroimaging studies reveal a close correspondence between activity in certain areas of one cerebral hemisphere and activity in the opposite cerebellar hemisphere (e.g., Junck *et al.*, 1988). Conceivably, therefore, some aspects of dyslexia are associated with disturbances of cortico-cerebellar circuitry as opposed to functional impairment of the cortex or cerebellum alone (see Eckert *et al.*, 2003). A sub-cortical lesion can affect the organisation of the cortex. Perhaps the minor focal cortical malformations said to be present in the brains of dyslexic adults (Galaburda *et al.*, 1985; Humphreys *et al.*, 1990) can be related to anomalies at the geniculate level. My guess is that in the not too distant future, our views of the biology of dyslexia will become much more sophisticated than at present, greater attention being paid to the integrated functions of a variety of neural systems rather than of specific brain parts.

REFERENCES

Ackermann, H., Gräber, S., Hertrich, I., & Daum, I. (1997). Categorical speech perception in cerebellar disorders. *Brain and Language, 60*, 323–331.

Adlard, A., & Hazan, V. (1998). Speech perception in children with specific reading difficulties (dyslexia). *The Quarterly Journal of Experimental Psychology, 51*, 153–177.

Allen, G., Buxton, R. B., Wong, E. C., & Courchesne, E. (1997). Attentional activation of the cerebellum independent of motor involvement. *Science, 275*, 1940–1943.

Annett, M. (1975). Hand preference and the laterality of cerebral speech. *Cortex, 11*, 305–328.

Annett, M. (1992). Parallels between asymmetries of planum temporale and of hand skill. *Neuropsychologia, 30*, 951–962.

Annett, M. (1996). Laterality and types of dyslexia. *Neuroscience and Behavioural Reviews, 20*, 631–636.

Annett, M. (2002). *Handedness and brain asymmetry: The right shift theory*. Hove, UK: Psychology Press.

Annett, M., & Alexander, M. P. (1996). Atypical cerebral dominance: Predictions and tests of the right shift theory. *Neuropsychologia, 34*, 1215–1227.

Annett, M., Eglinton, E., & Smythe, P. (1996). Types of dyslexia and the shift to dextrality. *Journal of Child Psychology and Psychiatry, 37*, 167–180.

Badian, N., & Wolff, P. H. (1977). Manual asymmetries of motor sequencing in boys with reading disability. *Cortex, 13*, 343–349.

Barta, P. E., Petty, R. G., McGilchrist, I., Lewis, R. W., Jerram, M., Casanova, M. F. *et al.* (1995). Asymmetry of the planum temporale: Methodological considerations and clinical associations. *Psychiatry Research Neuroimaging, 61*, 137–150.

Beaton, A. A. (1985). *Left side, right side: A review of laterality research.* London, Batsford and New Haven: Yale University Press.

Beaton, A. A. (1995). Hands, brains and lateral thinking: An overview of the right shift theory. *Cahiers de Psychologie Cognitive/Current Psychology of Cognition, 14*, 481–495.

Beaton, A. A. (1997). The relation of planum temporale asymmetry and morphology of the corpus callosum to handedness, gender, and dyslexia: A review of the evidence. *Brain and Language, 60*, 255–322.

Beaton, A. A. (2002). Dyslexia and the cerebellar deficit hypothesis. *Cortex, 38*, 479–490.

Beaton, A. A. (2003). The nature and determinants of handedness. In K. Hugdahl, & R. J. Davidson, (Eds.), *The asymmetrical brain* (Chap. 4, pp. 105–158). Cambridge, MA: MIT Press.

Beaton, A. A. (2004). *Dyslexia, reading and the brain: A sourcebook of psychological and biological research*. Hove: Psychology Press.

Beaton, A. A., Puddifer, A., & Edwards, R. (in preparation).

Best, M., & Demb, J. B. (1999). Normal planum temporale asymmetry in dyslexics with a magnocellular pathway deficit. *NeuroReport, 10*, 607–612.

Bishop, D. V. M. (2002). Cerebellar abnormalities in developmental dyslexia: Cause, correlate or consequences? *Cortex, 38*, 491–498.

Bishop, K. M., & Wahlsten, D. (1997). Sex differences in the human corpus callosum: Myth or reality? *Science and Behavioural Reviews, 21*, 581–601.

Borsting, E., Ridder, W. H., Dudeck, K., Kelley, C., Matsui, L., & Motoyama, J. (1996). The presence of a magnocellular defect depends on the type of dyslexia. *Vision Research, 36*, 1047–1053.

Brandt, J., & Rosen, J. J. (1980). Auditory phonemic perception in dyslexia: Categorical identification and discrimination of stop consonants. *Brain and Language, 9*, 324–337.

Breitmeyer, B. G. (1993). Sustained (P) and transient (M) channels in vision: A review and implications for reading. In D. M. Willows, R. S. Kruk, & E. Corcos (Eds.), *Visual processes in reading and reading disabilities*. Hillside, NJ, and Hove, England: Lawrence Erlbaum Associates.

Breitmeyer, B. G., & Ganz, L. (1976). Implications of sustained and transient channels for theories of visual pattern masking, saccadic suppression, and information processing. *Psychological Review, 83*, 1–36.

Brunswick, N., McCroy, E., Price, C. J., Frith, C. D., & Frith, U. (1999). Explicit and implicit processing of words and pseudowords by adult developmental dyslexics. *Brain, 122*, 1901–1917.

Bryden, M. P., McManus, I. C., & Bulman-Fleming, M. B. (1994). Evaluating the empirical support for the Geschwind-Behan–Galaburda model of cerebral lateralization. *Brain and Cognition, 26*, 103–167.

Bub, D. N. (2000). Methodological issues confronting PET and fMRI studies of cognitive function. *Cognitive Neuropsychology, 17*, 467–484.

Burr, D. C., Morrone, M. C., & Ross, J. (1994). Selective suppression of the magnocellular visual pathway during saccadic eye movements. *Nature, 371*, 511–513.

Castro-Caldas, A., Miranda, P., Carmo, I., Reis, A., Leote, F., Ribeiro, C. *et al.* (1999). The influence of learning to read and write on the morphology of the corpus callosum. *European Journal of Neurology, 6*, 23–28.

Cestnick, L., & Coltheart, M. (1999). The relationship between language-processing and visual-processing deficits in developmental dyslexia. *Cognition, 71*, 231–255.

Cornelissen, P. L., Hansen, P. C., Bradley, L., & Stein, J. F. (1996) Analysis of conceptual confusions between nine sets of consonant–vowel sounds in normal and dyslexic adults. *Cognition, 59*, 275–306.

Cornelissen, P. L., Hansen, P. C., Hutton, J. L., Evangelinou, V., & Stein, J. F. (1998). Magnocellular visual function and children's single word reading. *Vision Research, 38*, 471–482.

Cornelissen, P., Richardson, A., Mason, A., Fowler, S., & Stein, J. (1995). Contrast sensitivity and coherent motion detection measured at photopic luminance levels in dyslexics and controls. *Vision Research, 35*, 1483–1494.

Crosson, B. (1985). Subcortical language functions in language: A working model. *Brain and Language, 25*, 257–292.

Dejerine, M. J. (1891). Sur un cas de cécité verbale avec agraphie, suivi d'autopsie. *Mémoires de la Société de Biologie, 3*, 197–201.

Demb, J. B., Boynton, G. M., Best, M., & Heeger, D. J. (1998). Psychophysical evidence for a magnocellular pathway deficit in dyslexia. *Vision Research, 38*, 1555–1559.

Demb, J. B., Boynton, G. M., & Heeger, D. J. (1998). Functional magnetic resonance imaging of early visual pathways in dyslexia. *The Journal of Neuroscience, 18*, 6939–6951.

Denckla, M. B. (1985). Motor coordination in dyslexic children: Theoretical and clinical implications. In F. H. Duffy & N. Geschwind (Eds.), *Dyslexia: A neuroscientific approach to clinical evaluation* (pp. 187–195). Boston and Toronto: Little, Brown & Company.

Dennis, M. (1981). Language in a congenitally acallosal brain. *Brain and Language, 12*, 33–53.

De Schutter, E., & Maex, R. (1996). The cerebellum: Cortical processing and theory. *Current Opinion in Neurobiology, 6*, 759–764.

Drake, W. E. (1968). Clinical and pathological findings in a child with developmental learning disability. *Journal of Learning Disability, 1*, 486–502.

Driesen, N. R., & Raz, N. (1995). The influence of sex, age, and handedness on corpus callosum morphology: A meta analysis. *Psychobiology, 23*, 240–247.

Duara, B., Kushch, A., Gross-Glenn, K., Barker, W. W., Jallad, B., Pascal, S. *et al.* (1991). Neuroanatomic differences between dyslexic and normal readers on magnetic resonance imaging scans. *Archives of Neurology, 48*, 410–416.

Duffy, F. H., Denckla, M. B., Bartels, P. H., & Sandini, G. (1980). Dyslexia: Regional differences in brain electrical activity by topographic mapping. *Annals of Neurology, 7*, 412–420.

Eckert, M. A., Leonard, C. M., Richards, T. L., Aylward, E. H., Thomson, J., & Berninger, V. W. (2003). Anatomical correlates of dyslexia: Frontal and cerebellar findings. *Brain, 126*, 482–494.

Eden, G. F., Van Meter, J. W., Rumsey, J. M., Maisog, J. M., Woods, R. P., & Zeffiro, T. A. (1996). Abnormal processing of visual motion in dyslexia revealed by functional brain imaging. *Nature, 382*, 66–69.

Eglinton, E., & Annett, M. (1994). Handedness and dyslexia: A meta-analysis. *Perceptual and Motor Skills, 79*, 1611–1616.

Ehri, L. (1992). Reconceptualizing the development of sight word reading and its relationship to recoding. In P. B. Gough, L. C. Ehri, & R. Treiman (Eds.), *Reading acquisition* (Chap. 5, pp. 107–143). Hillsdale, NJ: Lawrence Erlbaum Associates.

Evans, B. J. W., Drasdo, N., & Richards, I. L. (1994). An investigation of some sensory and refractive visual factors in dyslexia. *Vision Research, 34*, 1913–1926.

Fabbro, F., Pesenti, S., Facoetti, A., Bonanomi, M., Libera, L., & Lorusso, M. L. (2001). Callosal transfer in different types of developmental dyslexia. *Cortex, 37*, 65–74.

Fabbro, F., Moretti, R., & Bava, A. (2000). Language impairments in patients with cerebellar lesions. *Journal of Neurolinguistics, 13*, 173–188.

Farmer, M. E., & Klein, R. M. (1995). The evidence for a temporal processing deficit linked to dyslexia: A review. *Psychonomic Bulletin and Review, 2*, 460–493.

Fawcett, A. J., & Nicolson, R. I. (1992). Automatisation deficits in balance for dyslexic children. *Perceptual and Motor Skills, 75*, 507–529.

Fawcett, A. J., & Nicolson, R. I. (1995). Persistent deficits in motor skill of children with dyslexia. *Journal of Motor Behaviour, 27*, 235–240.

Fawcett, A. J., & Nicolson, R. I. (1999). Performance of dyslexic children on cerebellar and cognitive tests. *Journal of Motor Behaviour, 31*, 68–78.

Felmingham, K. L., & Jakobson, L. S. (1995). Visual and visuomotor performance in dyslexic children. *Experimental Brain Research, 106*, 467–474.

Finch, A. J., Nicolson, R. I., & Fawcett, A. J. (2002). Evidence for a neuroanatomical difference within the olivo-cerebellar pathway of adults with dyslexia. *Cortex, 38*, 529–539.

Fletcher, J. M., Taylor, H. G., Morris, R., & Satz, P. (1982). Finger recognition skills and reading achievement: A developmental neuropsychological analysis. *Developmental Psychology, 18*, 124–132.

Flowers, D. L., Wood, F. B., & Naylor, C. E. (1991). Regional cerebral blood flow correlates of language processes in reading disability. *Archives of Neurology, 48*, 637–643.

Foundas, A. L., Eure, K. F., Luevano, L. F., & Weinberger, D. R. (1998). MRI asymmetries of Broca's area: The pars triangularis and pars opercularis. *Brain and Language, 64*, 282–296.

Frank, J., & Levinson, H. (1973). Dysmetric dyslexia and dyspraxia. *Journal of American Academy of Child Psychiatry, 12*, 690–701.

Franz, A. E., Waldie, K. E., & Smith, M. J. (2000). The effect of callosotomy on novel versus familiar bimanual actions: A neural dissociation between controlled and automatic processes? *Psychological Science, 11*, 82–85.

Fulbright, R. K., Jenner, A. R., Mencl, W. E., Pugh, K. R., Shaywitz, B. A., Shaywitz, S. E. *et al.* (1999). The cerebellum's role in reading. *American Journal of Neuroradiology, 20*, 1925–1930.

Galaburda, A. M., Menard, M. T., & Rosen, G. T. (1994). Evidence for aberrant auditory anatomy in developmental dyslexia. *Proceedings of the National Academy of Sciences of the USA, 91*, 8010–8013.

Galaburda, A. M., Sherman, G. F., Rosen, G. D., Aboitiz, F., & Geschwind, N. (1985). Developmental dyslexia: Four consecutive patients with cortical anomalies. *Annals of Neurology, 18*, 222–233.

Gallagher, A., Frith, U., & Snowling, M. (2000). Precursors of literacy delay among children at genetic risk of dyslexia. *Journal of Child Psychology and Psychiatry, 41*, 203–213.

Gardner, R. A., & Broman, M. (1979, Fall). The Purdue pegboard: Normative data on 1334 school children. *Journal of Clinical Child Psychology*, 156–162.

Gazzaniga, M. S. (2000). Cerebral specialization and interhemispheric communication: Does the corpus callosum enable the human condition? *Brain, 123*, 1293–1326.

Geffen, G., Nilsson, J., & Quinn, K. (1985). The effect of lesions of the corpus callosum on finger localization. *Neuropsychologia, 23*, 497–514.

Georgiewa, P., Rzanny, R., Hopf, J.-M., Knab, R., Glauche, V., Kaiser, W. *et al.* (1999). FMRI during word processing in dyslexic sand normal reading children. *NeuroReport, 10*, 3459–3465.

Geschwind & Galaburda. (1985). Cerebral lateralization. Biological mechanisms, associations, and pathology: 1. A hypothesis and program for research. *Archives of Neurology, 42*, 428–459.

Giedd, J. N., Rumsey, J. M., Castellanos, F. X., Rajapakse, J. C., Kaysen, D., Vaituzis, A. C. *et al.* (1996). A quantitative MRI study of the corpus callosum in children and adolescents. *Developmental Brain Research, 91*, 274–280.

Giedd, J. N., Blumenthal, J., Jeffries, N. O., Rajapakse, J., Vaituzis, A. C., Hong Liu *et al.* (1999). Development of the human corpus callosum during childhood and adolescence: A longitudinal MRI study. *Progress in Neuropharmacological and Biological Psychiatry, 23*, 571–588.

Gladstone, M., Best, C. T., & Davidson, R. J. (1989). Anomalous bimanual coordination among dyslexic boys. *Developmental Psychology, 25*, 236–246.

Glickstein, M. (1993). Motor skills but not cognitive tasks. *Trends in Neurosciences, 16*, 450–451.

Godfrey, J. J., Syrdal-Lasky, A. K., Millay, K. K., & Knox, C. M. (1981). Performance of dyslexic children on speech perception tests. *Journal of Experimental Child Psychology, 32*, 401–424.

Goswami, U., Thomson, J., Richards, U., Stainthorp, R., Hughes, D., Rosen, S., & Scott, S. K. (2002). Amplitude envelope onsets and developmental dyslexia: A new hypothesis. *Proceedings of the National Academy of Sciences of the USA, 99*, 10917–10922.

Griffiths, Y., & Snowling, M. (2002). Predictors of exception word and nonword reading in dyslexic children: The severity hypothesis. *Journal of Educational Psychology, 94*, 34–43.

Grimshaw, G. M., Bryden, M. P., & Finegan, J.-A. K. (1995). Relations between prenatal testosterone and cerebral lateralization in children. *Neuropsychology, 9*, 68–79.

Grosser, G. S., & Spafford, C. S. (1989). Perceptual evidence for an anomalous distribution of rods and cones in the retinas of dyslexics: A new hypothesis. *Perceptual and Motor Skills, 68*, 683–698.

Grosser, G. S., & Spafford, C. S. (1992) Reply to Stuart and Lovegrove's question, "Visual processing deficits in dyslexia: Receptors or neural mechanisms?". *Perceptual and Motor Skills, 75*, 115–120.

Gross-Glenn, K., & Rothenberg, S. (1984). Evidence for deficit in interhemispheric transfer of information in dyslexic boys. *International Journal of Neuroscience, 24*, 23–35.

Gross-Glenn, K., Skottun, B. C., Glenn, W., Kushch, A., Lingua, R., Dunbar, M. *et al.* (1995). Contrast sensitivity in dyslexia. *Visual Neuroscience, 12*, 153–163.

Habib, M. (2000). The neurological basis of developmental dyslexia. An overview and working hypothesis. *Brain, 123*, 2373–2399.

Hagman, J. O., Wood, F., Buchsbaum, M. S., Tallal, P., Flowers, L., & Katz, W. (1992). Cerebral brain metabolism in adult dyslexic subjects assessed with positron emission tomography during performance of an auditory task. *Archives of Neurology, 49*, 734–739.

Harris, L. J. (1991). Cerebral control for speech in right-handers and left-handers: An analysis of the views of Paul Broca, his contemporaries and his successors. *Brain and Language, 40*, 1–50.

Hayduk, S., Bruck, M., & Cavanagh, P. (1996). Low-level visual processing skills of adults and children with dyslexia. *Cognitive Neuropsychology, 13*, 975–1015.

Heim, S., Eulitz, C., Kaufmann, J., Füchter, I., Pantev, C., Lamprecht-Dinnesen, A. *et al.* (2000). Atypical organisation of the auditory cortex. *Neuropsychologia, 38*, 1749–1759.

Henderon, V. W. (1986). Anatomy of posterior pathways in reading: A reassessment. *Brain and Language, 29*, 119–133.

Holmes, G. (1917). Functional localization in the cerebellum. *Brain, 40*, 531–535.

Holmes, G. (1939). The cerebellum of man. *Brain, 62*, 1–30.

Howard, D., Patterson, K., Wise, R., Brown, W. D., Friston, K., Weiller, C. *et al.* (1992). The cortical localization of the lexicons. *Brain, 115*, 1769–1782.

Hulme, C. (1988). The implausibility of low-level visual deficits as a cause of children's reading difficulties. *Cognitive Neuropsychology, 5*, 369–374.

Humphreys, P., Kaufmann, W. E., & Galaburda, A. M. (1990). Developmental dyslexia in women: Neuropathological findings in three patients. *Annals of Neurology, 28*, 727–738.

Hurford, D. P., & Sanders, R. E. (1990). Assessment and remediation of a phonemic discrimination deficit in reading disabled second and fourth graders. *Journal of Experimental Child Psychology, 50*, 396–441.

Hynd, G. W., & Hiemenz, J. R. (1997). Dyslexia and gyral morphology variation. In C. Hulme & M. Snowling (Eds.), *Dyslexia: Biology, cognition and intervention* (Chap. 3., pp. 38–58). London: Whurr.

Hynd, G. W., & Semrud-Clikeman (1989). Dyslexia and brain morphology. *Psychological Bulletin, 106*, 447–482.

Hynd, G. W., Morgan, A. E., Edmonds, J. E., Black, K., Riccio, C. A., & Lombardino, L. (1995). Reading disabilities, cormorbid psychopathology, and the specificity of neurolinguistic deficits. *Developmental Neuropsychology, 11*, 311–322.

Indefrey, P., Kleinschmidt, A., Merboldt, K.-D., Krüger, G., Brown, C., Hagoort, P. *et al.* (1997). Equivalent responses to lexical and nonlexical visual stimuli in occipital cortex: A functional magnetic resonance imaging study. *NeuroImage, 5*, 78–81.

Ito, M. (1993). Movement and thought: Identical control mechanisms by the cerebellum. *Trends in Neurosciences, 16*, 448–450.

Ivry, M. (1997). Cerebellar timing systems. *International Review of Neurobiology, 41*, 555–573.

Ivry, R. B., & Deiner, H. C. (1991). Impaired velocity perception in patients with lesions of the cerebellum. *Journal of Cognitive Neuroscience, 3*, 355–366.

Ivry, R. B., & Keele, S. W. (1989) Timing functions of the cerebellum. *Journal of Cognitive Neuroscience, 1*, 136–152.

Jenner, A. R., Rosen, G. D., & Galaburda, A. M. (1999). Neuronal asymmetries in primary visual cortex of dyslexic and nondyslexic brains. *Annals of Neurology, 46*, 189–196.

Jeeves, M. A., & Temple, C. M. (1987). A further study of language function in callosal agenesis. *Brain and Language, 32*, 325–335.

Junck, L., Gilman, S., Rothley, J. R., Betley, A. T., Koeppe, R. A., & Hichwa, R. D. (1988). A relationship between metabolism in frontal lobes and cerebellum in normal subjects studied with PET. *Journal of Cerebral Blood Flow and Metabolism, 8*, 774–782.

Knecht, S., Deppe, M., Dräger, B., Bobe, L., Lohmann, H., Flöel, A. *et al.* (2000a). Language lateralization in healthy humans. *Brain, 123*, 74–81.

Knecht, S., Dräger, B., Deppe, M., Bobe, L., Lohmann, H., Flöel, A. *et al.* (2000b). Handedness and hemispheric language dominance in healthy humans. *Brain, 123*, 2512–2518.

Kujala, T., Myllyviita, K., Tervaniemi, M., Alho, K., Kallio, J., & Näätänen, R. (2000). Basic auditory dysfunction in dyslexia as demonstrated by brain activity measurements. *Psychophysiology, 37*, 262–266.

Larsen, J. P., Höien, T., Lundberg, I., & Ödegaard, H. (1990). MRI evaluation of the size and symmetry of the planum temporale in adolescents with developmental dyslexia. *Brain and Language, 39*, 289–301.

Larsen, J. P., Höien, T., Lundberg, I., & Ödegaard, H. (1992). Magnetic resonance imaging of the corpus callosum in developmental dyslexia. *Cognitive Neuropsychology, 9*, 123–134.

Leonard, C. M., Voeller, K. K. S., Lombardino, L. J., Morris, M. K., Hynd, G. W., Alexander, A. W. *et al.* (1993). Anomalous cerebral structure in dyslexia revealed with magnetic resonance imaging. *Archives of Neurology, 50*, 461–469.

Leonard, C. M., Lombardino, L. J., Mercado, L. R., Browd, S. R., Breier, J. I., & Agee, O. F. (1996). Cerebral asymmetry and cognitive development in children: A magnetic resonance imaging study. *Psychological Science, 7*, 89–95.

Lehmkuhle, S., Garzia, R. P., Turner, L., Hash, T., & Baro, J. A. (1993). A defective visual pathway in children with reading disability. *The New England Journal of Medicine, 328*, 989–996.

Leiner, H. C., Leiner, A. L., & Dow, R. S. (1989). Reappraising the cerebellum: What does the hindbrain contribute to the forebrain? *Behavioural Neuroscience, 103*, 998–1008.

Leiner, H. C., Leiner, A. L., & Dow, R. S. (1993). Cognitive and language functions of the human cerebellum. *Trends in Neurosciences, 16*, 444–447.

Leisman, G. (2002). Coherence of hemisphere function in developmental dyslexia. *Brain and Cognition, 48*, 425–431.

Levinson, H. N. (1988). The cerebellar-vestibular basis of learning disabilities in children, adolescents and adults: Hypothesis and study. *Perceptual and Motor Skills, 67*, 983–1006.

Lindgren, S. D. (1978). Finger localization and the prediction of reading disability. *Cortex, 14*, 87–101.

Livingstone, M. S., Rosen, G. D., Drislane, F. W., & Galaburda, A. M. (1991). Physiological and anatomical evidence for a magnocellular defect in developmental dyslexia. *Proceedings of the National Academy of Sciences of the USA, 88*, 7943–7947.

Logothetis, N. K., Pauls, J., Augath, M. A., Trinath, T., & Oeltermann, A. (2001). Neurophysiological investigation of the basis of the fMRI signal. *Nature, 412*, 150–157.

Lovegrove, W. J. (1991). Is the question of the role of visual deficits as a cause of reading disabilities a closed one? Comments on Hulme. *Cognitive Neuropsychology, 8*, 435–441.

Lovegrove, W. J. (B). (1996). Dyslexia and a transient/magnocellular pathway deficit: The current situations and future directions. *Australian Journal of Psychology, 48*, 167–171.

Lovegrove, W. J., Bowling, A., Badcock, D., & Blackwood, M. (1980). Specific reading disabilities: Differences in contrast sensitivity as a function of spatial frequency. *Science, 210*, 439–440.

Lovegrove, W., Martin, F., Bowling, A., Blackwood, M., Badcock, D., & Paxton, S. (1982). Contrast sensitivity functions and specific reading disability. *Neuropsychologia, 20*, 309–315.

Lovegrove, W., Martin, F., & Slaghuis, W. (1986). A theoretical and experimental case for a visual deficit in specific reading disability. *Cognitive Neuropsychology, 3*, 225–267.

McManus, I. C. (1985). Handedness, language dominance and aphasia: A genetic model. *Psychological Medicine, Monograph Supplement 8*. Cambridge University Press.

McManus, I. C. (1991). The inheritance of left-handedness. In G. R. Bock, & J. Marsh (Eds.), *CIBA Foundation Symposium 162: Biological asymmetry and Handedness* (pp. 251–267). Chichester: Wiley.

McManus, I. C., Murray, B., Doyle, K., & Baron-Cohen, S. (1992). Handedness in childhood autism shows a dissociation of skill and preference. *Cortex, 28*, 373–381.

McManus, I. C., Shergill, S., & Bryden, M. P. (1993). Annett's theory that individuals heterozygous for the right shift gene are intellectually advantaged: Theoretical and empirical problems. *British Journal of Psychology, 84*, 517–537.

McPhillips, M., Hepper, P. G., & Mulhern, G. (2000). Effects of replicating primary-reflex movements on specific reading difficulties in children: A randomised, double-blind, controlled trial. *The Lancet, 335*, 537–541.

Manis, F. R., McBride-Chang, C., Seidenberg, M. S., Doi, L. M., Munson, B., & Peterson, A. (1997). Are speech perception deficits associated with developmental dyslexia? *Journal of Experimental Child Psychology, 66*, 211–235.

Martin, F., & Lovegrove, W. (1984). The effects of field size and luminance on contrast sensitivity differences between specifically reading disabled and normal children. *Neuropsychologia, 22*, 73–77.

Masterson, J., Hazan, V., & Wijayatilake, L. (1995). Phonemic processing problems in developmental phonological dyslexia. *Cognitive Neuropsychology, 12*, 233–259.

Maunsell, J. H. R., & Van Essen, D. C. (1983). The connections of the middle temporal visual area (MT) and their relationship to a cortical hierarchy in the macaque monkey. *The Journal of Neuroscience, 3*, 2563–2586.

Menell, P., McAnally, K. I., & Stein, J. F. (1999). Psychophysical sensitivity and physiological response to amplitude modulation in adult dyslexic listeners. *Journal of Speech, Language, and Hearing Research, 42*, 797–803.

Merigan, W. H., & Maunsell, J. H. R. (1993). How parallel are the primate visual pathways? *Annual Review of Neuroscience, 16*, 369–402.

Metsala, J. (1997). Spoken word recognition in reading disabled children. *Journal of Educational Psychology*, 89, 159–169.

Miles, T. R. (1993) *Dyslexia: The pattern of difficulties* (2nd ed.). London: Whurr.

Mody, M., Studdert-Kennedy, M., & Brady, S. (1997). Speech perception deficits in poor readers: Auditory processing or phonological coding? *Journal of Experimental Child Psychology, 64*, 199–231.

Moffat, S. D., & Hampson, E. (1996). Salivary testosterone levels in left- and right-handed adults. *Neuropsychologia, 34*, 225–233.

Moffat, S. D., Hampson, E., Wickett, J. C., Vernon, P. A., & Lee, D. H. (1997). Testosterone is correlated with regional morphology of the human corpus callosum. *Brain Research, 767*, 297–304.

Moore, L. H., Brown, W. S., Markee, T. E., Theberge, D. C., & Zvi, J. C. (1995). Bimanual coordination in dyslexic adults. *Neuropsychologia, 33*, 781–793.

Moore, L. H., Brown, W. S., Markee, T. E., Theberge, D. C., & Zvi, J. C. (1996). Callosal transfer of finger localization information in phonologically dyslexic adults. *Cortex, 32*, 311–322.

Nawrot, M., & Rizzo, M. (1998). Chronic motion perception deficits from midline cerebellar lesions in human. *Vision Research, 38*, 2219–2224.

Nicolson, R. I., & Fawcett, A. J. (1990). Automaticity: A new framework for dyslexia research? *Cognition, 35*, 159–182.

Nicolson, R. I., & Fawcett, A. J. (1994). Comparison of deficits in cognitive and motor skills among children with dyslexia. *Annals of Dyslexia, 44*, 147–163.

Nicolson, R. I., Fawcett, A. J., & Dean, P. (1995). Time estimation deficits in developmental dyslexia: Evidence of cerebellar involvement. *Proceedings of the Royal Society, 259*, 43–47.

Nicolson, R. I., Fawcett, A. J., & Dean, P. (2001). Developmental dyslexia: The cerebellar deficit hypothesis. *Trends in Neurosciences, 24*, 508–511.

Nicolson, R. I., Fawcett, A. J., Berry, E. L., Jenkins, I. H., Dean, P., & Brooks, D. J. (1999). Association of abnormal cerebellar activation with motor learning difficulties in dyslexic adults. *The Lancet, 353*, 1662–1667.

Orton, S. T. (1925). "Word-blindness" in school children. *Archives of Neurology and Psychiatry, 14*, 581–615.

Orton, S. T. (1928). A physiological theory of reading disability and stuttering in children. *The New England Journal of Medicine, 199*, 1046–1052.

Parashos, I. A., Wilkinson, W. E., & Coffey, C. E. (1995). Magnetic resonance imaging of the corpus callosum: Predictors of size in normal adults. *The Journal of Neuropsychiatry and Clinical Neurosciences, 7*, 35–41.

Paulesu, E., Frith, U., Snowling, M., Gallagher, A., Morton, J., Frackowiak, R. S. J. *et al.* (1996). Is developmental dyslexia a disconnection syndrome? Evidence from PET scanning. *Brain, 119*, 143–157.

Pennington, B. F., Filipek, P. A., Lefly, D., Churchwell, J., Kennedy, D. N., Simon, J. H. *et al.* (1999). Brain morphometry in reading-disabled twins. *Neurology, 53*, 723–729.

Petersen, S. E., Fox, P. T., Snyder, A. Z., & Raichle, M. E. (1990). Activation of extrastriate and frontal cortical areas by visual words and word-like stimuli. *Science, 249*, 1041–1044.

Pipe, M.-E. (1991). Developmental changes in finger localization. *Neuropsychologia, 29*, 339–342.

Poldrack, R. A., Wagner, A. D., Prull, M. W., Desmond, J. E., Glover, G. H., & Gabrieli, J. D. E. (1999). Functional specialization for semantic and phonological processing in the left inferior prefrontal cortex. *NeuroImage, 10*, 15–35.

Preilowski, B. (1972). Possible contribution of the anterior forebrain commissures to bilateral motor co-ordination. *Neuropsychologia, 10*, 267–277.

Price, C. J., Moore, C. J., & Frackowiak, R. S. (1996). The effect of varying stimulus rate and duration on brain activity during reading. *NeuroImage, 3*, 40–52.

Pugh, K. R., Shaywitz, B. A., Shaywitz, S. E., Constable, R. T., Skudlarski, P., Fulbright, R. K. *et al.* (1996). Cerebral organization of component processes in reading. *Brain, 119*, 1221–1238.

Pugh, K. R., Mencl, W. E., Shaywitz, B. A., Shaywitz, S. E., Fulbright, R. K., Constable, R. *et al.* (2000). The angular gyrus in developmental dyslexia: Task-specific differences in functional connectivity within posterior cortex. *Psychological Science, 11*, 51–56.

Pujol, J., Vendrell, P., Junqué, C., Marti-Vilalta, J. L., & Capdevila, A. (1993). When does human brain development end? Evidence of corpus callosum growth up to adulthood. *Annals of Neurology, 34*, 71–75.

Quinn, K., & Geffen, G. (1986). The development of tactile transfer of information. *Neuropsychologia, 24*, 793–804.

Rae, C., Lee, M. A., Dixon, R. M., Blamire, A. M., Thompson, C. H., Styles, P. *et al.* (1998). Metabolic abnormalities in developmental dyslexia detected by 1H magnetic resonance spectroscopy. *The Lancet, 351*, 1849–1852.

Rae, C., Harasty, J., Dzendrowskyj, T. E., Talcott, J. B., Simpson, J. M., Blamire, A. M. *et al.* (2002). Cerebellar morphology in developmental dyslexia. *Neuropsychologia, 46*, 1285–1292.

Ramus, F., Rosen, S., Dakin, S. C., Day, B. L., Castellotte, J. M., & Frith, U. (2003). Theories of developmental dyslexia: Insights from a multiple study of dyslexic adults. *Brian, 126*, 841–865.

Reed, M. (1989). Speech perception and discrimination of brief auditory cues in reading disabled children. *Journal of Experimental Child Psychology, 48*, 270–292.

Reynolds, D., Nicolson, R., & Hambly, H. (2003). Evaluation of an exercise-based treatment for children with reading difficulties. *Dyslexia, 9*, 48–71.

Ridder, W. H., Borsting, E., Cooper, M., McNeel, B., & Huang, E. (1997). Not all dyslexics are created equal. *Optometry and Vision Science, 74*, 99–104.

Robichon, F., & Habib, M. (1998). Abnormal callosal morphology in male adult dyslexics: Relationships to handedness and phonological abilities. *Brain and Language, 62*, 127–146.

Ross, J., Burr, D., & Morrone, C. (1996). Suppression of the magnocellular pathways during saccades. *Behavioural Brain Research, 80*, 1–8.

Rousselle, C., & Wolff, P. H. (1991). The dynamics of bimanual coordination in developmental dyslexia. *Neuropsychologia, 29*, 907–924.

Roush, W. (1995). Arguing over why Johnny can't read. *Science, 267*, 1896–1898.

Rumsey, J. M., Andreason, P., Zametkin, A. J., Aquino, T., King, A. C., Hamburger, S. D. *et al.* (1992). Failure to activate the left temporoparietal cortex in dyslexia. *Archives of Neurology, 49*, 527–534.

Rumsey, J. M., Donohue, B. C., Brady, D. R., Nace, K., Geidd, J. N., & Andreason, P. A. (1997). Magnetic resonance imaging study of planum temporale asymmetry in men with developmental dyslexia. *Archives of Neurology, 54*, 1481–1489.

Rumsey, J. M., Dorwart, R., Vermess, M., Denckla, M. B., Kruesi, M. J. P., & Rapoport, J. L. (1986). Magnetic resonance imaging of brain anatomy in severe developmental dyslexia. *Archives of Neurology, 43*, 1045–1046.

Rumsey, J. M., Horwitz, B., Donohue, B. C., Nace, K. L., Maisog, J. M., & Andreason, P. (1999). A functional lesion in developmental dyslexia: Left angular gyral blood flow predicts severity. *Brain and Language, 70*, 187–204.

Rumsey, J. M., Horwitz, B., Donohue, B. C., Nace, K., Maisog, J. M., & Andreason, P. (1997a). Phonological and orthographic components of word recognition: A PET–rCBF study. *Brain, 120*, 739–759.

Rumsey, J. M., Nace, K., Donohue, B., Wise, D., Maisog, J. M., & Andreason, P. (1997b). A positron emission tomographic study of impaired word recognition and phonological processing in dyslexic men. *Archives of Neurology, 54*, 562–573.

Rumsey, J. M., Zametkin, A. J., Andreason, P. A., Hanahan, A. P., Hamburger, S. D., Aquino, T. *et al.* (1994). Normal activation of frontotemporal language cortex in dyslexia, as measured with oxygen 15 positron emission tomography. *Archives of Neurology, 51*, 27–38.

Salmelin, R., Service, E., Kiesilä, Uutela, K., & Salonen, O. (1996). Impaired visual word processing in dyslexia revealed with magnetoencephalography. *Annals of Neurology, 40*, 157–162.

Schmahmann, J. D., & Sherman, J. D. (1998). The cerebellar cognitive affective syndrome. *Brain, 121*, 561–579.

Schulte-Körne, G., Deimel, W., Bartling, J., & Remschmidt, H. (1998). Auditory processing and dyslexia: Evidence for a specific speech processing deficit. *NeuroReport, 9*, 337–340.

Schultz, R. T., Cho, N. K., Staib, L. H., Kier, L. E., Fletcher, J. M., Shaywitz, S. E. *et al.* (1994). Brain morphology in normal and dyslexic children: The influence of age and sex. *Annals of Neurology, 35*, 732–739.

Shapleske, J., Rossell, S. L., Woodruff, P. W. R., & David, A. S. (1999). The planum temporale: A systematic, quantitative review of its structural, functional and clinical significance. *Brain Research Reviews, 29*, 26–49.

Shaywitz, S. E., Shaywitz, B. A., Pugh, K. R., Fulbright, R. K., Constable, R. T., Mencl, W. E. *et al.* (1998). Functional disruption in the organisation of the brain for reading in dyslexia. *Proceedings of the National Academy of Sciences, 95*, 2636–2641.

Skottun, B. C. (1997). Some remarks on the magnocellular deficit theory of dyslexia. *Vision Research, 37*, 965–966.

Skottun, B. C. (2000). The magnocellular deficit theory of dyslexia: The evidence from contrast sensitivity. *Vision Research, 40*, 111–127.

Slaghuis, W. L., Lovegrove, W. J., & Davidson, J. A. (1993). Visual and language processing deficits are concurrent in dyslexia. *Cortex, 29*, 601–615.

Slaghuis, W. L., Twell, A. J., & Kingston, K. R. (1996). Visual and language processing disorders are concurrent in dyslexia and continue into adulthood. *Cortex, 32*, 413–438.

Snowling, M., Bishop, D. V. M., & Stothard, S. E. (2000). Is preschool language impairment a risk factor for dyslexia in adolescence? *Journal of Child Psychology and Psychiatry, 41*, 587–600.

Spafford, C. S., Grosser, G. S., Donatelle, J. R., Squillace, S. R., & Dana, J. P. (1995). Contrast sensitivity differences between proficient and disabled readers using colored lenses. *Journal of Learning Disabilities, 28*, 240–252.

Spinelli, D., Angelelli, P., De Luca, M., Di Pace, E., Judica, A., & Zoccolotti, P. (1997). Developmental surface dyslexia is not associated with deficits in the transient visual system. *NeuroReport, 8*, 1807–1812.

Steffens, M. L., Eilers, R. E., Gross-Glenn, K., & Jallad, B. (1992). Speech perception in adult subjects with familial dyslexia. *Journal of Speech and Hearing Research, 35*, 192–200.

Stein, J. (1994). Developmental dyslexia, neural timing and hemispheric lateralisation. *International Journal of Psychophysiology, 18*, 241–249.

Stein, J. (2000). The neurobiology of reading difficulties. *Prostaglandins, Leukotrienes and Essential Fatty Acids, 63*, 109–116.

Stein, J., & Walsh, V. (1997). To see but not to read: The magnocellular theory of dyslexia. *Trends in Neurosciences, 20*, 147–152.

Stordy, J. (2000) Dark adaptation, motor skills, docosahexaenoic acid, and dyslexia. *American Journal of Clinical Nutrition, 71*, 323S–326S.

Studdert-Kennedy, M., & Mody, M. (1995). Auditory temporal perception deficits in the reading-impaired: A critical review of the evidence. *Psychonomic Bulletin and Review, 2*, 508–514.

Tallal, P., Miller, S., & Fitch, R. H. (1993). Neurobiological basis of speech: A case for the preeminence of temporal processing. In P. Tallal, A. M. Galaburda, R. R. Llinas, & C. von Euler (Eds.), *Temporal information processing in the nervous system (Annals of the New York Academy of Sciences, 682*, 27–47). New York: New York Academy of Sciences.

Talcott, J. B., Hansen, C., Willis-Owen, C., McKinnell, I. W., Richardson, A. J., & Stein, J. F. (1998). Visual magnocellular impairment in adult developmental dyslexics. *Neuro-ophthalmology, 20*, 187–201.

Tan, U., & Tan, M. (1999). Incidences of asymmetries for the palmar grasp reflex in neonates and hand preference in adults. *NeuroReport, 10*, 3253–3256.

Tan, U., & Tan, M. (2001). Testosterone and grasp-reflex differences in human neonates. *Laterality, 6*, 181–192.

Temple, C., Jeeves, M., & Villaroya, O. (1989). Ten pen men: Rhyming skills in two children with callosal agenesis. *Brain and Language, 37*, 548–564.

Temple, C., Jeeves, M., & Villaroya, O. (1990). Reading in callosal agenesis. *Brain and Language, 39*, 235–253.

Temple, C., & Ilsley, J. (1993). Phonemic discrimination in callosal agenesis. *Cortex, 29*, 341–348.

Thach, W. T. (1996). On the specific role of the cerebellum in motor learning and cognition: Clues from PET activation and lesion studies in man. *Behavioural and Brain Sciences, 19*, 411–431.

Tootell, R. B. H., Reppas, J. B., Kwong, K. K., Malach, R., Born, R. T., Brady, T. J. *et al.* (1995). Functional analysis of human MT and relayed visual cortical areas using magnetic resonance imaging. *The Journal of Neuroscience, 15*, 3215–3230.

Tzeng, O. J. L., & Wang, W. S.-Y. (1984). Search for a common neurocognitive mechanism for language and movements. *American Journal of Physiology, 246*, R904–R911.

Vanni, S., Uusitalo, M. A., Kiesila, P., & Hari, R. (1997). Visual motion activates V5 in dyslexics. *NeuroReport, 8*, 1939–1942.

Velay, J.-L., Daffaure, V., Giraud, K., & Habib, M. (2002). Interhemispheric sensorimotor integration in pointing movements: A study on dyslexic adults. *Neuropsychologia, 40*, 827–834.

Victor, J. D., Conte, M. M., Burton, L., & Nass, R. D. (1993). Visual evoked potentials in dyslexics and normals: Failure to find a difference in transient or steady-state responses. *Visual Neuroscience, 10*, 939–946.

Volpe, B., Sidtis, J., Holtzman, J., Wilson, D., & Gazzaniga, M. (1982). Cortical mechanisms involved in praxis: Observations following partial and complete section of the corpus callosum in man. *Neurology, 32*, 645–650.

Von Plessen, K., Lundervold, A., Duta, N., Heiervang, E., Klauschen, F., Smievoll, A. I. *et al.* (2002). Less developed corpus callosum in dyslexic subjects—a structural MRI study. *Neuropsychologia, 40*, 1035–1044.

Walther-Müller, P. U. (1995). Is there a deficit of early vision in dyslexia? *Perception, 24*, 919–936.

Watson, J. D. G., Myers, R., Frackowiak, R. S. J., Hajnal, J. V., Woods, R. P., Mazziotta, J. C. *et al.* (1993). Area V5 of the human brain: Evidence from a combined study using positron emission tomography and magnetic resonance imaging. *Cerebral Cortex, 3*, 79–94.

Wimmer, H., Mayringer, H., & Raberger, T. (1999). Reading and dual-task balancing: Evidence against the automatization deficit explanation of developmental dyslexia. *Journal of Learning Disabilities, 32*, 473–478.

Wolff, P. H. (1993). Impaired temporal resolution in developmental dyslexia. *Annals of the New York Academy of Sciences, 682*, 87–103.

Wolff, P. H., Cohen, C., & Drake, C. (1984). Impaired motor timing control in specific reading retardation. *Neuropsychologia, 22*, 587–600.

Werker, J. F., & Tees, R. C. (1987). Speech perception in severely disabled and average reading children. *Canadian Journal of Psychology, 41*, 48–61.

Young, A. W., & Ellis, A. W. (1981). Asymmetry of cerebral hemispheric function in normal and poor readers. *Psychological Bulletin, 89*, 183–190.

Yabuta, N. H., Sawatari, A., & Callaway, E. M. (2001). Two functional channels from primary visual cortex to dorsal visual cortical areas. *Science, 292*, 297–300.

Yap, R. L., & van der Leij, A. (1994). Testing the automization deficit hypothesis of dyslexia via a dual task paradigm. *Journal of Learning Disabilities, 27*, 660–665.

Zeki, S. (1991). Cerebral akinetopsia (visual motion blindness): A review. *Brain, 114*, 811–824.

Zeki, S., Watson, J. D. G., Lueck, C. J., Friston, K. J., Kennard, C., & Frackowiak, R. S. J. (1991). A direct demonstration of functional specialization in human visual cortex. *The Journal of Neuroscience, 11*, 641–649.

Zetzche, T., Meisenzahl, E. M., Preuss, U. W., Holder, J. J., Kathmann, N., Leinsinger, G. *et al.* (2001). In-vivo analysis of the human planum temporale (PT): Does the definition of PT borders influence the results with regard to cerebral asymmetry and correlation with handedness? *Psychiatry Research: Neuroimaging, 107*, 99–115.

Zihl, J., Von Cramon, D., & Mai, N. (1983). Selective disturbance of movement after bilateral brain damage. *Brain, 106*, 313–340.

Zihl, J., Von Cramon, D., Mai, N., & Schmid, Ch. (1991). Disturbance of movement vision after bilateral posterior brain damage: Further evidence and follow up observations. *Brain, 114*, 2235–2252.

4

The Science of Dyslexia: A Review of Contemporary Approaches*

Margaret J. Snowling

Up until the 1960s, the study of dyslexia was primarily the domain of medical specialists. Foremost amongst them was Samuel T. Orton, whose insights into the condition that he described as "strephosymbolia" presaged much present thinking. Orton (1925) considered dyslexia to be a brain-based disorder with a hereditary component, with affected family members often reporting associated speech or language difficulties. Orton considered that the problems of dyslexia were amenable to intervention and, together with his colleagues Gillingham and Stillman, he pioneered a highly structured, multisensory approach to the remediation of reading and spelling difficulties that is at the root of most contemporary approaches (Snowling, 1996). As we shall see, although details of Orton's theory were incorrect, his characterization of the problem was broadly similar to that which is held today.

A SHORT HISTORY OF THE CONCEPT OF DYSLEXIA

Medical approaches to dyslexia, such as that exemplified by the work of pioneers such as Orton and Macdonald Critchley (Critchley, 1970) fell from favor mostly

*A more detailed version of the argument in this chapter can be found in Snowling (2000).

Margaret J. Snowling, Department of Psychology, University of York, Heslington, York Y01 5DD, UK.

The Study of Dyslexia, edited by Turner and Rack.
Kluwer Academic Publishers, New York, 2004.

because they were difficult to put into practice. The dyslexic child was described as having severe difficulty in learning to read and write, despite adequate IQ, normal sensory function, and adequate opportunity both at home and at school. It is perhaps worth taking time to consider these criteria. What is adequate IQ? What is adequate opportunity? Why should a sensory impairment rule out the possibility of dyslexia? It will be clear that many of the terms in this kind of a definition are difficult to operationalize. It is a "definition by exclusion" and does not list any of the positive signs of dyslexia.

In fairness, the medical model of dyslexia also carried with it a list of behavioral symptoms, such as the tendency to reverse letters and numbers, directional confusion, clumsiness, and delays and difficulties with language. However, it was never made clear which were the critical symptoms, how many of them needed to be observed or at what degree of severity. While checklists for dyslexia have their place, such instruments do not fare well in discriminating dyslexia from other types of learning disorder. In fact, they can be most misleading to parents whose knowledge of *normal* development may be poor.

A more scientific approach to dyslexia emerged in the late 1960s when one of the main issues of debate was whether "dyslexia" was different from plain poor reading. Studies of whole child populations, notably the epidemiological studies of Rutter and his colleagues, provided data about what differentiated children with *specific* reading problems (dyslexia) from those who were slow in reading but for whom reading was in line with general cognitive ability (Rutter & Yule, 1975). The results of these studies were salutary for proponents of dyslexia. In fact, there were relatively few differences in etiology between children with specific reading difficulty and the group they described as backward readers (often referred to in the US literature as "*garden-variety poor readers*"). The group differences that were found included a higher preponderance of males among children with specific reading difficulties, and more specific delays and difficulties with speech and language development. On the other side of the coin, the generally backward group showed more hard signs of brain damage, for example, cerebral palsy and epilepsy.

Importantly, the two groups differed in the progress they had made at a 2-year follow-up. Contrary to what might have been expected on the basis of their IQ, the children with specific reading difficulties (who had higher IQ) made less progress in reading than the generally backward readers. This finding suggested that their problems were intransigent, perhaps because of some rather specific cognitive deficit, at the time unspecified. It is perhaps worth noting here, that this differential progress rate has not been replicated in more recent studies (Shaywitz *et al.*, 1992), perhaps because advances in knowledge have led to better focused remedial approaches.

Following on from these large-scale studies, the use of the term "dyslexia" became something of a taboo in UK schools. Rather, children were described as having specific reading difficulties (SRD) or specific learning disability (SpLD) if there was a discrepancy between their expected attainment in reading, as predicted by their age and IQ, and their actual reading attainment. The use of IQ as part of the definition has attracted a lot of criticism. First, IQ is not strongly related to reading. Indeed, many children with low IQ can read perfectly well even though they may encounter reading comprehension difficulties. Second, and perhaps most important, there is suggestive evidence that verbal IQ may decline over time among poor readers. To some extent, this decline might be as a consequence of limited access to knowledge in books. However, there are other reasons too. Difficulties in retrieving verbal information and problems of verbal short-term memory can influence test performance, as can low self-esteem. Whatever the reason, the discrepancy definition of dyslexia may disadvantage those children with the most severe problems whose apparently low verbal IQ may obscure the "specificity" of the reading problem.

Another problem with the discrepancy definition of dyslexia is that it does not work well for younger children, who may as yet not have failed sufficiently to fulfill its criteria! Indeed there is a considerable lack of stability at the individual level in children who obtain the classification during their early school years. Moreover, the definition is silent with regard to at-risk signs of dyslexia, and to how to diagnose dyslexia in young people who may have overcome basic literacy difficulties. This limitation relates to the same problem that was mentioned above, the lack of positive diagnostic criteria for dyslexia.

WHAT IS THE COGNITIVE DEFICIT IN DYSLEXIA?

At about the same time as the first epidemiological studies were being conducted, cognitive psychologists began comparing groups of normal and dyslexic readers in a range of experimental paradigms. These researchers pursued the then popular idea that dyslexia was a perceptual deficit and studies investigated visual perception, visual memory, cross-modal transfer between visual and verbal codes and perceptual learning, and other skills in relation to reading ability. These early studies were fraught with methodological problems as Vellutino's (1979) ground-breaking review makes explicit. The nub of his argument was that, when it is appropriate, children recruit verbal codes to support perceptual performance. Imagine a visual memory task in which you have to recall a string of letters. Would you do this visually or verbally? In some elegant experiments, Vellutino and his colleagues showed that when it was not possible to use verbal codes (e.g., if you had to remember strings of unfamiliar symbols), dyslexics performed as well as normal readers. However, when it was helpful to recode visual stimuli verbally, they had problems. Thus, Vellutino (1979) reinterpreted a large body of evidence suggesting there are perceptual problems in dyslexia as consistent with a *verbal deficit hypothesis*.

PHONOLOGICAL DEFICITS IN DYSLEXIA

Since the 1980s, the most widely accepted view of dyslexia has been that it can be considered part of the continuum of language disorders. However, there has been a gradual shift from the verbal deficit hypothesis to a more specific theory, that dyslexia is characterized by *phonological processing difficulties*. What is the difference between these views?

The language processing system that the child brings to the task of learning to read can be thought of as comprising different subsystems. While the phonological processing system (phonology) is concerned with how speech sounds are perceived, coded, and produced, the semantic system is concerned with the meanings of words. The other two subsystems have attracted much less interest in dyslexia research. These are the syntactic processing system concerned with the grammatical structures of sentences, and the pragmatic system, concerned with the use of language for communication (Bishop, 1997). Typically, dyslexic children have been found to have difficulties that primarily affect the phonological domain, though as we shall see later, this may turn out to be an overly narrow view.

The most consistently reported phonological difficulties found in dyslexia are limitations of verbal short-term memory and, more directly related to their reading problems, problems with phonological awareness. It is interesting to note here, however, that the persistent difficulties with phonological awareness associated with dyslexia are not a universal phenomenon. Rather, they appear to be specific to children learning to read in irregular or "opaque" orthographies, such as English. In more regular or transparent orthographies where the relationships between spellings and their sounds are consistent (e.g., German, Italian, Spanish, and Greek), children learn to decode quickly and at the same time they rapidly acquire awareness of the phonemic structure of spoken words (Cossu, 1999). It follows that, in these languages, deficits in phonological awareness are not a good marker of dyslexia. It is indeed, the other impairments of phonological processing, such as rapid naming that are the landmark of dyslexia in these writing systems (e.g., Landerl & Wimmer, 2000).

There is also evidence that dyslexic children have trouble with long-term verbal learning. This problem may account for many classroom difficulties, including problems memorizing the days of the week or the months of the year, mastering multiplication tables, and learning a foreign language. In a similar vein, this problem may be responsible for the poor vocabulary development often observed in dyslexic children.

Dyslexic children also have difficulties with the retrieval of phonological information from long-term memory. Word finding difficulties are often seen clinically and experimental studies using both rapid naming and object naming tasks report deficiencies. It is possible that these problems stem from more basic deficits in speech perception, speech production, or temporal processing (Klein & Farmer, 1995). The strength of the evidence pointing to the phonological deficits associated with dyslexia led Stanovich and his colleagues to propose that dyslexia should be defined as a core phonological deficit. Importantly, within the *phonological core-variable difference* model of dyslexia (Stanovich & Siegel, 1994), poor phonology is related to poor reading performance, irrespective of IQ.

PHONOLOGICAL REPRESENTATIONS, LEARNING TO READ, AND DYSLEXIA

Although the role of visual deficits in dyslexia continues to be debated (Stein & Talcott, 1999; Willows, Kruk, & Corcos, 1993), the best candidate for the cause of dyslexia is an underlying phonological deficit. A useful way in which to think about this is that dyslexic individuals come to the task of learning to read with poorly specified phonological representations—the way in which their brain codes phonology is less efficient than that of normally developing children (Snowling & Hulme, 1994). As we have seen, this problem at the level of phonological representation causes a range of typical symptoms which can be observed in the overt behavior of dyslexic individuals. However, it is reasonable to ask why a deficit in *spoken* language should affect the acquisition of *written* language.

Studies of normal reading development offer a framework for considering the role of phonological representations in learning to read, and for understanding the problems of dyslexia. At the basic level, learning to read requires the child to establish a set of mappings between the letters (graphemes) of printed words and the speech sounds (phonemes) of spoken words. These mappings between orthography and phonology allow novel words to be decoded and provide a foundation for the acquisition of later and more automatic reading skills. In English, they also provide a scaffold for learning multi-letter (e.g., *ough*, *igh*), morphemic (-tion, cian), and inconsistent (-ea) spelling–sound correspondences. Indeed, the early developing ability of the child to "invent" spellings that are primitive phonetic transcriptions of spoken words (e.g., LEVNT for elephant) is one of the best predictors of later reading and spelling success. More broadly, there are strong relationships between phonological skills and reading ability throughout development and into adulthood, where the phonological deficits of dyslexics persist (Bruck, 1992; Snowling, Nation, Moxham, Gallagher, & Frith, 1997).

However, there is more to reading than decoding and indeed, more than one way in which letters strings can be mapped to sound. A subsystem of reading that involves the creation of mappings between orthography, semantics, and phonology is important in English for reading exception words (e.g., sign, guest) that do not conform to letter–sound rules (Plaut, McClelland, Seidenberg, & Patterson, 1996). This system may also play a role in the development of reading fluency as it is primed by the influence of semantic and syntactic context (Nation & Snowling, 1998). Children with poor language resources have specific difficulty reading exception words that cannot be dealt with adequately by the orthography–phonology system because of their semantic processing weaknesses.

Within this model of reading development (Figure 1), deficits at the level of phonological representation constrain the reading development of dyslexic children (Snowling, Hulme, & Nation, 1997a). A consequence is that although dyslexic children can learn to read words (possibly with heavy reliance on context), they have difficulty generalizing this knowledge. For dyslexic readers of English, a notable consequence is poor nonword reading (Rack, Snowling, & Olson, 1992).

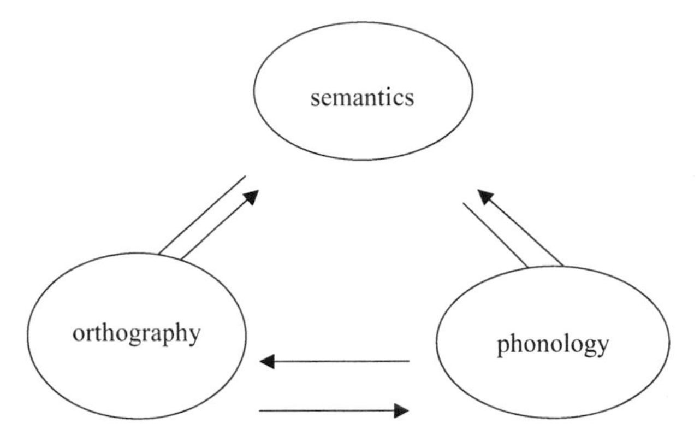

Figure 1. Interactive framework of reading development (after Seidenberg & McClelland, 1989).

In contrast, the semantic skills of dyslexic readers are, by definition, within the normal range, and these can be used to facilitate the development of exception word reading in some cases.

In short, learning to read is an interactive process to which the child brings all of his or her linguistic resources. However, it is phonological processing that is most strongly related to the development of reading, and the source of most dyslexic problems in reading and spelling. The phonological representations hypothesis, therefore, provides a parsimonious explanation of the disparate symptoms of dyslexia that persist through school to adulthood. It also makes contact with theories of normal reading development and with scientific studies of intervention. Here the consensus view is that interventions that promote phonological awareness in the context of a highly structured approach to the teaching of reading have a positive effect both in preventing reading failure and ameliorating dyslexic reading difficulties (Troia, 1999). There is also, as we shall see below, biological evidence in support of the theory.

BIOLOGICAL EVIDENCE IN SUPPORT OF THE PHONOLOGICAL DEFICIT HYPOTHESIS

It has been known for many years that poor reading tends to run in families and there is now conclusive evidence that dyslexia is heritable. Gene markers have been identified on chromosomes 6 and 15 but we are still a long way from understanding the precise genetic mechanisms that are involved. What we do know is there is as much as a 50% probability of a boy becoming dyslexic if his father is dyslexic (about 40% if his mother is affected), and a somewhat lower probability of a girl developing dyslexia. What is inherited is not of course reading disability *per se*, but aspects of language processing. Results of large-scale twin studies suggest there is heritability of phonological ("phonic") aspects of reading and that phonological awareness shares heritable variance with this (Olson, Forsberg, & Wise, 1994).

Studies of dyslexic readers using brain imaging techniques also supply a piece in the jigsaw. In one such study, we investigated differences in brain function between dyslexic and normal readers while they performed two phonological processing tasks (Paulesu *et al.*, 1996). This study involved five young adults with a well-documented history of dyslexia; all of these dyslexics had overcome their reading difficulties but they had residual problems with phonological awareness. Under positron emission tomography (PET scan), they completed two sets of parallel tasks. The phonological tasks were a rhyme judgment and a verbal short-term memory task; the visual tasks were visual similarity judgment and visual short-term memory.

Although these dyslexic adults had compensated well for their developmental difficulties, they showed different patterns of left hemisphere brain activation from controls during performance on the phonological processing tasks. The brain regions associated with reduced activity in the dyslexics were those involved in the transmission of language and plausibly, allow the translation between the perception and production of speech. It is therefore possible to speculate that this area may be the "seat" of the problems viewed at the cognitive level, as a difficulty in setting up phonological representations. Interestingly, recent studies that have measured the gray matter in the brain in these regions, suggest a reduction in its volume among dyslexic readers.

ISSUES FOR PHONOLOGICAL REPRESENTATIONS THEORY OF DYSLEXIA

The concept of dyslexia as a deficiency in the way in which the brain codes phonology is appealing but the idea of "fuzzy" phonological representations is vague. What exactly is meant by phonological representation? How are phonological representations created? Does each word have one (a localist account) or are these really distributed patterns of activation across a set of yet to be specified processing units? We do not have answers to these questions. However, one influential theory proposes that problems in basic auditory processing (of rapidly presented sounds) cause deficits in speech perception that lead on to problems of phonological awareness and reading development (Tallal, Miller, Jenkins, & Merzenich, 1997).

Rapid Auditory Processing Deficits in Dyslexia

The rapid auditory processing (RAP) theory is very persuasive; moreover, interventions that train RAP are advocated as a "cure" for language learning impaired children including dyslexics. But where is the evidence? Tallal's original studies which are the foundation of the RAP theory were with language impaired children, not with dyslexics. Since then, there have been failures to replicate the findings with reading disabled children (Mody, Studdert-Kennedy, & Brady, 1997; McArthur, Hogben, Edwards, Heath, & Mengler, 2000) and to confirm predictions of the theory (Nittrouer, 1999). Studies from our own laboratory involving dyslexic children and dyslexic adults, have typically found only a small subset of dyslexic readers who display low-level auditory processing impairments. Moreover, it is important to note that these participants have *not* been those with the more severe phonological deficits. Thus, the relationship between RAP and phonological processing remains an open question. It seems to us, that dyslexic children may have difficulty with the paradigms used to test auditory processing for a number of reasons; some have attentional difficulties and these are demanding paradigms; some have verbal labeling difficulties and verbal codes may be used to mediate perceptual judgments, as Vellutino proposed. Finally, many of these paradigms carry working memory demands (Marshall, 2000). This field of research highlights a very important issue for the field of dyslexia research, that of cause and effect. What might at first seem a cause of phonological processing impairment (RAP) may turn out to be its consequence.

Individual Differences in Dyslexia

A more serious issue for the phonological representation view is that of individual differences. The phonological deficit theory has no difficulty explaining the problems of a child with poor word attack skills, who cannot read nonwords and whose spelling is dysphonetic (Snowling, Stackhouse, & Rack, 1986). However, there are also dyslexic children who appear to have mastered alphabetic skills. Such children have been referred to as developmental "surface" dyslexics. The classic characteristic of these children is that, in single-word reading, they rely heavily upon a phonological strategy. Thus, they tend to pronounce irregular words as though they were regular (e.g., glove → gloave; island → izland), they have particular difficulty distinguishing between homophones like *pear–pair* and *leek–leak*, and their spelling is usually phonetic (e.g., Biscuit → BISKT; pharmacist → FARMASIST).

While evidence in favor of distinct subtypes is lacking (Seymour, 1986), most systematic studies of individual differences among dyslexics have revealed variations in their reading skills (Castles & Coltheart, 1993). One way of characterizing a child's reading strategies is to assess how well they can decode words they have not seen before (e.g., using a nonword reading test) and to compare this to how well they recognize words that they cannot "sound-out," such as irregular or exception words that do not conform to English spelling rules (e.g., *pint*, *yacht*). A number of studies have now shown that dyslexic children who have relatively more difficulty in reading nonwords than exception words (phonological dyslexia) perform significantly less well than younger, reading age-matched controls on tests of phonological awareness. In contrast, dyslexic children who have more difficulty with exception words than nonwords (classified as surface dyslexic) perform at a similar level to controls on these tests.

More generally, however, it does not seem useful to classify dyslexic children into subtypes because all taxonomies leave a substantial number of children unclassified (Griffiths, 1999). Rather, in studies from our own laboratory, we have found that individual differences in phonological processing, as measured by performance on tests of phonological awareness and phonological memory, predict individual differences in nonword reading, even when reading age has already been taken into account. In essence, the more severe a child's phonological deficit, the greater is its impairment in nonword reading. In contrast, we found that variation in exception word reading was tied to reading experience, reflecting the fact that print exposure is required to learn about the inconsistencies of the English orthographic system. As we saw earlier, exception word reading builds on a foundation of grapheme–phoneme mappings, but it is also supported by semantics. To this extent, exception word reading may develop independently of decoding skill in some dyslexic children, forging a pattern of "phonological dyslexia" at the behavioral level.

The Relationship between Language and Phonology

Surprisingly, given the strength of the phonological representational account of dyslexia, rather little is known about the relationship between phonological skills and other language abilities. While it is clear that semantic and phonological skills can be selectively impaired in developmental reading disorders (Nation, Marshall, & Snowling, 2001), how the phonological deficits that are causally related to reading problems in dyslexia relate to these children's oral language development is less well understood.

We have been exploring the early language precursors of dyslexia in a longitudinal study of children at genetic risk by virtue of having a first degree affected relative, from 4 to 8 years (Gallagher, Frith, & Snowling, 2000). Our preliminary findings indicate that more than 60% of the high-risk group develop literacy problems. These children showed a pattern of oral speech and language delay in the preschool years and poor phonological awareness at age 6 with concomitant difficulty in literacy tasks that require the use of letter–sound relationships. In contrast, the high-risk children who went on to be normal readers were no different to a socio-economically matched control group in their early language and metaphonological development. A surprising finding, however, was that this high-risk group, who turned out to be normal readers, was as impaired as the dyslexic children in their ability to use letter–sound knowledge in reading and spelling at 6 years. We must, therefore, conclude that they can compensate for this "grapheme–phoneme" deficit, perhaps by drawing on their strong general language resources, to become normal readers by the age of 8 years.

The findings of our at-risk study are in line with a biological explanation of dyslexia. The rate of dyslexia among the high-risk children was directly proportional to the severity (and persistence) of their parents' literacy problems. The first behavioral manifestation that we detected among 4-year-olds was poor letter knowledge. In turn, letter knowledge and speech development drove phonological awareness at 6, an important predictor of later word level reading and decoding skill. The critical role of grapheme–phoneme knowledge at 6–8 years literacy development was in line with studies pointing to nonword reading deficits in dyslexia. Importantly, however, children with a smaller "dose" of dyslexia were able to compensate for impairments of grapheme–phoneme knowledge. It remains an issue whether these unaffected children will develop literacy problems at a later stage in their development, for instance, in developing reading fluency. At least some of them might be among the main "hidden" dyslexics who are first diagnosed as university students who have difficulty coping with the increased literacy demands of their advanced programs of study.

WHY DOESN'T EVERYONE AGREE?

You may at this point be wondering, why there is disagreement among dyslexia specialists. First, it may be that disagreement is more apparent than real. Some theories that I have not dealt with here, such as the magnocellular deficit or the cerebellar deficit theories do not reject the evidence for the phonological processing problems of dyslexia. Rather, they seek a biological cause for these deficits. It falls to these theorists to demonstrate that their theories explain both a necessary and a sufficient cause of dyslexia. Arguably, at the present time, the phonological deficit hypothesis provides the only clear-cut evidence of causal links with reading failure.

Second, it may be that different researchers are working with rather different groups of children, and selection biases may be operating to distort results. Researchers working in close liaison with speech and language therapists are far more likely to find children with broader language impairments in their samples than those working with optometrists, who may be referred a sizeable proportion of children with associated visual defects. It falls to researchers to be clear in their descriptions of how research samples were selected. In turn, consumers of research must be clear as to how far it is possible to generalize from highly selected, discrepancy-defined populations of dyslexics to those with less specific reading difficulties.

The Issue of Comorbidity

Some of the apparent differences among researchers of different orientations hinges on the issue of what is known as *comorbidity*. Comorbidity refers to the fact that, given any developmental disorder, there is a high probability that it will co-occur with at least one other disorder. Commonly co-occurring with dyslexia are difficulties with coordination (dyspraxia) or with attention control (ADHD). The cause of this comorbidity may be the sharing of brain mechanisms involved in the two disorders or because they share similar risk factors (e.g., family adversity).

In cases of children with comorbid disorders, it is easy to mistake a behavioral symptom of one disorder for that of the other. Many dyslexic children are clumsy, but not all are by any means. It is important therefore, not to build a theory of dyslexia on the assumption that motor impairments play a causal role. Similarly, one of the key cognitive features of ADHD is a difficulty in inhibiting automatic responses, an aspect of executive function. However, there is evidence to suggest that children with comorbid ADHD and dyslexia show phonological deficits rather than executive deficits (Pennington, Grossier, & Welsh, 1993). The implication of this finding is that the ADHD symptoms have different roots in comorbid cases to those seen in pure cases. Indeed, it is possible that they arise as a secondary consequence of reading problems.

Finally, and of considerable theoretical importance, the behavioral profile of dyslexic children may change with age. Studies of the early language development of dyslexic children point to language impairments beyond the phonological module, encompassing slow vocabulary development and grammatical delays (Scarborough, 1990). In the same way, children who have specific difficulties in reading comprehension may develop decoding problems at a later stage in their development because their language skills are not sufficiently strong to bootstrap word recognition (Snowling, Bishop, & Stothard, 2000).

WHERE DO WE GO FROM HERE?

The field of dyslexia research has burgeoned in the last 20 years, and the problem is now well understood. Equally, public awareness of dyslexia has increased to the benefit of dyslexic children. But there remain some challenges. The first of these is to clarify how dyslexia manifests itself in different languages (Goulandris, 2003). In turn, this research must build on a solid foundation of knowledge about the normal development of reading in readers of different orthographies (Harris & Hatano, 1999). The cross-linguistic study of dyslexia is really just opening up and there are many questions to be answered. Already, there is good evidence that the language in which a child learns has a strong effect on the development of their decoding and metaphonological skills. It follows that more sensitive tests will be needed to detect dyslexia in readers of transparent (e.g., Dutch, Spanish) than of opaque orthographies, such as English and Danish.

A second issue to be grappled with is that of dyslexic children who do not respond to teaching techniques that have been demonstrated to be otherwise effective. This chapter has not focused on intervention studies. However, there is a reasonable consensus that a child's phonological skill at the start of intervention is one of the best predictors of their progress, while their IQ is not. But, even in the most effective interventions, there is a proportion of children who resist treatment (Torgesen *et al.*, 1999). Much more needs to be known about why this is so. Generally, it is the children with the most severe phonological deficits. Do these children have co-occurring speech–language difficulties? Or ADHD? Perhaps some suffer from some kind of psychological problem, such as low self-esteem, or are they subject to social and family adversity? We simply don't have the answers at the present time. A related issue is what accounts for the maintenance of treatment gains when specialist teaching stops. A critical factor seems to be levels of print exposure—but how do you encourage a dyslexic who dislikes reading to read? Such questions turn on the talent of teachers who can assess the needs of their dyslexic students with a high degree of precision.

What are the implications of our discussion for research and practice? Many of these are dealt with in other chapters. At the risk of being prescriptive, it seems useful at this point to draw to a close with a list of ways to avoid some obvious pitfalls.

1. Don't assume every child that is underachieving to be dyslexic. Now that the cognitive deficits in dyslexia are well understood, research should turn to clarify other causes of underachievement, for instance, those grouped under the umbrella term "nonverbal learning disabilities."
2. Take care to assess the severity of a child's phonological difficulty and to hone teaching carefully.
3. Make sure to take account of the amount of practice children are having in reading. Low levels of print exposure will have disastrous effects on a child's ability to learn to read exception words that cannot be decoded.
4. Beware of comorbidity! If the dyslexic children you are working with have other problems, these may cloud your data or your therapy.
5. Don't assume that every difficulty experienced by a dyslexic child is the consequence of dyslexia; antisocial behavior is more likely a consequence of dyslexia + ADHD rather than dyslexia *per se*.
6. Don't assume that every child will improve with a phonological approach. If they do not respond, or indeed if they have plateaued, ask yourself about their language development more broadly. Maybe they could benefit from direct help with vocabulary development, expressive language, or comprehension strategies. Alternatively, maybe they need some classroom survival strategies.

In short, it is important to take a scientific stance to the field of dyslexia. There is nothing bad in allowing your research and practice to be theory-led. You must, however, be willing to accept the evidence when theories do not stand up to empirical test.

REFERENCES

Bishop, D. V. M. (1997). *Uncommon understanding*. Hove: Psychology Press.

Bruck, M. (1992). Persistence of dyslexics' phonological awareness deficits. *Development Psychology, 28*, 874–886.

Castles, A., & Coltheart, M. (1993). Varieties of developmental dyslexia. *Cognition, 47*, 149–180.

Cossu, G. (1999). Biological constraints on literacy acquisition. *Reading and Writing, 11*(3), 213–237.

Critchley, M. (1970). *The dyslexic child*. London: Heinemann Medical Books.

Gallagher, A., Frith, U., & Snowling, M. J. (2000). Precursors of literacy-delay among children at genetic risk of dyslexia. *Journal of Child Psychology and Psychiatry, 41*(2), 203–213.

Goulandris, N. (Ed.). (2003). *Dyslexia in different languages: Cross-linguistic comparisons*. London: Whurr.

Griffiths, Y. M. (1999). *Individual differences in developmental dyslexia*. York: University of York.

Harris, M., & Hatano, G. (Eds.). (1999). *Learning to read and write: A cross-linguistic perspective*. Cambridge, UK: Cambridge University Press.

Klein, R. M., & Farmer, M. E. (1995). Dyslexia and a temporal processing deficit: A reply to the commentaries. *Psychonomic Bulletin and Review, 2*(4), 515–526.

Landerl, K., & Wimmer, H. (2000). Deficits in phoneme segmentation are not the core problem of dyslexia: Evidence from German and English children. *Applied Psycholinguistics, 21*, 243–262.

Marshall, C. M. (2000). *The relationship between rapid auditory processing and phonological skill in reading development and dyslexia*. Unpublished DPhil, York, York.

McArthur, G. M., Hogben, J. H., Edwards, V. T., Heath, S. M., & Mengler, E. D. (2000). On the "specifics" of specific reading disability and specific language impairment. *Journal of Child Psychology and Psychiatry, 41*(7), 869–874.

Mody, M., Studdert-Kennedy, M., & Brady, S. (1997). Speech perception deficits in poor readers: Auditory processing or phonological coding? *Journal of Experimental Child Psychology, 58*, 112–123.

Nation, K., Marshall, C., & Snowling, M. J. (2001). Phonological and semantic contributions to children's picture naming skills: Evidence from children with developmental reading disorders. *Language and Cognitive Processes, 16*, 241–259.

Nation, K., & Snowling, M. J. (1998). Semantic processing and the development of word recognition skills: Evidence from children with reading comprehension difficulties. *Journal of Memory and Language, 39*, 85–101.

Nittrouer, S. (1999). Do temporal processing deficits cause phonological processing problems? *Journal of Speech, Language and Hearing Research, 42*, 925–942.

Olson, R. K., Forsberg, H., & Wise, B. (1994). Genes, environment, and the development of orthographic skills. In V. W. Berninger (Ed.), *The varieties of orthographic knowledge I: Theoretical and development issues* (pp. 27–71). Dordrecht, The Netherlands: Kluwer Academic.

Orton, S. T. (1925). "Word blindness" in schoolchildren. *Archives of Neurology and Psychiatry, 14*, 581–615.

Paulesu, E., Frith, U., Snowling, M., Gallagher, A., Morton, J., Frackowiak, F. S. J. *et al.* (1996). Is developmental dyslexia a disconnection syndrome? Evidence from PET scanning. *Brain, 119*, 143–157.

Pennington, B. F., Grossier, D., & Welsh, M. C. (1993). Contrasting cognitive defects in attention deficit hyperactivity disorder vs. reading disability. *Developmental Psychology, 29*, 511–523.

Plaut, D. C., McClelland, J. L., Seidenberg, M. S., & Patterson, K. (1996). Understanding normal and impaired word reading: Computational principles in quasi-regular domains. *Psychological Review, 103*, 56–115.

Rack, J. P., Snowling, M. J., & Olson, R. K. (1992). The nonword reading deficit in developmental dyslexia: A review. *Reading Research Quarterly, 27*, 29–53.

Rutter, M., & Yule, W. (1975). The concept of specific reading retardation. *Journal of Child Psychology and Psychiatry, 16*, 181–197.

Scarborough, H. S. (1990). Very early language deficits in dyslexic children. *Child Development, 61*, 1728–1743.

Seidenberg, M. S., & McClelland, J. (1989). A distributed, developmental model of word recognition. *Psychological Review, 96*, 523–568.

Seymour, P. H. K. (1986). *A cognitive analysis of dyslexia*. London: Routledge & Kegan Paul.

Shaywitz, B. A., Fletcher, J. M., Holahan, J. M., & Shaywitz, S. E. (1992). Discrepancy compared to low achievement definitions of reading disability: Results from the Connecticut longitudinal study. *Journal of Learning Disabilities, 25*(10), 639–648.

Snowling, M., Bishop, D. V. M., & Stothard, S. E. (2000). Is pre-school language impairment a risk factor for dyslexia in adolescence? *Journal of Child Psychology and Psychiatry, 41*(5), 587–600.

Snowling, M. J. (1996). Annotation: Contemporary approaches to the teaching of reading. *Journal of Child Psychology and Psychiatry, 37*(2), 139–148.

Snowling, M. J. (2000). *Dyslexia*. Oxford: Blackwell.

Snowling, M. J., & Hulme, C. (1994). The development of phonological skills. *Philosophical Transactions of the Royal Society B, 346*, 21–28.

Snowling, M. J., Hulme, C., & Nation, K. (1997a). A connectionist perspective on the development of reading skills in children. *Trends in Cognitive Science, 1*, 88–91.

Snowling, M. J., Nation, K., Moxham, P., Gallagher, A., & Frith, U. (1997b). Phonological processing deficits in dyslexic students: A preliminary account. *Journal of Research in Reading, 20*, 31–34.

Snowling, M. J., Stackhouse, J., & Rack, J. (1986). Phonological dyslexia and dysgraphia: A developmental analysis. *Cognitive Neuropsychology, 3*, 309–339.

Stanovich, K. E., & Siegel, L. S. (1994). The phenotypic performance profile or reading-disabled children: A regression-based test of the phonological-core variable-difference model. *Journal of Educational Psychology, 86*, 24–53.

Stein, J., & Talcott, J. (1999). Impaired neuronal timing in developmental dyslexia—the magnocellular hypothesis. *Dyslexia, 5*, 59–77.

Tallal, P., Miller, S. L., Jenkins, W. M., & Merzenich, M. M. (1997). The role of temporal processing in developmental language-based learning disorders: Research and clinical implications. In B. A. Blachman (Ed.), *Foundations of reading acquisition and dyslexia* (pp. 49–66). Hillsdale, NJ: Laurence Erlbaum Associates.

Torgesen, J. K., Wagner, R. K., Rashotte, C. A., Rose, E., Lindamood, P., Conway, T. *et al.* (1999). Preventing reading failure in young children with phonological processing disabilities: Group and individual responses to instruction. *Journal of Educational Psychology, 91*, 579–593.

Troia, G. A. (1999). Phonological awareness intervention research: A critical review of the experimental methodology. *Reading Research Quarterly, 34*(1), 28–52.

Vellutino, F. R. (1979). *Dyslexia: Research and theory*. Cambridge, MA: MIT Press.

Willows, D. M., Kruk, R. S., & Corcos, E. (Eds.). (1993). *Visual processes in reading and reading disabilities*. Hillsdale, NJ: Lawrence Erlbaum Associates.

5

Phonological Skills, Learning to Read, and Dyslexia

Valerie Muter

INTRODUCTION

During the last 20 years, extensive research has been conducted into individual differences in children's phonological abilities, that is, their sensitivity to the speech sound structure of words. The findings have consistently converged on the conclusion that these are strongly causally related to the normal acquisition of beginning reading skill. There is also considerable evidence that children whose phonological abilities are deficient are likely to be delayed or deviant in their reading development. In response to the overwhelming evidence that phonological processing plays such a central role in the reading acquisition of normal and disabled readers, there has been a strong impetus to develop both assessment measures of phonological skill and also intervention procedures that specifically train up phonological awareness.

This chapter will examine the contribution of phonological processing skills to learning to read, emphasizing at the same time the connection between phonological deficiencies and dyslexia. Findings from research into the phonological abilities of normal children will be discussed and then related to what we know about patterns of phonological deficiency in children whose reading is delayed or deviant. The interaction of phonological processes with other reading-related skills, such as letter knowledge acquisition, verbal memory, and the ability to use

Valerie Muter, Department of Psychology, University of York, Heslington, York Y01 5DD, UK.

The Study of Dyslexia, edited by Turner and Rack.
Kluwer Academic Publishers, New York, 2004.

syntactic or semantic cues will also be considered. Finally, from a practitioner perspective, the measurement, assessment, and teaching of phonological skills will be critically evaluated.

PHONOLOGICAL AWARENESS, ITS NATURE, AND MEASUREMENT

It is possible to measure phonological awareness, that is, the ability to identify, segment, and manipulate speech segments of words in a variety of ways. Adams (1990) divides phonological awareness tasks that successfully predict reading skills into four main types. First are syllable and phoneme segmentation tasks in which the child taps, counts out, or identifies the constituent syllables and/or phonemes within words, for example, for the word "cat," the child taps three times to indicate the three constituent phonemes within the word (Liberman, Shankweiler, Fischer, & Carter, 1974). Second are phoneme manipulation tasks which require the child to delete, add, substitute, or transpose phonemes within words, for example, in a consonant deletion task, "cat" without the "c" says "at" (Bruce, 1964). Third come sound blending tasks in which the examiner provides the phonemes of a word and the child is asked to put them together, for example, "c-a-t" blends to yield "cat" (Perfetti, Beck, Bell, & Hughes, 1987). Finally, there are rhyming tasks that include knowledge of nursery rhymes (MacLean, Bryant, & Bradley, 1987), and the identification of the "odd word out" (non-rhyming word) in a sequence of three or four words as in the sound categorization task of Bradley and Bryant (1983), for example, the odd word out in the series "cat, pat, fan" is "fan." Some phonological awareness tasks are clearly easier than others and may be demonstrated in children as young as 2 and 3 years, for example, syllable blending, syllable segmentation, and some aspects of rhyming skill. Other skills do not emerge until later in development, and may depend on exposure to print and even explicit reading instruction, for example, phoneme segmentation and manipulation tasks. This would seem to suggest that there is a two-way interactive process between phonological skills and learning to read. Indeed, there is clear evidence that reading plays a reciprocal role in promoting more advanced phonological awareness (Cataldo & Ellis, 1988; Perfetti *et al.*, 1987).

While some phonological skills are available at a rudimentary level to children as young as 2 and 3 years of age, they are not necessarily stable abilities at that age. However, as demonstrated in a study of children from 2 to 5 years (Lonigan, Burgess, Anthony, & Baker, 1998), phonological sensitivity becomes increasingly stable during the preschool period. In addition, a number of long-term longitudinal studies have shown that phonological processing abilities are remarkably stable during the elementary school years (Muter & Snowling, 1998; Wagner, Torgesen, & Rashotte, 1994; Wagner *et al.*, 1997). In other words, level of phonemic skill at an early age is significantly predictive of later phonemic ability. As Wagner and his associates conclude, phonological processing abilities should be viewed as stable and coherent individual difference variables akin to other cognitive abilities, as opposed to more transitory indices of reading-related knowledge that might vary from year to year. This observed stability is essential for any variable that might be used to assess and identify children who have significant and, if untreated, persistent reading problems.

While it is well accepted that phonological skills are causally related to the development of reading, the nature of these skills and their precise bearing on reading requires elucidation, and is indeed an area of some controversy. How many sorts of phonological skills are there, and if there are different sorts, which play the most important role in the beginning stages of learning to read? Do different phonological skills influence reading at different stages of development? In general, early correlational/factor analytic studies of phonological awareness suggested that a single factor accounted for individual variance in phonological ability (Stanovich, Cunningham, & Cramer, 1984; Wagner & Torgesen, 1987). More recent studies have indicated that it may be possible to derive multiple, admittedly intercorrelated, factors within the phonological domain. Yopp (1988) uncovered two factors from a principal components analysis of 10 phonological awareness tests given to 96 kindergarten children. The factors were highly correlated and appeared to reflect two levels of difficulty rather than two qualitatively different kinds of skill—simple or implicit phonemic awareness required only one cognitive operation (e.g., blending), while complex or explicit phonemic awareness involved more than one cognitive operation and placed a heavier burden on memory (e.g., phoneme deletion). Other studies have claimed to identify differing types of phonological ability. For instance, the longitudinal study by Wagner *et al.* (1994) uncovered two nonredundant factors in first- and second-grade readers, namely phonological "analysis" (breaking words down into sounds) and phonological "synthesis" (blending sounds).

Clearly, the usefulness of making distinctions between highly correlated phonological ability factors really depends on the nature of their predictive relationships with reading. If factors differ in their predictive power, then it may be important to distinguish between them. If they do not, then it is simpler merely to consider differences in phonological skill as a unitary construct. Also, taking a developmental perspective, it is important to emphasize that although different phonological skills may be highly intercorrelated, this is not incompatible with a model in which these skills develop at different times and have different causal relationships with reading ability that are dependent on the age of the child and the stage he or she has reached in the reading process.

Relevant to the theoretical position promoted in this chapter is the influential model developed by Goswami and Bryant (1990), who maintain that phonological skills relate to different levels of analysis for spoken words, and that awareness of these develops at different rates. They propose that tasks of rhyming skill tap the child's understanding of onset and rime units within words, the onset being the consonant or the consonant cluster that precedes the vowel, and the rime being the vowel and any succeeding consonants. So for instance, in the single syllable word "trip," the onset is "tr" and the rime is "ip." Rhyming tasks may be performed by children as young as 4- and 5-years old and are highly predictive of later success in reading (Bradley & Bryant, 1983). Goswami and Bryant (1990) propose that the ability to segment phonemes within words is a very different skill, develops later partly as a consequence of learning to read, and is thought to be particularly closely related to learning to spell.

Partial support for the Goswami and Bryant model comes from a study by Muter, Hulme, Snowling, and Taylor (1998) who followed the development of 38 normal children during their first 2 years of learning to read. Principal components analysis of a battery of phonological measures, initially administered when the children were prereaders, identified two distinct and relatively independent factors: rhyming (defined by measures of rhyme detection and rhyme production) and segmentation (defined by measures of phoneme identification, phoneme blending, and phoneme deletion). Segmentation was strongly correlated with attainment in both reading and spelling at the end of the first year of learning to read, while rhyming was not. By the end of the second year, however, rhyming had started to exert a predictive effect on spelling, though not on reading. The relative importance of rhyming and segmentation skill to beginning literacy development has generated heated debate in the scientific literature (see Bryant, 1998; Hulme, Muter, & Snowling, 1998). While Bryant (1998) maintains a strong stance as to the important predictive relationship between rhyming and reading, there are other studies which like that of Muter *et al.* have also suggested that segmentation may be a more influential phonological skill in the beginning stages of learning to read than rhyming (Duncan, Seymour, & Hill, 1997; Seymour & Duncan, 1997; Wimmer, Landerl, & Schneider, 1994). The studies by Seymour and his colleagues have suggested that even when children start school with well-established rhyming skills, they are not necessarily disposed to use this knowledge in their initial attempts at reading. In their studies, the introduction of letters and the alphabetic principle (learning to connect letters with their sounds) resulted in the children becoming more proficient in identifying smaller phonemic rather than larger rime units within words. Wimmer *et al.* (1994) found that rime awareness was only minimally predictive of reading and spelling achievement at the end of Grade 1 in German-speaking children, but that it gained substantially in predictive importance for reading and spelling achievement in Grades 3 and 4. These findings taken together suggest that rhyming skill has relevance at a later stage in reading development, and that rhyming ability may have a particular bearing on children's ability to apply analogical strategies (Goswami, 1990; Muter, Snowling, & Taylor, 1994). Clearly, the resolution of this controversy is crucial not only from a theoretical standpoint, but also in respect of knowing which phonological skills to prioritize in the assessment of children with reading difficulties, in the teaching of reading to children in their early years at school, and in the remediation of reading difficulties.

In achieving a better understanding of how phonological abilities influence reading skill, and in particular how phonological deficits might be avoided through intervention, we need to know more about the early sources of phonological awareness word. Fowler (1991) maintains that phonological representations are initially of whole words. During the preschool years, they are gradually reorganized into increasingly smaller segments: syllables, then onsets and rimes, and eventually phonemes. So in essence, phonological representations become increasingly fine grained. Awareness of segmental size, which is assumed to be driven by earlier vocabulary growth, is viewed as an important predictor of the ease with which children develop phonological processing skill (Walley, 1993). Evidence for the relationship between vocabulary growth and phonological development comes from a correlational study by Metsala (1999) that showed that performance on phonological awareness tasks was related to vocabulary size. Byrne (1998) has suggested that vocabulary growth contributes to the early stages of phonological awareness development, but that it does not play a continuing role. It could be that once children have developed rudimentary phonological skills (syllables and onsets and rimes), which may in themselves be driven by vocabulary, the further growth of refined phonemic awareness may simply depend on those earlier established phonological skills.

Relatively few studies have reported on the influence of early language (especially phonological) skills on much later reading development. Those that have show a strong relationship between phonological abilities at kindergarten and Grade 1 and word level reading skill toward the end of elementary school (Klicpera & Schabmann, 1993; Schneider & Naslund, 1993; Stuart, 1990; Stuart & Masterson, 1992; Wagner *et al.*, 1997). In a long-term longitudinal study of 34 children, Muter and Snowling (1998) found that one of the best predictors of reading skill at age 9 was the phoneme deletion measure obtained at ages 5 and 6 years. Consistent with the results of our earlier longitudinal study (Muter *et al.*, 1998), it was the segmentation (in this case deletion) measure that was predictive of reading skill in the later primary years; rhyming scores obtained at ages 5 and 6 did not predict reading skill at age 9. Bearing in mind the observed stability and the established long-term predictive power of phonological skills, it is evident that tasks that tap these abilities are prime candidates for inclusion in any early screening or assessment battery of reading achievement and in teaching programs designed to remediate reading problems.

PHONOLOGICAL SKILLS AND THEIR RELATION TO LETTER KNOWLEDGE ACQUISITION

While phonological skills are possibly the most stable and best long-term determiners of literacy skill outcome, they do not always emerge as the most powerful predictor of reading skill at the beginning stages of literacy development. There is strong evidence that children's knowledge of letter names (and probably also sounds) is a very strong predictor of early reading skill, and that its interaction with emerging phonological skills is the major driving force behind reading progress in the first year at school.

Early large-scale studies found prereaders' letter knowledge to be a very good predictor of first-grade reading skills (Bond & Dykstra, 1967; Chall, 1967).

More recently, Byrne, Fielding-Barnsley, Ashley, and Larsen (1997) found that letter knowledge accounted for more variance in a decoding task within a teaching experiment with preschool and kindergarten children than did a measure of phonemic awareness. In a longitudinal study of nearly 100 normal children, conducted by the author and colleagues, Margaret Snowling, Charles Hulme, and Jim Stevenson (in press), letter knowledge skill on entry to school was found to be the best single predictor of word recognition 1 year later. It is clear that knowledge of letter names or sounds is an important prerequisite for children learning to read and spell in an alphabetic orthography such as English. There are a number of possible reasons for this. Most obviously, knowledge of letter sounds is necessary for children to understand that there is a systematic relationship between the spelling patterns of words and their pronunciations. Alternatively, it can be argued that from the very earliest stages of learning to read, young children are creating mappings between the orthographic representations of words and their phonological forms (Ehri, 1992; Rack, Hulme, Snowling, & Wightman, 1994). The creation of such mappings, however, depends upon knowledge of the phonetic characteristics of the sounds for which letters stand. In the longitudinal study conducted by Muter *et al.* (1998), letter knowledge predicted both reading and spelling during the first year at school. In fact, not only did segmentation and letter knowledge make separate and specific contributions to beginning literacy, but there was also an additional significant contribution from the product term "Letter knowledge × Segmentation," which reflected the interaction of the two component skills. The product term exerted a small additional influence on reading, and a massive effect on spelling. We may relate this finding to the results of experimental and field training studies that have shown that in order to progress in reading, children need to forge meaningful connections between their developing phonological skills and their appreciation of print (Ball & Blachman, 1988; Bradley & Bryant, 1983; Byrne & Fielding-Barnsley, 1989; Iverson & Tunmer, 1993; Tunmer & Hoover, 1992)—what Hatcher, Hulme, and Ellis (1994) refer to as "phonological linkage." It is thus not simply having adequate phonological awareness and letter knowledge that permits good progress in learning to read, rather, both these factors are important and they interact.

Given the emerging evidence that ease of acquiring letter identities is a very powerful predictor of beginning reading success, an important question we need to ask is what are the determinants of the ease with which letter–sound connections are acquired? It is tempting to think that the relationships between phonemes and graphemes are learned in a rote, paired associate fashion. It is, therefore, assumed that given equal exposure, /w/-w and /t/-t are learned equally quickly. However, as Treiman, Weatherston, and Berch (1994) point out, it is easier for kindergartners to learn letter–sound correspondences when the sound of the letter is contained in its name (e.g., p, b, k) than when the letter name does not help to specify the sound (e.g., h, w). They also suggest that, not only do children bring their prior knowledge of letter names to their learning of alphabetic sound to letter relationships, but they also use their ability to segment letter names into their component phonemes. Thus, children's phonological awareness skills are intimately involved in their learning of phoneme–grapheme correspondences. The authors go on to suggest that more extensive teaching is needed for some phoneme–grapheme correspondences than for others. Consequently, teachers may need to spend more time on the sounds of letters such as /y/ and /w/ than on the sounds of letters such as /b/ and /v/. However, the direction of influence between phonological segmentation and letter knowledge acquisition remains an area of dispute. While Treiman *et al.* and indeed others (Rack *et al.*, 1994) have suggested that phonological awareness skills help to promote letter–sound knowledge acquisition, Barron (1991) and Johnston, Anderson, and Holligan (1996) have proposed that learning the alphabet may cause children to become explicitly aware of the phonological structure of words. Johnston *et al.* found that letter knowledge accounted for unique variance in phoneme awareness in 4-year-old nonreaders. Clearly, there is an interactive process between children's emerging letter–sound knowledge and their phonological awareness. The bidirectional nature of this relationship has been explored by Burgess and Lonigan (1998) in a 1-year longitudinal study of children aged 4 and 5 years. Multiple regressions revealed that phonological sensitivity predicted growth in letter knowledge, and that letter knowledge predicted growth in phonological sensitivity when controlling for children's age and oral language skills. However, the size of letter knowledge's effect on phonological sensitivity was smaller than the effect of phonological sensitivity on growth in letter knowledge.

PHONOLOGICAL SKILLS AND SHORT-TERM VERBAL MEMORY

The role of verbal short-term memory in early reading development is dealt with by Pickering (Chapter 6, this volume). However, a discussion of the importance of phonological skills in reading development would be incomplete without a consideration of its interaction with verbal memory processes. Indeed, the relative importance of verbal memory and phonology in learning to read is an issue that has been nearly as controversial as that of the contribution made by specific phonological processes to reading. It is well documented in the scientific literature that short-term verbal memory is closely related to level of reading skill, whether the materials are digits and letters (Katz, Healy, & Shankweiler, 1983), words (Brady, Shankweiler, & Mann, 1983) or sentences (Mann, Liberman, & Shankweiler, 1980). However, whether verbal memory span is an important predictor of reading skill independent of children's phonological abilities is not entirely clear. Hansen and Bowey (1994) found in their correlational study of 7-year-olds that both phonological analysis and verbal working memory accounted for unique variance in three reading measures. However, other studies have suggested that short-term verbal memory does not significantly predict reading skills after controlling for phonological abilities (Rohl & Pratt, 1995; Wagner *et al.*, 1994). Margaret Snowling has suggested that short-term verbal memory tasks may essentially be tapping the completeness or intactness of the child's phonological representation (Snowling, 1998). Wagner *et al.* (1994) would agree with this view; they propose that various phonological processing tasks, including those of short-term verbal or working memory, are tapping the quality of underlying phonological representations, and it is the quality of these representations which in turn affects children's ability to learn to read. With words only coarsely represented at the phonological level, there is a restriction in the number of verbal items that can be retained in memory. Another way of looking at this is to suggest that children who are able to process phonetic information only slowly and inefficiently suffer a constriction in their working memory system. Slow phonological processing might result in a bottleneck that impedes transfer of information to higher levels of processing within the system (Shankweiler, Crain, Brady, & Macaruso, 1992). The interactive relationship between phonology and short-term verbal memory that is embodied in these two views may explain why in some studies short-term verbal memory does not uniquely predict reading skill once phonological abilities have been taken into account. Both perspectives see the core problem for poor readers being a deficit in phonological processing, whether degraded phonological representations or slow processing speed, which then impacts on their short-term verbal memory.

Gathercole and Baddeley (1989) have conceptualized nonword repetition tasks as measures of phonological working memory that may relate to reading skill (see Chapter 6 by Pickering, this volume, for a fuller discussion). Gathercole, Baddeley, and Willis (1991) failed to uncover a relationship between nonword repetition and reading ability in the first year at school, although the correlation between these skills was statistically significant by year 2. Furthermore, Muter and Snowling (1998) found that, in a discriminant function analysis for good versus poor reading accuracy skill at age 9, the best long-term predictor set consisted of the phoneme deletion and nonword repetition scores at ages 5 and 6, with a successful classification rate of 80%. While Gathercole and her colleagues regard a test of nonword repetition as a pure and direct measure of phonological memory, Snowling, Chiat, and Hulme (1991) have argued that this task assesses not only memory processes, but also phonological segmentation and possibly assembly of articulatory instructions. They showed that there is not a systematic increase in repetition difficulty as length of nonword increases, a finding which is inconsistent with a memory interpretation of the test. The ease of repetition of a given nonword may be as strongly related to its morphemic and phonological similarity to real words as it is to its length. Also, even the effects of length cannot simply be attributed to the influence of memory. Longer items are not only harder to remember, but they also place greater demands upon other phonological processes such as segmentation skill and the assembly of articulatory motor programs. Hulme and MacKenzie (1992) have argued that it may be very difficult to separate phonological storage and retrieval mechanisms from other phonological mechanisms. It is clear that some phonological awareness tests (such as the Bradley odd word out task) place a considerable load on the child's phonological working memory, while tasks purporting to assess phonological memory (such as nonword repetition) are often simultaneously tapping other phonological processes.

The speed and accuracy with which pictures, digits, and letters can be named is a well-documented linguistic correlate of reading ability that is thought by some to reflect phonological memory or retrieval processes (see Wolf & O'Brien, 2001, for a recent review). Individual differences in what is referred to as rapid automatized naming have been shown to predict reading development in the first grade (Felton & Brown, 1990) and in the third and fourth grades (Badian, McAnulty, Duffy, & Als, 1990). Naming speed with digits and letters appears to be more predictive than with pictures, not a surprising finding given the more reading-related symbolic nature of these stimuli (Badian, 2000). Whether naming skill is related to ease of accessing phonological representations in long-term memory or whether it merely reflects the distinctness of those representations has been an area of debate. Most recently, Manis, Seidenberg, and Doi (1999) have attempted to elucidate the empirical relations between naming speed and other reading-related tasks within a computational model of word recognition. They examined the contributions of naming, verbal ability, and phonological awareness to the prediction of phonological and orthographic skills from the first to the second grade. Both naming (of letters and numbers) and phonological awareness accounted for unique variance in later reading scores. Naming was a stronger predictor than phoneme awareness for three tasks in which orthographic information was critical whereas the opposite held true for nonword reading and paragraph comprehension. Their model of reading suggests that naming tasks account for distinct variance in reading when compared with phoneme awareness because naming involves arbitrary associations between print and sound, whereas phoneme awareness is more related to the learning of systematic spelling–sound correspondences. Learning arbitrary associations (between sounds and letters) probably plays a central role in the development of early reading skill, whereas knowledge of segmental phonology is relevant to both the earlier and later phases of learning to read. This suggestion fits in with the observations of Wagner *et al.* (1997) who found that individual differences in naming speed influenced subsequent individual differences in word level reading initially, but that these influences faded with development; in contrast, phonological awareness was determined to have both a powerful and long-term bearing on reading progress and outcome. One might hypothesize that naming speed is one of the predictors or determiners, alongside phonological awareness, of letter knowledge acquisition in young children.

Hulme and his colleagues have explored the relationship between short-term memory and phonological processing in a rather different and less obvious way by examining speech rate; that is, the speed with which a specified word or words can be spoken (see Hulme & Roodenrys, 1995, for a review). They present evidence from developmental studies that demonstrates a close relationship between changes in speech rate and increases in memory span. With regard to reading, McDougall, Hulme, Ellis, and Monk (1994) studied the relationship between memory span, speech rate, and phonological awareness in good, average, and poor readers.

They were able to demonstrate that differences in reading ability are associated with differences in the efficiency of the speech-based rehearsal component of short-term memory span as measured by a speech rate task. Children's scores on the test of speech rate made a significant contribution to reading skill independent of that made by two measures of phonological awareness. Furthermore, after accounting for speech rate, memory span made no contribution to reading skill. Conversely, however, speech rate did predict reading skill even after controlling for the effects of memory span. Muter and Snowling (1998) obtained similar results as to the relative salience of speech rate and memory span to progress in reading in normal 9-year-old children. Hulme and his colleagues suggest that speech rate provides an index of the speed and efficiency with which phonological representations of words in long-term memory can be activated, and that this is a skill separate from the phonological processes tapped by measures of phonological awareness. Thus, in speech rate, we appear to have a quick and easy-to-implement reading-relevant assessment technique that can complement and supplement the more extensively documented measures of phonological awareness and letter knowledge.

To what extent phonological analysis ability and phonological memory/access skills are inextricably linked within the reading process and to what extent they function independently is clearly an issue in need of further clarification. One view is that tasks of phonological awareness, verbal memory, naming speed, and speech rate are merely indirect tests of children's underlying phonological representations (Snowling & Hulme, 1994; Wagner *et al.*, 1994). Consequently, as I discussed earlier, these tests are tapping the specificity of underlying phonological representation, and it is the quality of these representations which in turn affects children's ability to learn to read. The alternative view is that naming speed, measures of working memory, and speech rate may be tapping processes separate from phoneme awareness, though still within the phonological domain, and that consequently each accounts for unique variance in reading skill.

PHONOLOGICAL SKILLS AND THEIR RELATION TO SEMANTIC AND GRAMMATIC COMPONENTS OF READING DEVELOPMENT

The foregoing discussion has highlighted the importance of measures of phonological processing ability, together with short-term verbal memory and letter knowledge, in influencing literacy development. Whitehurst and Lonigan (1998) refer to these as "inside-in" processes that determine children's knowledge of the rules for translating the written word into sounds. Increasing attention has been paid of late to other non-phonological language skills, including syntax and semantics, that might have a bearing on how easily children learn to read. These skills, viewed by Whitehurst and Lonigan as "outside-in" processes, relate to children's understanding of the context in which they are reading.

Tunmer (1989) emphasized the importance of syntactic awareness as an independent contributor to early reading skill. In a longitudinal study, he administered tests of verbal ability, phonological awareness, syntactic awareness, and reading to 100 children at the end of first grade and again 1 year later. The results clearly demonstrated that both phonological and syntactic awareness influenced reading comprehension through phonological recoding (as measured by a nonword reading test). Further to this, Tunmer and Chapman (1998) have suggested that young readers often combine incomplete phoneme–grapheme information with their knowledge of sentence constraints in order to identify unfamiliar words. This increases their word-specific knowledge (including irregular words) and their knowledge of grapheme–phoneme correspondences. The ability to use contextual information allows young readers to monitor their accuracy of word identification by providing them with immediate feedback when their attempted readings of unfamiliar words fail to conform to the surrounding grammatical context. They can then make the necessary adjustments to their subsequent attempts at reading the unfamiliar word up to the point where their response satisfies both the phoneme–grapheme relations within the word and its surrounding sentence constraints.

Further support for the combined phonology-context view of reading comes from our recent study (Muter & Snowling, 1998) in which an important concurrent predictor of reading accuracy at age 9 (alongside that of phonological segmentation ability) was found to be grammatical awareness. Indeed, it is important to note that grammatical awareness in this study was a predictor of reading accuracy in context, as measured by a prose reading test, but not of a pure measure of decoding skill, namely nonword reading. Thus, grammatical knowledge interacts with decoding ability to increase word identification skills. This may be particularly true of older children who have moved beyond single word decoding of simple text toward an increasing appreciation of the value of context and content cues contained in more complex reading materials. Of relevance to the age at which children might take advantage of contextual information is the finding from the recent longitudinal study of Muter, Hulme, Snowling, and Stevenson (in press) that measures of syntactic or grammatic awareness failed to predict reading accuracy skill in the first year of learning to read. However, as Willows and Ryan (1986) have shown, children become increasingly sensitive to semantic and syntactic features in reading tasks from Grades 1 through 3. Thus, syntactic and semantic abilities are likely to gather in predictive importance later on in elementary schooling when children read in context and when there is greater emphasis on comprehension skills. Indeed, children need to develop syntactic and semantic skills that are sufficiently refined to enable them to make sense of contextual cues as they proceed from early single word decoding to the mastery of more complex text. Consistent with this idea, Nation and Snowling (1998) found that the extent to which children can use a spoken sentence context to facilitate decoding depends on their verbal skills. They presented children with a printed word, either in isolation or following a spoken context. Not surprisingly, the children's reading accuracy increased when context was made available to them. However, what was of particular interest was that the children who benefited most were those with poor phonological skills but good verbal ability, while the children who benefited least were those with relatively weak verbal skills. Clearly, there is a strong interaction between children's phonological skills, their verbal capability, and their capacity to take advantage of context-based information.

Another way of viewing the interaction of semantic and phonological information in influencing reading skill is to draw on computer-simulated connectionist models of reading. The earliest most influential word recognition model of this sort, proposed by Seidenberg and McClelland (1989) consists of only two types of representational units—a set of input units coding the letters present in printed words (orthographic units) and a set of output units coding the pronunciation of words (phonological units). There is additionally a set of hidden units that connect the orthographic and phonological units. These connections, together with the hidden units, carry weights that govern the spread of activation across the units. It is these weights that encode the model's knowledge about written English. During the training procedure, a learning algorithm adjusts the weights of the connections in the network in proportion to the extent that this will reduce the mismatch between the actual and "correct" pattern of excitation that represents the word being read correctly. The Seidenberg and McClelland model has been successful in simulating a number of different aspects of reading performance, and has been found to generate acceptable pronunciations for many novel items, including nonwords. A more recent model by Plaut, McClelland, Seidenberg, and Patterson (1996) has introduced a semantic mechanism in order to more comprehensively describe the reading process. There are two pathways by which orthographic information can influence phonological information: a phonological pathway and a semantic pathway. Thus, from this connectionist perspective, the task when learning to read is to learn the mappings between the representations of written words (orthographic units), spoken words (phonological units), and their meanings (semantic units). As the training proceeds, the semantic pathway becomes increasingly specialized for the pronunciation of exception words, while the phonological pathway becomes more specialized for the pronunciation of words with consistent spelling patterns.

ENVIRONMENTAL INFLUENCES ON PHONOLOGICAL AWARENESS

It is well documented that young children's level of language proficiency and their reading skill are closely correlated with socioeconomic status (SES) of the parents, with middle-class children attaining higher levels of language and literacy than their lower-class or deprived peers (Feagans & Farran, 1982; White, 1982). Of particular relevance to the present discussion is the research that addresses the relationship between SES, reading progress, and phonological sensitivity. Raz and Bryant (1990) and Bowey (1995) have suggested that SES differences in word level reading in young children are mediated partly through preexisting differences in phonological sensitivity. Raz and Bryant reported strong SES differences in reading performance even when IQ score effects were covaried. Furthermore, when phonological sensitivity scores were covaried, SES differences were no longer significant in the tests of reading accuracy. This suggests that SES differences in phonological sensitivity may mediate SES differences in word level reading achievement. Bowey conducted a longitudinal study of 5-year-olds differing in SES during their first year of learning to read. Marked differences on a wide variety of measures emerged when the children were grouped according to high or low SES. When general verbal ability effects were controlled, differences in phonological sensitivity and word level reading remained. Thus, many children from lower socioeconomic groups may be arriving at school with underdeveloped phonological awareness which of course then seriously disadvantages them in acquiring early reading skills. These findings point to the importance of preschool and kindergarten programs, especially in disadvantaged areas, focusing strongly on language activities that foster children's sensitivity to sound patterns in words and which link these to their experience of print in terms of alphabet and story books.

It has been suggested that home environmental differences may be more important than demographic indices such as parental occupational status in accounting for variation in children's literacy development, and that parent-preschooler reading experiences could contribute substantially to the overall literacy environment of the young child's home. Scarborough and Dobrich (1994) reviewed 31 studies that looked at the impact of parents' reading to their preschoolers on the children's oral language and literacy development; 20 of these were correlational in their methodology and 11 were intervention programs. Frequency of reading to preschoolers seemed to be the important factor. In general, the correlational studies showed that frequency of reading was associated with growth in lexical and semantic content of language and in developing literacy (but not in syntactic or phonological structures). In the intervention studies, more positive results emerged for oral language than for literacy outcome. However, while there appears to be a significant association between parent-preschool reading and language/literacy outcome, this is of modest proportions, with most correlations at or less than 0.28 (accounting for no more than about 8% of the overall variance in language or reading achievement). A recent study by Senechal, LeFevre, Thomas, and Daley (1998) contrasted frequency of storybook reading and frequency of parents teaching the reading and writing of words in their impact on the children's oral and written language skills. The children who were from middle-class biased schools were studied while in kindergarten and Grade 1. After controlling for parents' print exposure and the children's IQ, storybook exposure explained statistically significant unique variance in children's oral language skills, but not in their written language. In contrast, parent teaching of reading explained unique variance in the children's written language skills, but not in their oral language. The authors conclude that these findings are consistent with the hypothesis that storybook exposure may enhance children's oral language skills, whereas additional support in the form of reading teaching may be necessary to enhance written language skills. There seems to be little evidence that parent-preschooler storybook experiences have an impact on children's reading via the promotion of their phonological skills; Lonigan, Anthony, and Dyer (1996) showed that growth in phonological sensitivity was not linked to shared reading experience. However, a number of the studies reviewed by Scarborough and Dobrich (1994) have suggested that early shared storybook experiences at home may enhance children's vocabulary development, which is thought by some to be an important precursor of emerging phonological segmentation skill (Fowler, 1991; Walley, 1993), that is, the effect of storybook reading on phonological ability may be indirect, being mediated through earlier established language skills.

PHONOLOGICAL DEFICITS IN DYSLEXIA

We have seen that current theories of early reading development propose that children set up direct mappings between printed words and representations of spoken words in the child's language systems (Ehri, 1992; Rack *et al.*, 1994). It follows from this that the status of a child's underlying phonological representations determines the ease with which they learn to read. Since 1980, there have been many studies that point to language difficulties in the dyslexic child, specifically at the level of phonology. Dyslexic children typically perform poorly on a wide range of measures of phonological awareness, verbal short-term memory, rapid naming, speech perception, and verbal representation, tasks which essentially tap children's representation of, access to, and recall of phonological information (see Snowling, 2000, for a recent review). From this extensive research on phonological deficits in dyslexic children, Snowling has formulated the following definition of dyslexia:

> Dyslexia is a specific form of language impairment that affects the way in which the brain encodes the phonological features of spoken words. The core deficit is in phonological processing and stems from poorly specified phonological representations. Dyslexia specifically affects the development of reading and spelling skills but its effects can be modified through development leading to a variety of behavioural manifestations (Snowling, 2000: 213–214).

Note that Snowling's definition makes it clear that it is a deficit in phonology that results in the reading and related difficulties that dyslexics experience.

How do phonological deficits that are evident in dyslexic children manifest themselves in reading failure? In an earlier section of this chapter, I discussed how young children's phonological awareness interacts with their emerging letter knowledge acquisition in order to promote reading development. Brian Byrne has described this process in terms of the acquisition of the *alphabetic principle*, that is, the child's realization that particular speech sounds are systematically associated with printed letters (Byrne, 1998). Byrne has specified three conditions that need to be met for children to acquire the alphabetic principle. First, they need to know the letters of the alphabet. Second, they require a minimum level of phonological segmentation skill that enables them to split words into sounds. Finally, they need to connect or link their emerging speech sound sensitivity with their experience of print—what Hatcher *et al.* (1994) refer to as "phonological linkage." In dyslexic children, these three precursors of the alphabetic principle are slow to develop, with the result that the child has problems in mapping alphabetic symbols to speech sounds. This inability to abstract letter–sound correspondences from experience with printed words results in a failure to develop phonic reading strategies; so, not surprisingly, the dyslexic child soon falls further behind his or her peers in reading.

One direct way of tapping the level of phonic reading skill is to ask children to attempt to read nonsense words that cannot be recognized from previous experience nor accessed through semantic clues. Examples of nonwords taken from the Graded Nonword Reading Test (Snowling, Stothard, & MacLean, 1996) are "twesk, nolcrid, chamgalp." There is a great deal of evidence that dyslexic children's nonword reading skill is significantly worse than their other reading abilities (Rack, Snowling, & Olson, 1992).

Although dyslexic children invariably have phonological and phonic decoding deficits, most do eventually acquire a reading vocabulary. This may be achieved partly through the implementation of compensatory strategies that develop in the individual in response to, not only teaching and experiential factors, but also the availability of specific cognitive strengths. For instance, reading disabled children who are particularly verbally able may be able to compensate for their phonological and decoding deficiencies by paying attention to context cues within prose reading (Nation & Snowling, 1998); thus, combining their good vocabularies and syntactic/grammatic skills with their fragmented phonological/phonic knowledge permits them eventually to arrive at the correct reading of a given word which they might not have been able to achieve on the basis of their incomplete decoding abilities alone.

Children's phonology-based reading difficulties may also be considered within a connectionist framework. In a computer-simulated model of reading difficulty, it is necessary to constrain the mappings that are established between orthography and phonology during training. In line with the extensive research on the contribution of phonological processing to reading, the most obvious modification to the training procedure would be to coarsen or underspecify the units in the phonological store. Just such a simulation has been conducted by Seidenberg and McClelland (1989) who showed that the model not only learned less effectively, but that there was poor generalization to novel items (equivalent to the poor nonword reading shown by many dyslexic youngsters). If we accept that the majority of severely disabled readers have problems in mapping alphabetic symbols to sounds (83% in Vellutino and Scanlon's 1991 analysis of hundreds of impaired readers), then this "coarsened phonological representation" view would seem to be the most plausible way of representing dyslexia within a connectionist model. However, Snowling (1998) has suggested that there are two other strategies which might be used to interfere with the normal "running" of a connectionist model. One would be to decrease the number of hidden units so that the activation between the phonological and orthographic units is "lost," perhaps by limitations in verbal short-term memory. Alternatively, the connectionist framework could be compromised due to insufficiencies in the coding of the orthographic representations. This might result from a visual perceptual problem, a view that has been held in disfavor since Vellutino's attack on visual models of reading difficulty (Vellutino, 1979), but which has recently generated a resurgence of interest within the reading research field (see Hogben, 1997, and Chapter 1 by Evans, this volume). Thus, short-term memory limitations or visual processing deficits might account for the problems of a minority of reading disabled children whose phonological representations are relatively unimpaired. Further to this, Manis *et al.* (1999) have proposed that, consistent with their computerized model of word recognition, a deficit in rapid naming may be a more predictive measure of reading for those individuals who have "surface (or orthographic) dyslexia," that is, a delay in the development of word recognition skills and problems in reading exception words, but with lesser impairment in phonological skill. In contrast, naming may be less predictive than a phonological segmentation task of the reading skill in a phonological dyslexic who has difficulties in decoding (as reflected in nonword reading). Wolf (1997) has described children who have both phonological awareness and naming speed deficits as having in effect a double deficit that makes their reading (including comprehension) problems more severe than if only one of these skills was affected; the literacy outcome for such children is further worsened by the fact that they also have fewer compensatory mechanisms on which they can draw.

PHONOLOGICAL DIFFICULTIES AND THE DYSLEXIC CHILD—A CASE STUDY

To illustrate the progression of phonological abilities and their relationship to long-term reading development in relation to a single child who proved to be dyslexic, it is proposed to describe the case of Nicholas who was a subject in the original longitudinal and the long-term follow-up studies of Muter *et al.* (1998) and Muter and Snowling (1998). Nicholas was first assessed at age 4 when he recorded a WPPSI IQ of 110 (Wechsler, 1967). At that time, he scored zero on two measures of phoneme segmentation (phoneme identification and phoneme deletion), and he identified just one letter of the alphabet correctly. However, there was nothing striking in Nicholas's profile at age 4; true, he had not developed any meaningful phonological skills, but then neither had most of the other children in the study.

When seen 1 year later at his state elementary school, he scored zero on the phoneme identification test and one out of ten on the phoneme deletion test. By that time, he could read five letters of the alphabet. At this stage, there was emerging evidence of a gap between Nicholas and his peers. The study group as a whole obtained a score of three on both of the phonological tests and they could read on average 12 letters of the alphabet. He was able to read four words correctly from the British

Abilities Scales Reading Test (Elliott, Murray, & Pearson, 1983), but he was unable to spell any words from the Schonell spelling test (Schonell & Goodacre, 1971).

The gap had widened further by the time Nicholas was seen in the following year, now aged 6 years. His phoneme identification score had risen to five (out of eight), but he scored zero on the phoneme deletion test (the group mean was five), and he was still reading only five letters of the alphabet correctly (in contrast to the mean of nineteen for the group as a whole). He read five words from the reading test and spelled one word from the spelling test (the mean scores for the whole group were 17 and 8, respectively). However, his oral and written arithmetic skills at age 6 were well up to the standard of the rest of the sample.

When Nicholas was followed up at age 10, his reading and spelling attainments fell at the third and eighth centiles, respectively, although his written arithmetic was of a higher standard, that is, thirty-ninth centile (British Abilities Scales attainment tests, Elliott *et al.*, 1983). He read only two words correctly from the Graded Nonword Reading Test (Snowling, Stothard, & MacLean, 1996) in contrast to the mean score for the follow-up sample of 14.56. Nicholas continued to display substantial problems in phonological processing skill; on a complex phoneme deletion task developed by McDougall *et al.* (1994), he scored only 4 out of 24, while the mean for the sample was 17.24. On a speech rate measure, he scored 3.39 words per second, nearly two standard deviations below the sample mean of 4.82. By the age of 10, Nicholas's frustration over his severe and persisting reading failure was now beginning to affect his behavioral adjustment. His scores on the "anxious/depressed" and "attention problem" scales of the Achenbach Child Behavior Checklist (Achenbach, 1991) fell within the clinical problem range. This was in contrast to his excellent examiner-observed and teacher-judged behavioral adjustment when younger.

Nicholas's story gives a clear picture of a child of normal, indeed above average, intelligence who, having had a relatively happy and uneventful nursery school career, had encountered enormous problems in learning to read and spell. By the time he was reaching the end of his elementary schooling, he was failing within the educational system and was becoming increasingly anxious, demotivated, and negative toward learning. Yet it is evident that the warning signs that Nicholas was at risk for significant reading problems were emerging quite clearly by the age of 5 years. This clinical vignette illustrates how measures of phonological awareness and letter knowledge might serve as potentially powerful, early indicators and diagnostic tools to assess and to identify those children who are likely to display persisting reading failure.

GENETICS AND THE NEUROBIOLOGICAL BASIS OF PHONOLOGICAL AWARENESS AND DYSLEXIA

There is accumulating evidence that dyslexia is an inherited condition—not just the reading difficulty *per se*, but in particular, the underlying phonological deficit (for a fuller discussion see Chapter 2 by Olson, this volume). Results of large-scale twin studies have shown that there is greater heritability of phonologically based reading skills (e.g., as measured by a nonword reading test) than of orthographic skills (e.g., selecting the correct spelling from phonologically viable pairs, e.g., train–trane) (Olson, Wise, Connors, Rack, & Fulker, 1989). In a recent twin study conducted by Gayan and Olson (1999), heritability of phonological awareness was estimated at 0.89, which is significantly higher than for other components of the reading process (word recognition 0.68, phonic decoding 0.77, and orthographic decoding 0.76).

Additionally there have been recent attempts to relate phonological deficits experienced by dyslexics to specifics of brain function. The phonological processing skills essential to reading development are thought to be associated with regions in the left temporal lobe of the brain. As a rule, these left temporal regions are larger in volume than those homologous areas in the right temporal lobe (Geschwind & Levitsky, 1968). However, when the temporal lobes of dyslexic individuals were studied at postmortem by these authors, it was found that this characteristic asymmetry favoring the left hemisphere structures was absent. Clearly, this phenomenon of planum symmetry in dyslexia is an important brain-behavior finding, but how does it relate to the specific phonologically based reading difficulties that dyslexics experience? A study that has attempted to answer this question was carried out by Larsen, Hoien, Lundberg, and Odegaard (1990). They studied nineteen 15-year-old dyslexics who were individually matched to controls for age, gender, IQ, socioeconomic class, and educational environment. Magnetic resonance imaging (MRI) analyses of the subjects' brain structures revealed symmetry of the planum for the majority of the dyslexics while asymmetry was evident for the majority of the normals. The children were then classified into dyslexic subgroups of "pure phonological," "pure orthographic," and "mixed phonological–orthographic." Accepting that the sample size was small and the classification crude, the results nonetheless are interesting. All five of the pure phonological dyslexics and seven out of the nine mixed dyslexics showed symmetry of the planum. One child who had a pure orthographic profile and three out of the four who were deemed "unclassifiable" showed the normal pattern of asymmetry.

Paulesu *et al.* (1996) studied the positron emission tomography (PET) scans of five adult dyslexics who had demonstrable problems on phonological tasks. This procedure permits the analysis of brain activity and function during ongoing cognitive processing. Major sites of the phonological system (Broca's area and temporo-parietal cortex) were activated in the dyslexic subjects, but in contrast to the controls, this activation was not in concert. The authors propose that the defective phonological system of dyslexics is due to weak connectivity between central and posterior language areas. Consequently, segmented and unsegmented phonological codes, which are represented separately, both neuroanatomically and neurofunctionally, are not activated simultaneously in dyslexics as they are in normal readers. The simultaneous holding of segmented and unsegmented phonological codes in working memory is essential to the successful completion of complex phonological tasks.

Brunswick, McCrory, Price, Frith, and Frith (1999) conducted a PET scan of six adult dyslexics that demonstrated that, when the dyslexics read real or non-words, they showed decreased activation (when compared with matched controls) in the following brain areas—the left posterior inferior temporal cortex (also known as Brodman area 37, BA37), the left and midline cerebellum, the left thalamus and the medial extrastriate cortex. It has been found that BA37 is critical in the specification and retrieval of phonological information. In fact, it may function as a sort of thesaurus, that is, a mental lexicon (or dictionary) that provides an access facility to the names of words. Of course, it is extensively documented that dyslexics have specific difficulties in retrieving phonological codes that correspond to items to be named. In keeping with the findings of Paulesu *et al.*, the present authors propose that no single component of the dyslexics' phonological reading system may be defective *per se*, but that the system malfunctions as a whole because the components do not work in concert.

Studies of children with known dyslexic problems have their limitations. Since it is well established that the relationship between phonological skills and reading is a reciprocal one, it follows that the observed phonological deficits in dyslexics could have been influenced by the reading impairment itself. To circumvent this problem, and to complement the retrospective studies, some researchers have concentrated on looking at young children who are highly at risk for developing dyslexia. It follows from the genetic research on dyslexia that this condition must show a tendency to run in families. One way of identifying children at risk for dyslexia before they have had the opportunity to learn to read is to select children from families where there is high incidence of dyslexia and to then contrast their performance on language tasks with children from families in which there is no dyslexia.

Hollis Scarborough (1990, 1991) compared the performance of reading disabled children from dyslexic families on language, phonological, and preliteracy measures at several different ages with that of normal reading children from non-dyslexic families, all of whom had been recruited for the study at age 2. The children in the reading disabled group were significantly poorer than those in the control group on tests of expressive syntax and sentence comprehension at 30–48 months, and on tests of phonological awareness, letter–sound knowledge, letter identification, and object naming at 60 months. Scarborough suggested that dyslexic children have a broader language disorder that is not simply reflected in reading failure. This disorder is expressed as different observable weaknesses at different ages; first, syntax problems, then weaknesses in phonological awareness, naming and other preliteracy skills; and finally, difficulties with reading and spelling during the school years. The fact that different cognitive skills have differing predictive power according to the age at which they are assessed is an important consideration when devising predictor and early assessment instruments; a series of tests relevant for 3-year-olds may have a very different language content, beyond that of difficulty level, from that devised for 5-year-olds.

More recent studies of children at risk for dyslexia have also highlighted differences between high-risk and low-risk children in respect of phonological awareness tasks and letter knowledge/identification (Byrne *et al.*, 1997; Elbro, Borstrom, & Petersen, 1998; Lefly & Pennington, 1996). Gallagher, Frith, and Snowling (2000) looked at reading outcome at age 6 from language scores obtained at 45 months in a group of reading "at-risk" children. Half the children were delayed in their early literacy development when compared with low-risk children. Gallagher *et al.* found that the strongest predictor was letter knowledge. The literacy delayed at-risk children were also subject to mild delays in all aspects of spoken language, semantic and syntactic as well as phonological. The involvement of semantic and syntactic, as well as phonological, skills in these children fits in well with the view that semantic skills primarily determine the acquisition of an initial sight vocabulary, while phonological skills become increasingly important in setting up a set of mappings between orthography and phonology (Snowling, 1998). The literacy delayed children in this study who were making the best progress were those with better vocabulary and expressive language skills. Thus, eventual success in overcoming early reading delay may depend not only on the status of the children's phonological deficit, but also upon how successfully they can use syntactic and semantic skills to compensate for their likely decoding inadequacies.

SCREENING AND ASSESSMENT OF PHONOLOGICAL SKILLS

The foregoing discussion has highlighted the importance of "prediction" studies that tell us a great deal about the knowledge and skills young children bring to bear on the task of learning to read. Research-generated phonological tasks might, if given to sufficiently large samples of children, provide norms for the purposes of screening young children, or against which individual "slow starter" youngsters might be compared. However, whether such studies can suggest a strategy for reliably identifying those specific children who go on to have persisting phonologically based reading problems that necessitate special needs intervention is a rather more complex issue. When considering individual children, it is not always possible to conclude confidently that a child who obtains a low score on a measure of phonological awareness will necessarily go on to have significant reading difficulties. In Bradley and Bryant's (1983) longitudinal study of early readers, only 30% of those children who initially produced good sound categorization scores went on to become exceptionally good readers. Of greater relevance to the early identification issue is the finding that just 28% of those who initially produced poor sound categorization scores became exceptionally poor readers. These authors suggest that a phonological awareness test on its own might not be a particularly effective way of predicting persisting reading problems. More recent studies that have used a number of independent predictors have reported higher sensitivity ratings of 48% (Schneider & Naslund, 1993), 42% (Badian, 1988), and 56% (Butler, Marsh,

Sheppard, & Sheppard, 1985). An even higher sensitivity was achieved in a study by Badian (1994) in which she was able to predict the problems of 14 out of 15 poor readers. However, a high sensitivity such as this may be at the cost of a relatively high number of false alarms, depending on where the cutoff point for group inclusion is placed. So, in the Badian' (1994) study, 10 out of 24 children did not develop reading difficulties as predicted. At the extreme, adopting a large number of predictors in order to achieve higher prediction rates also has its disadvantages. Using a large number of independent measures will provide an almost perfect, or even a perfect, prediction (Elbro *et al.*, 1998). Beyond the practical limitations of time and cost, this is not in itself desirable theoretically or methodologically. Prediction is best when each measure is strongly correlated with the outcome measure of reading, but uncorrelated with the other measures (Tabachnick & Fidell, 1989). Thus, the goal of any screening or early assessment measure is to select the fewest independent measures necessary to provide a good prediction of reading outcome where each measure predicts a substantial and independent proportion of the variability in reading outcome.

There are a number of tests that have employed a phonological awareness measure as a screening or early diagnostic instrument. In the United States, Diane Sawyer's (1987) Test of Awareness of Language Segments (TALS) can be used as a screen for children at the kindergarten or Grade 1 level, but is also recommended for use diagnostically and prescriptively for older children who are already exhibiting delay in learning to read. In this test, children are required to segment language, first from sentences to words, then words to syllables, and finally from words to sounds. Predictive validation studies have shown that the TALS administered at age 4 years 6 months to 5 years 6 months significantly predicted reading right through to Grade 3. Also in the United States, Torgesen and Bryant's (1994) Test of Phonological Awareness (TOPA) is a group administered test in which children use pictorial material to demonstrate their ability to identify initial sounds (kindergarten version) or end sounds (elementary version) within words. The TOPA–Kindergarten was normed on 857 children aged 5 and 6, while the TOPA–Early Elementary was normed on 3,654 children aged 6 through 8 years. Predictive validity from kindergarten TOPA scores showed that the correlation with reading analysis skill 1 year later was 0.55.

Bearing in mind Bradley and Bryant's (1983) caution against relying on a single measure to predict reading outcome reliably, the present author, together with Charles Hulme and Margaret Snowling developed a multi-measure screening and early diagnostic instrument aimed at being simple to use, economic, and able to provide good prediction, while at the same time being congruent with recent research findings and with current theoretical perspectives on early reading development. The Phonological Abilities Test (PAT) was published by the Psychological Corporation UK in 1997 (Muter, Hulme, & Snowling, 1997). Tests were selected that tap the skills young children bring to bear on their earliest reading experiences and which reflect the authors' theoretical position on the important relationship between phonological abilities and reading. Since phonological awareness skills are the most stable and robust of the available predictor skills, and in view of their ability to predict reading outcome over long periods of time, they were given selection priority over more transitory predictors like vocabulary or naming skill. The PAT contains two measures of children's ability to segment words into syllables or phonemes—a test of syllable and phoneme completion, and a test of phoneme deletion. While current evidence seems to favor segmentation over rhyming as being the better predictor of beginning literacy, measures of onset-rime awareness were not excluded. There is the well-documented view that rhyming skill may have a bearing on later stages of learning to read, and, in particular, on children's ability to adopt analogical strategies. We have seen that there is increasingly strong evidence emerging as to the paramount importance of letter knowledge acquisition in early reading development. Thus, a test of letter knowledge as an indicator of the thoroughness and possibly ease with which the letter identities have been learned was viewed as an obligatory component of an early screening battery. A measure of children's phonological memory and access, speech rate completes the battery.

The PAT comprises four phonological awareness subtests (Rhyme Detection; Rhyme Production; Word Completion—Syllables and Phonemes; Beginning and End Phoneme Deletion), a Speech Rate test (timed repeating of the word "buttercup"), and a test of Letter Knowledge. The test was standardized on 826 children aged 4–7 years, and norms provided in 6-month age bands between 5 and 7 years (and in a 12-month age band for the 4-year-olds). When given in full, the PAT takes approximately 25–30 min to administer, with each subtest varying in administration time from around 3 min (Letter Knowledge and Speech Rate) to up to about 8 min (Beginning and End Phoneme Deletion). The individual subtests demonstrated good internal and test–retest reliability. Criterion-related validity studies have shown that the individual subtests correlate significantly with a concurrently administered standardized reading test (British Abilities Scales Reading Test, Elliott *et al.*, 1983). Multiple regression analyses have also demonstrated that the subtests are significant predictors of concurrent reading skill in the age range 5–7 years, with the best and most consistent predictor subtests being Phoneme Deletion and Letter Knowledge. The predictive validity of the PAT has been recently evaluated in a longitudinal study of almost 100 children. The present author administered the PAT to rising 5-year-olds who had had only very minimal exposure to formal reading instruction. The measures of Letter Knowledge, Phoneme Completion, and Beginning and End Phoneme Deletion together accounted for 55% of the variance in children's reading skills 1 year later. In effect, they predicted, at an accuracy rate of 90%, whether the children would be categorized as good or poor readers 1 year later. The PAT, like the TALS, may be used as a screening measure in the age range 5–7 years (either when given in its entirety or using a subset of tests), or it may be administered to selected older children who are already experiencing reading problems and for whom a diagnostic and prescriptive phonological profile is required.

For older children for whom the PAT may be used diagnostically, reference is made not only to the norms (given in centiles), but also to the graphically represented PAT Profile. The following single case study illustrates how the PAT might be used diagnostically and as a prescription for teaching. Andrew aged 7 years 11 months had a WISC III Verbal IQ of 101 (Wechsler, 1992), and could thus be regarded as being of average ability. He scored at barely the 6-year level on a standardized test of single word reading (Wechsler Objective Reading Dimensions, WORD, Wechsler, 1993), and he was unable to read any nonwords from the Graded Nonword Reading Test (Snowling *et al.*, 1996). Andrew's scores on the Rhyme Detection, Rhyme Production, Beginning and End Phoneme Deletion, and Letter Knowledge subtests were all at or under the tenth centile. He had no difficulty with Sentence or Phoneme Completion (fiftieth centile) or with Speech Rate (seventy-fifth centile). His PAT profile is shown in Figure 1. While Andrew clearly has no problems in respect of phonological short-term memory, and his simple segmentation abilities are established, he is nonetheless experiencing great difficulty in many other aspects of phonological skill and in acquiring letter knowledge. His pattern of difficulty, taken together with his marked reading underachievement and his total lack of decoding ability, is clearly indicative of his having a specific literacy difficulty that has the hallmarks of a dyslexic problem. He needs to embark on a systematic literacy training program which emphasizes phonological awareness and related skills. Of the phonological abilities in which Andrew is deficient, rhyming is the ability which appears earliest in the developmental progression of phonological skills. It is, therefore, recommended that this be trained first. Later on, Andrew needs to work on his phonological manipulation skills through exercises which teach him to add, delete, substitute, or transpose phonemes within words. Andrew also needs to be trained in his letter knowledge, with a teaching approach that emphasizes multi-sensory learning (feeling, writing, naming) of both letter names and sounds and where there is the opportunity for a lot of practice and reinforcement. When Andrew is more phonologically aware, and when he knows all his letters, he should be exposed to "linkage" exercises which help him make important connections between his improving speech sound sensitivity and his experience of print. After Andrew has worked through a program such as this, he should be ready to embark on a structured phonic-based program that teaches him about grapheme-to-phoneme consistencies and about sequential decoding skill.

The PAT is an instrument to be used for assessing phonological skills in children in the early stages of learning to read. However, as we have already seen, phonological skills continue to have a significant long-term bearing on children's reading progress and their ultimate literacy outcome. A UK-based phonological test battery geared toward somewhat older children is the Phonological Assessment Battery (PhAB) (Frederickson, Frith, & Reason, 1997). The PhAB provides norms for children aged 6 years to 14 years 11 months and consists of six subtests: alliteration, naming speed (for pictures and digits), rhyme, spoonerisms, fluency, and nonword reading. The PhAB is intended not so much as a screening

pat record form

Phonological Abilities Test	
Name	ANDREW
School	
Examiner	
Gender	

	Year	Month	Day
Date Tested			
Date of Birth			
Age	7	11	

Test Scores		Raw Score	Centiles
Rhyme Detection	Number correct - (max 10)	4	<10
Rhyme Production	Number of words - day	1	
	Number of words - bell	1	
	Total:	2	10
Word Completion - Syllables and Phonemes	Number correct - syllables - (max 8)	8	50
	Number correct - phonemes - (max 8)	8	50
	Total: (max 16)	16	
Phoneme Deletion Beginning and End sounds	Number correct beginning sounds - (max 8)	0	<10
	Number correct end sounds - (max 8)	1	<10
	Total correct: (max 16)	1	
Speech Rate	A. Buttercup Time 1	1.7"	
	B. Buttercup Time 2	2.7"	
	C. Buttercup Time 3	3.7"	
	D. Mean Time ((A+B+C) / 3)		
	Words per second (10 / D)	1.43"	75
Letter Knowledge	Letters correctly identified - (max 26)	23	<10

PAT Profile

Centiles 100 90 80 70 60 50 40 30 20 10 0

Rhyme Detection; Rhyme Production; Word Completion-Syllables; Word Completion-Phonemes; Phoneme Deletion-Beginning Sounds; Phoneme Deletion-End Sounds; Speech Rate; Letter Knowledge

Figure 1. Phonological Abilities Test (PAT) record form and PAT profile for Andrew.

instrument for early identification (the main purpose of the PAT), but as a diagnostic and prescriptive instrument "in particular... for children whose literacy progress is causing concern" (p. 3). Criterion-related validation studies have confirmed that the PhAB tests show significant correlations with a standardized reading measure, the Neale Analysis of Reading Ability (Neale, 1989). The clinical applicability of the PhAB was demonstrated by comparing the PhAB scores of children with significant reading underachievement with those in the standardization sample; the specific reading disabled group obtained significantly lower scores on all the PhAB subtests apart from the fluency measures.

A recently available, and highly recommended, phonological assessment instrument is the Comprehensive Test of Phonological Processing (CTOPP), developed by Wagner, Torgesen, and Rashotte (1999). It is based on their model of phonological processing ability which comprises three separate, though related, skills: phonological awareness tapped by tests of phonological blending and segmentation and the like; phonological working memory assessed by tests of memory span or nonword repetition; and phonological access in long-term memory evaluated by naming speed measures. There are two versions of the CTOPP, one for use in the 5–6 year age range and the other developed for use with older individuals, aged from 7 to 24 years. The CTOPP was normed on 1,656 individuals across the United States, and its manual reports extensive data on the impressive reliability and validity of the instrument. The core phonological awareness tests are measures of phoneme blending and phoneme deletion (referred to as "elision"); these measures in the 5–7 year age range correlate with reading 1 year later with correlations as high as 0.7 and 0.8. The CTOPP has measures of both memory for digits and nonword repetition which tap phonological working memory skills, together with color, picture, letter, and digit naming tasks that are sensitive to phonological access in long-term memory. Rapid naming and working memory tasks correlate significantly with later reading skills but the correlations are not as impressive as for phonological awareness with later reading; rapid naming tasks in 5–6-year-olds have correlations with later reading in the 0.6–0.7 region, while working memory tasks given at the same age correlate with later reading at a rather lower level (0.4–0.5). Children with severe dyslexia are likely to experience problems on all three types of phonological measure. However, there are some children who may have difficulty with one or two of the phonological tasks, but not all three. Phonological awareness tests are particularly strongly diagnostic of dyslexic difficulties but it is not uncommon to assess children who are able to blend and segment phonemes within words, but who exhibit short-term verbal memory limitations or naming speed problems or both. Whether these apparent individual differences in phonological skill reflect varying degrees of severity or complexity in dyslexia or whether there are qualitatively different subtypes of dyslexia is still an area of some debate.

PREVENTING READING FAILURE THROUGH EARLY PHONOLOGICAL TRAINING

Once a young child has been identified as at risk for reading failure, and assuming that he or she has been found to have weaknesses in phonological processing, the next step is to provide a systematic intervention program that will improve the child's phonological skills and so ameliorate his or her impending reading problem.

There are now several curricula aimed at developing phonological awareness in young children which have been empirically validated in training studies. Lundberg, Frost, and Petersen (1988) developed a program that comprised daily metaphonological games that were given to 235 preschool children in a group context over a period of 1 year. At school entry, the trained children out-performed untreated control children on tests of phonemic awareness. More importantly, the children who had had phonological awareness training obtained significantly higher reading scores than those in the control group when they were tested at 7-monthly intervals through to Grade 3. Lundberg (1994) evaluated the predictive power of this preschool intervention by randomly allocating 50 "at-risk" children (i.e., those performing poorly on phonological awareness tests) to either of two groups: a training group which participated in phonological awareness training activities during their preschool year and an untrained control group. The control group made slow progress in reading over the next 3 years at school; in contrast, the trained group performed at a similar level to a not "at-risk" control group, and by Grade 3 were said to be achieving normal levels of attainment. The results of the Lundberg studies have had far-reaching effects on educational policy within Scandinavia, to such an extent that group-based phonological awareness activities conducted for 15–20 min per day are now routinely included in kindergarten classrooms throughout that country. The Lundberg program has been translated into English by Adams and Huggins (1993). There is a further English adaptation by David Fielker in 1997 which was used as the basis for a training study by Brennan and Ireson (1997) and which proved to be similarly effective with English mother tongue children.

Byrne and Fielding-Barnsley (1991) have developed a training program called "Sound Foundations." This is designed for 4-year-olds and uses colorful illustrations of words that either begin or end with the same sound. Preschoolers trained on this phoneme identification program showed benefits in their reading evident even at 3-year follow-up (Byrne & Fielding-Barnsley, 1995). The program developed by Blachman, Ball, Black, and Tangel (1994) uses a simple task they refer to as "say-it-and-move-it" to teach children to move discs to represent the sounds in words. The children begin by learning to move one disc to represent one sound. Eventually, they learn to represent two-phoneme and three-phoneme items by moving discs to represent individual sounds as they say each word slowly. After segmenting an item into its constituent phonemes, the children repeat the original word as a whole unit. Blachman *et al.* have demonstrated that children's phonological awareness can be heightened using this structured phoneme segmentation task. There is recent evidence that the benefits of phonological awareness training for beginner readers can be even further enhanced when the phonological training is actively and explicitly linked either to the learning of letter names/sounds (Ball & Bryant, 1983; Bradley & Blachman, 1988) or to direct reading experience (Cunningham, 1990). In each of these three studies, young children who received phonological awareness training combined with letter knowledge or reading instruction made greater gains in later reading than those children who had only "pure" phonological training.

It is clear that the above early phonological training programs have considerable similarities, though it is acknowledged that there are some differences in terms of methodology and emphasis. What might the necessary components of a successful phonological training program for beginning readers be? Spector (1995) has suggested seven components to effective early programs for training phonological awareness:

1. Engaging children in activities that direct their attention to the sounds in words; in particular, children need to come to appreciate that words share common sounds.
2. Encouraging children to segment and blend sounds in words.
3. Including training in letter–sound relationships.
4. Teaching segmentation and blending as complementary skills.
5. Sequencing activities systematically (for instance, introducing words with simple consonants before consonant blends, e.g., "cat" before "clap").
6. Encouraging transfer to novel tasks; children need to practice segmentation and blending with words that have different sound structures and within a variety of contexts.
7. Ensuring that teachers have a clear understanding of the rationale for a phonological awareness training program.

The evidence presented thus far may well give the impression that explicit phonological awareness training is critical to the development of good reading skills—and that children who do not receive this training in their early school years are likely to be markedly disadvantaged. While there is very clear evidence for phonological awareness skills being a powerful predictor of reading ability, and bearing in mind the findings of the above training studies, it is tempting to conclude that all children should be given phonological training as part of their classroom learning routine from an early age. However, it is important not to lose sight of the fact that phonological training can be time consuming, demanding on teachers, and, in theory, could take time away from other important classroom activities. If some children do not need phonological training in order to promote literacy development, then clearly we need to know this, given the cost in terms of time and energy involved in delivering such programs.

A very recent study by Hatcher, Hulme, and Snowling (2001) has thrown light on this important issue. One thousand children in the north of England were recruited at the age of 4 years for an evaluation of *whole class teaching* of phonological training programs. One of the treatment groups received a structured phonic reading program while three other treatment groups received the phonic reading program with additional phonological training exercises. There was also an unseen control group. All four experimental groups showed greater improvements in literacy skill development than those in the unseen control group. However, there were no differences between the treatment groups; the additional phonological training did not bestow benefits beyond those achieved in the phonic-reading-only program.

These results were clearly very disappointing, leading the authors to look at their data again, but in a rather different way. First, they selected 137 children from the intervention cohort who had obtained very low scores on measures of phonological skill and letter knowledge at the outset of the study. They then looked at the performance of these "at-risk" children according to the teaching group to which they had been allocated. The children who had the best later reading outcomes were those who received phonic reading, together with phoneme awareness training. Finally, Hatcher and his colleagues charted the progress over time of the reading skills in the children in the at-risk cohort. The children in the phonic-reading-only group tended to show a relative decline over time. However, this decline was effectively arrested in those children who had received the phonic reading and additional phonological training intervention. Hatcher *et al.* have concluded that the optimal early phonological training strategy should begin by screening at-risk children at ages 4–5 years using standardized tests of phonological awareness and letter knowledge. These at-risk children should then have access to individual or small group training programs that particularly emphasize phoneme awareness activities that are linked to literacy instruction. It would seem unnecessary, and indeed wasteful of resources, to provide phonological training for all children. The majority of 4–5-year-old children have established, through normal language learning experiences, phonological awareness skills sufficient to support the early stages of literacy development. Screening and targeting selected at-risk preschoolers makes far better use of teacher time and resources, and enables children to be taught individually or in small groups, a procedure that may reap greater and faster benefits than whole class teaching.

PHONOLOGICAL SKILLS AND READING INSTRUCTION IN CHILDREN WITH DYSLEXIA

The final section of this chapter will look at evidence for phonological training programs improving the literacy achievements of known dyslexic children. Two excellent training studies (one British and one American) will be described in some detail as they stand as models of good clinically relevant experimental research.

The seminal study by Hatcher *et al.* (1994) empirically demonstrated the efficacy of a phonological training program conducted within the context of learning to read. They carried out a training study with 128 poor readers aged 7 years who were allocated to one of four groups matched for age, IQ, and reading age. The reading + phonology group received phonological awareness training, reading experience, and activities that linked the two components. The phonology alone program experienced the same phonological training given the reading + phonology group, but had no explicit reading instruction or phonology linkage exercises. The reading alone group read books, had multisensory training, and learned letter names, but had no phonological training. A control group received conventional classroom instruction. After pretesting on cognitive, phonological awareness, and educational measures, the children were subjected to 40 sessions of individual instruction over a 20-week period. At posttest, the reading + phonology group scored significantly higher than the other groups on measures of reading; these improvements were sustained at follow-up 9 months later. Thus, phonological training is effective in promoting the literacy skills of delayed readers, though training in phonological skill may be less effective than training that forms explicit connections or links between children's underlying phonological skills and their experiences in learning to read.

The second study by Torgesen *et al.* (1999) contrasted two phonologically based training programs that were delivered to young poor readers. They selected 180 kindergartners of normal verbal ability who had obtained low scores on both a letter knowledge and a phoneme deletion task. The children were then randomly assigned to one of four groups: a no-treatment control (NTC) condition; regular classroom support (RCS) condition; the embedded phonics (EP) condition; and the phonological awareness + synthetic phonics (PASP) condition. At pretest, the children were subjected to an extensive battery of tests of language, nonverbal ability, phonological skill, and reading attainment. The children in the two treatment conditions were provided with four 20-min sessions of one-to-one instruction per week for 2½ years. In the PASP teaching condition, the emphasis was largely on word level learning; this strongly emphasized phonological awareness training, teaching letter–sound correspondences, and the decoding of single words. Teachers in the PASP condition spent 80% of the teaching time on word level instruction and only 20% on text level activities. In contrast, the children in the EP condition spent 43% of their training time on word level activities and 57% on text level activities (the latter involved reading or writing connected text). When the children were reassessed on completion of the training study, the authors found that the children in the PASP condition had significantly stronger skills than those in the EP condition in phonological awareness, phonemic decoding, and context-free word reading. Indeed, the word level skills of the children in the PASP condition were in the middle of the average range. In contrast, the children in the EP and RCS conditions demonstrated no significant improvements in word level skill when compared with the control group. The findings of this study echo those of Hatcher *et al.* in that they highlight the importance of phonological and decoding training in effecting significant improvements in young "at-risk" poor readers.

The above studies provide compelling evidence that phonologically based reading programs can facilitate reading progress in young at-risk 5-year-old children (Torgesen *et al.*) and older children who have fallen significantly behind in their reading development (Hatcher *et al.*). The programs adopted by Hatcher and Torgesen are available commercially for use by learning support teachers in their work with reading delayed children.

The study by Hatcher *et al.* produced such encouraging results that Peter Hatcher decided to develop and publish the materials that had been adopted in the study in the form of a phonological training package, called Sound Linkage, now in its second edition (Hatcher, 2000). The phonological activities are divided into nine sections and are graded in the order of difficulty that determines the sequence or order in which they are taught—from word and syllable identification activities through to phoneme manipulation activities. Sound Linkage may be inserted into any of a number of existing structured reading support programs. In Hatcher's own work, he has incorporated Sound Linkage into the Marie Clay Reading Recovery framework (Clay, 1985). This has required the insertion of an additional section that centers around letter identification, phonological training, and phonological linkage activities into the traditional Clay framework of "reading text and recording reading behavior, writing a story, cutting up a story and reading a new book" (for a fuller description, see Hatcher, 1996). A field study of this method revealed encouraging results (Hatcher, 1996). Seven-year-old children, who were about 18 months delayed in their reading at the start of the intervention, made average gains of 8 months in word reading and 11 months in spelling over the 3 months of intervention. Thus, incorporating phonological training within an existing well-tried structured reading program shows considerable promise for tackling the reading deficits shown by dyslexic or reading delayed children.

A popular phonological awareness training program in the United States is the Auditory Discrimination in Depth program developed by Lindamood and Lindamood (1984). Indeed, it was the Lindamood program that formed the core of the PASP teaching condition in the Torgesen *et al.* study. This program begins with instruction designed to make children aware of the specific mouth movements required to produce each phoneme. As part of this instruction, the children also learn labels for each phoneme that are descriptive of these mouth movements and positions, that is, the label "lip popper" is used for the phonemes /b/ and /p/, the label "tip tapper" is assigned to the phonemes /t/ and /d/, while the phonemes /k/ and /g/ are labeled "scrapers." Once the children attain a high level of knowledge in oral awareness, they engage in an extensive series of exercises in which they represent sequences of phonemes with either mouth-form pictures or colored blocks. This training aims to enable them to acquire sensitivity to the sequences of sounds in syllables, and it also enables them to learn to represent these sequences with concrete visual objects. As the children learn to label each phoneme with a descriptive name, they are also taught to associate specific letters with each phoneme. Once they become facile at representing sequences of letters with concrete objects, it is a natural transition to begin to represent them with letters. Like Sound Linkage, the Lindamood program aims to improve children's sensitivity to the sound structure of words while at the same time forging connections between their knowledge of phonemes and other concrete representations of these, be they visual objects or of course ultimately letters. What makes the Lindamood program different from other phonological training packages is the emphasis on articulatory gestures. To what extent these enhance a program and whether they make a bigger difference for the child with particularly severe problems are questions for further research. The Lindamood–Bell Company is developing computer-based programs that help reinforce and extend the teaching principles inherent in their auditory training. These computer programs are being adopted within the computer training model of Olson, Wise, and their colleagues at the University of Colorado. They have developed a talking computer system to provide speech support for children who have decoding difficulties—"ROSS," reading with orthographic and segmented speech feedback (Olson, Foltz, & Wise, 1986). ROSS has been demonstrated to be a successful adjunct to direct teacher–child methods in the promotion of segmentation and decoding skills in reading disabled children. Olson and Wise (1992) noted that subjects who had the lowest initial levels of phoneme awareness tended to gain significantly less from reading with speech feedback. In a training study conducted by Wise and Olson (1995), children with low levels of phonemic awareness were first taught to become aware of the distinctive articulatory movements associated with different speech sounds, using the Lindamood Auditory Discrimination training program (and reinforcer computer materials). They were then switched to the ROSS program. The results following training showed dramatically improved performance (relative to a comprehension training group and the findings of previous ROSS studies) in phoneme awareness and, importantly, in decoding skill.

While most phonological/phonic programs have taught at the level of the smallest unit, that is, the phoneme, there has been an upsurge of interest in larger sublexical units, specifically at the level of onset and rime (what teachers have usually termed "word family" learning, e.g., learning related groups of words like "light, might, fight" etc.). Goswami and her colleagues (see Goswami, 1994, for a review) have argued that children as young as 5 years can benefit significantly from instruction in onset-rime awareness and specific instruction in orthographic analogies. As well as the empirical evidence for teaching analogies derived from Goswami's elegant experimental studies, there are two further arguments as to why this approach may be particularly helpful for the dyslexic child. First, decoding using larger chunks at the level of the rime rather than the phoneme acts to reduce the load on memory when phonological segments need to be synthesized to form pronunciations. Second, larger sublexical units show greater consistency and regularity than do smaller unit graphemes within the English language; consequently, it should be easier to learn and to apply generalizations about these larger units without being confused and confounded by so many exceptions that occur at the level of smaller units. However, on a cautionary note, some experimental studies have shown that rime training may not reap such long-term benefits as teaching smaller phoneme–grapheme correspondences (Bruck & Treiman, 1992), and that it may be necessary for children to have some decoding skill before they can make sense of an analogical approach to reading (Ehri & Robbins, 1992). Furthermore, while Goswami has demonstrated powerful analogy effects in experimental studies, there is recent evidence that children may not use analogies spontaneously, at any rate not until they have had the opportunity to develop a sight vocabulary on which to base their analogical inferences (Muter *et al.*, 1994;

Savage, 1997). Goswami (1994) has herself suggested that very young children need to have established onset-rime awareness, if necessary achieved through rhyming-oriented phonological training, to take advantage of rime-based reading approaches. On an encouraging note, a series of training studies with disabled readers conducted by Greany, Tunmer, and Chapman (1997) have demonstrated that children with reading ages as low as 7 years are able to profit from instruction in rime analogies when compared with reading-age matched controls. In response to 3–4 times weekly individual instruction over a 12-week term, the children taught rime analogies not only significantly improved on experimental measures of rime awareness and word recognition at postterm, but importantly they also showed a 6-month gain in reading age on a standardized test.

CONCLUSION

Much has been learned of the important contribution of phonological awareness to reading development in both normal and disabled readers. The implications for early reading practices within the classroom and in particular for children whose phonological deficits are preventing them from becoming good readers, are not inconsiderable. However, there are still many unanswered questions. The age at which children should be screened for phonological difficulties and the stage at which intervention should begin are important pedagogical issues. Phonological skills are less stable in young children, a factor which can impair their predictive relationship to later reading skill; consequently, it may be advisable not to attempt large-scale screening of preschool children, but instead concentrate screening and early identification around the first year of formal schooling. There is also the question of children's developmental readiness to profit from a systematic phonological training program and whether such programs meet the needs of all preschoolers. We have come a long way in respect of developing measures of phonological awareness, but there remains the controversy as to which skills (for instance, rhyming vs segmentation) best predict reading development. This is an important issue because it has a bearing on which tasks are to be selected for screening and assessment procedures and which skills are prioritized in training programs. Further research needs to be carried out into how long and how intensive phonological awareness training programs should be, not just from the crucial economic point of view, but also because this factor is likely to have a considerable bearing on the long-term efficacy and generalizability of phonological training programs. We need to know more about how phonological awareness relates to other phonological processing, and reading-related skills such as naming speed and phonological working memory; it may become increasingly relevant not only to measure these related abilities during assessment, but also to find ways of programming them into intervention packages. Finally, there are crucial issues of individual responsivity to phonological awareness training programs. We need to know which children are resistant to such training programs and what factors predict responsivity. Do resistant children merely require more intensive, lengthier programs or is it a question of looking at different types of teaching techniques? The individual differences variable also draws attention to children's capacity for developing compensatory strategies; is it possible in intervention programs to promote children's syntactic, visual, or other strengths in order to enable them to find alternative routes to word recognition beyond those of relying on their depleted phonological abilities? Phonological awareness training and alphabetic coding skills are now becoming a feature of classroom teaching in many countries. To ensure that we optimize the benefits to beginner readers and children with reading difficulties, it will be necessary to refine further and develop our phonologically based assessment and intervention procedures. These in turn need to be grounded in empirical research and sound theory, a challenge that will hopefully be met as we advance into the 21st century, with the goal of making long-term reading disability a rarity.

REFERENCES

Achenbach, T. M. (1991). *Child behavior checklist.* Burlington, VT: University of Vermont, Department of Psychiatry.

Adams, M. J. (1990). *Beginning to read: Learning and thinking about print.* Cambridge, MA: MIT Press.

Adams, M. J., & Huggins, A. (1993). *Lundberg, Frost and Petersen's program for stimulating phonological awareness among kindergarten, first-grade and special education students.* Unpublished Manuscript, BBN Labs, Cambridge, MA.

Badian, N. A. (1988). Predicting dyslexia in a preschool population. In R. L. Maslund, & M. W. Maslund (Eds.), *Preschool prevention of reading failure* (pp. 78–103). Parkton, MD: York Press.

Badian, N. (1994). Preschool prediction: Orthographic and phonological skills, and reading. *Annals of Dyslexia, 44,* 3–25.

Badian, N. (2000). Do orthographic skills contribute to prediction in reading? In N. Badian (Ed.), *Prediction and prevention of reading failure* (pp. 31–56). Baltimore, MD: York Press.

Badian, N. A., McAnulty, G. B., Duffy, F., & Als, H. (1990). Prediction of dyslexia in preschool boys. *Annals of Dyslexia, 40,* 152–169.

Ball, E. W., & Blachman, B. A. (1988). Phoneme segmentation training: Effect on reading readiness. *Annals of Dyslexia, 38,* 208–225.

Barron, R. (1991). Protoliteracy, literacy and the acquisition of phonological awareness. *Learning and Individual Differences, 3,* 243–255.

Blachman, B. A., Ball, E. W., Black, R. S., & Tangel, D. M. (1994). Kindergarten teachers develop phoneme awareness in low-income, inner city classrooms. Does it make a difference? *Reading and Writing: An Interdisciplinary Journal, 6,* 1–18.

Bond, G. L., & Dykstra, R. (1967). The cooperative research program on first grade reading instruction. *Reading Research Quarterly, 2,* 5–142.

Bowey, J. (1995). Socioeconomic status differences in preschool phonological sensitivity and first-grade reading achievement. *Journal of Educational Psychology, 87,* 476–487.

Brady, S., Shankweiler, D., & Mann, V. (1983). Speech perception and memory coding in relation to reading ability. *Journal of Experimental Psychology, 35,* 345–367.

Bradley, L., & Bryant, P. E. (1983). Categorising sounds and learning to read: A causal connection. *Nature, 301,* 419–521.

Brennan, F., & Ireson, J. (1997). Training phonological awareness: A study to evaluate the effects of a program of metalinguistic games in kindergarten. *Reading and Writing: An Interdisciplinary Journal, 9*, 241–263.

Bruce, L. J. (1964). The analysis of word sounds by young children. *British Journal of Educational Psychology, 34*, 158–174.

Bruck, M., & Treiman, R. (1992). Learning to pronounce words: The limitations of analogies. *Reading Research Quarterly, 27*, 375–388.

Brunswick, N., McCrory, E., Price, C. J., Frith, C. D., & Frith, U. (1999). Explicit and implicit processing of words and pseudowords by adult developmental dyslexics: A search for Wernicke's Worschatz? *Brain, 122*, 1901–1917.

Bryant, P. E. (1998). Sensitivity to onset and rime does predict young children's reading: A comment on Muter, Hulme, Snowling and Taylor. *Journal of Experimental Child Psychology, 71*, 29–37.

Burgess, S. R., & Lonigan, C. J. (1998). Bidirectional relations of phonological sensitivity and pre-reading abilities: Evidence from a preschool sample. *Journal of Experimental Child Psychology, 70*, 117–141.

Butler, S. R., Marsh, H. W., Sheppard, M. J., & Sheppard, J. L. (1985). Seven year longitudinal study of the early prediction of reading achievement. *Journal of Educational Psychology, 77*, 349–361.

Byrne, B. (1998). *The foundation of literacy: The child's acquisition of the alphabetic principle*. Hove, UK: Psychology Press.

Byrne, B., & Fielding-Barnsley, R. (1989). Phonemic awareness and letter knowledge in the child's acquisition of the alphabetic principal. *Journal of Educational Psychology, 82*, 805–812.

Byrne, B., & Fielding-Barnsley, R. (1991). Evaluation of a program to teach phonemic awareness to young children. *Journal of Educational Psychology, 83*, 451–455.

Byrne, B., & Fielding-Barnsley, R. (1995). Evaluation of a program to teach phoneme awareness to young children: A 2- and 3-year follow-up and a new pre-school trial. *Journal of Educational Psychology, 87*, 488–503.

Byrne, B., Fielding-Barnsley, R., Ashley, L., & Larsen, K. (1997). Assessing the child's and the environment's contribution to reading acquisition: What we know and what we don't know. In B. Blachman (Ed.), *Foundations of dyslexia and early reading acquisition* (pp. 265–285). Hillsdale, NJ: Erlbaum & Associates.

Cataldo, S., & Ellis, N. (1988). Interactions in the development of spelling, reading and phonological skills. *Journal of Research in Reading, 11*, 86–109.

Chall, J. (1967). *Learning to read: The great debate*. New York: McGraw-Hill.

Clay, M. (1985). *The early detection of reading difficulties* (3rd ed.). Oxford: Heinemann Educational.

Cunningham, A. E. (1990). Explicit versus implicit instruction in phonemic awareness. *Journal of Experimental Child Psychology, 50*, 429–444.

Duncan, L. G., Seymour, P. H. K., & Hill, S. (1997). How important are rhyme and analogy in beginning reading? *Cognition, 63*, 171–208.

Ehri, L. C. (1992). Reconceptualising the development of sight word reading and its relationship to recoding. In P. B. Gough, L. C. Ehri, & R. Treiman (Eds.), *Reading acquisition* (pp. 107–143). Hillsdale, NJ: Erlbaum Associates.

Ehri, L. C., & Robbins, C. (1992). Beginners need some decoding skills to read words by analogy. *Reading Research Quarterly, 27*, 13–25.

Elbro, C., Borstrom, I., & Petersen, D. (1998). Predicting dyslexia from kindergarten: The importance of distinctness of phonological representations of lexical items. *Reading Research Quarterly, 33*, 39–60.

Elliott, C. D., Murray, D. J., & Pearson, L. S. (1983). *British abilities scales*. Windsor, Berkshire, UK: NFER-Nelson Press.

Feagans, L., & Farran, D. C. (1982). *The language of children reared in poverty: Implications for evaluation and intervention*. New York: Academic Press.

Felton, R. H., & Brown, I. S. (1990). Phonological processes as predictors of specific reading skills in children at risk for reading failure. *Reading and Writing: An Interdisciplinary Journal, 2*, 39–59.

Fielker, D. (1997). *Language games for the reinforcement of language awareness*. Translation from the program developed by J. Frost & A. Lonnegaard (1995). Unpublished Manuscript.

Fowler, A. F. (1991). How early phonological development might set the stage for phoneme awareness. In S. Brady & D. Shankweiler (Eds.), *Phonological processes in literacy: A tribute to Isabelle Y. Liberman* (pp. 97–118). Hillsdale, NJ: Erlbaum & Associates.

Frederickson, N., Frith, U., & Reason, R. (1997). *Phonological assessment battery (PhAB)*. Windsor, UK: NFER-Nelson.

Gallagher, A., Frith, U., & Snowling, M. J. (2000). Precursors of literacy delay among children at risk of dyslexia. *Journal of Child Psychology and Psychiatry, 41*, 203–213.

Gathercole, S., & Baddeley, A. (1989). Development of vocabulary in children and short-term phonological memory. *Journal of Memory and Language, 28*, 200–213.

Gathercole, S., Baddeley, A., & Willis, C. (1991). Differentiating phonological memory and awareness of rhyme: Reading and vocabulary development in children. *British Journal of Psychology, 82*, 387–406.

Gayan, J., & Olson, R. (1999). Genetic and environmental influences on individual differences in IQ, phonological awareness, word recognition, phonemic and orthographic decoding. Poster presented at the meeting of the *Society for the Scientific Study of Reading*, Montreal, Canada.

Geschwind, N., & Levitsky (1968). Human brain left–right asymmetries in the temporal speech region. *Science, 161*, 186–187.

Goswami, U. (1990). A special link between rhyming skills and the use of orthographic analogies by beginning readers. *Journal of Child Psychology and Psychiatry, 31*, 301–311.

Goswami, U. (1994). Reading by analogy: Theoretical and practical perspectives. In C. Hulme & M. Snowling (Eds.), *Reading development and dyslexia* (pp. 18–30). London: Whurr.

Goswami, U., & Bryant, P. E. (1990). *Phonological skills and learning to read*. London: Erlbaum & Associates.

Greany, K., Tunmer, W. E., & Chapman, J. (1997). The use of rime-based analogy training as an intervention strategy for reading-disabled children. In B. Blachman (Ed.), *Foundations of reading acquisition and dyslexia* (pp. 327–345). Hillsdale, NJ: Erlbaum & Associates.

Hansen, J., & Bowey, J. A. (1994). Phonological analysis skills, verbal working memory and reading ability in second grade children. *Child Development, 65*, 938–950.

Hatcher, P. (1996). Practising sound links in reading intervention with children. In M. Snowling, & J. Stackhouse (Eds.), *Dyslexia, speech and language* (pp. 146–170). London: Whurr.

Hatcher, P. (2000). *Sound linkage* (2nd ed.). London: Whurr.

Hatcher, P., Hulme, C., & Ellis, A. W. (1994). Ameliorating early reading failure by integrating the teaching of reading and phonological skills: The Phonological Linkage hypothesis. *Child Development, 65*, 41–57.

Hatcher, P., Hulme, C., & Snowling M. J. (2001). Training rhyme and phoneme skills facilitates reading through phoneme awareness. Paper presented at the Society for Scientific Study of Reading, SSSR. Boulder, CO.

Hogben, J. (1997). How does a visual transient deficit affect reading? In C. Hulme & M. Snowling, (Eds.), *Dyslexia: Biology, cognition and intervention* (pp. 59–71). London: Whurr.

Hulme, C., & MacKenzie, S. (1992). *Working memory and severe learning difficulties*. Hove, UK: Erlbaum & Associates.

Hulme, C., & Roodenrys, S. (1995). Verbal working memory development and its disorders. *Journal of Child Psychology and Psychiatry, 36*, 373–398.

Hulme, C., Muter, V., & Snowling, M. (1998). Segmentation does predict early progress in learning to read better than rhyme: A reply to Bryant. *Journal of Experimental Child Psychology, 71*, 39–44.

Iverson, S., & Tunmer, W. E. (1993). Phonological processing and the reading recovery program. *Journal of Educational Psychology, 85*, 112–126.

Johnston, R., Anderson, M., & Holligan, C. (1996). Knowledge of the alphabet and explicit awareness of phonemes in pre-readers: The nature of the relationship. *Reading and Writing: An Interdisciplinary Journal, 8*, 217–234.

Katz, R. B., Healy, A. F., & Shankweiler, D. (1983). Phonetic coding and order memory in relation to reading proficiency: A comparison of short-term memory for temporal and spatial order information. *Applied Psycholinguistics, 4*, 229–250.

Klicpera, C., & Schabmann, A. (1993). Do German-speaking children have a chance to overcome reading and spelling difficulties? A longitudinal survey from second until eighth grade. *European Journal of Psychology of Education, 8*, 307–323.

Larsen, J. P., Hoien, T., Lundberg, I., & Odegaard, H. (1990). MRI evaluation of the size and planum temporale in adults with developmental dyslexia. *Brain and Language, 39*, 289–301.

Lefly, D. L., & Pennington, B. F. (1996). Spelling errors and reading fluency in compensated adult dyslexics. *Annals of Dyslexia, 41*, 143–162.

Liberman, I. Y., Shankweiler, D., Fischer, F. W., & Carter, B. (1974). Reading and the awareness of language segments. *Journal of Experimental Child Psychology, 18*, 201–212.

Lindamood, C., & Lindamood, P. (1984). *Auditory discrimination in depth.* Allen, TX: DLM/Teaching Resources.

Lonigan, C. J., Burgess, S., Anthony, J. L., & Barker, T. A. (1998). Development of phonological sensitivity in 2- to 5-year old children. *Journal of Educational Psychology, 90*, 294–311.

Lonigan, C. J., Anthony, J. L., & Dyer, S. (1996). The influence on the home literacy environment on the development of literacy skills in children from diverse economic backgrounds. Paper presented at the Annual Convention of the American Educational Research Institution, New York.

Lundberg, I. (1994). Reading difficulties can be predicted and prevented: A Scandinavian perspective on phonological awareness and reading. In C. Hulme, & M. Snowling (Eds.), *Reading development and dyslexia* (pp. 180–199). London: Whurr.

Lundberg, I., Frost, J., & Petersen, O.-P. (1988). Effects of an extensive program for stimulating phonological awareness in preschool children. *Reading Research Quarterly, 23*, 264–284.

Manis, F., Seidenberg, M., & Doi, L. (1999). See Dick RAN: Rapid naming and the longitudinal prediction of reading subskills in first and second graders. *Scientific Studies of Reading, 3*, 129–157.

Mann, V. A., Liberman, I. Y., & Shankweiler, D. (1980). Children's memory for sentences and word strings in relation to reading ability. *Memory and Cognition, 8*, 329–335.

MacLean, M., Bryant, P. E., & Bradley, P. E. (1987). Rhymes, nursery rhymes and reading in early childhood. *Merrill–Palmer Quarterly, 33*, 255–281.

Metsala, J. L. (1999). Young children's phonological awareness and nonword repetition as a function of vocabulary development. *Journal of Educational Psychology, 91*, 3–19.

Muter, V., Hulme, C., & Snowling, M. (1997). *The Phonological Abilities Test, PAT.* London: The Psychological Corporation.

Muter, V., Hulme, C., Snowling, M., & Taylor, S. (1998). Segmentation, not rhyming predicts early progress in learning to read. *Journal of Experimental Child Psychology, 71*, 3–27.

Muter, V., & Snowling, M. (1998). Concurrent and longitudinal predictors of reading: The role of metalinguistic and short-term memory skills. *Reading Research Quarterly, 33*, 320–337.

Muter, V., Snowling, M., & Taylor, S. (1994). Orthographic analogies and phonological awareness: Their role and significance in early reading development. *Journal of Child Psychology and Psychiatry, 35*, 293–310.

Muter, V., Hulme, C., Snowling, M., Stevenson, J. (in press). Phonemes, rimes, vocabulary and grammatical skills as foundations of early reading development: Evidence from a longitudinal study. *Developmental Psychology.*

McDougall, S., Hulme, C., Ellis, A. W., & Monk, A. (1994). Learning to read: The role of short-term memory and phonological skills. *Journal of Experimental Child Psychology, 58*, 112–133.

Nation, K., & Snowling, M. (1998). Contextual facilitation of word recognition: Evidence from dyslexia and poor reading comprehension. *Child Development, 69*, 996–1011.

Neale, M. (1989). *Neale analysis of reading ability—revised.* Windsor, UK: NFER-Nelson.

Olson, R., Foltz, G., & Wise, B. (1986). Reading instruction and remediation with the aid of computer speech. *Behavior Research Methods, Instruments and Computers, 18*, 93–99.

Olson, R., & Wise, B. (1992). Reading on the computer with orthographic and speech feedback: An overview of the Colorado Remedial Reading Project. *Reading and Writing: An Interdisciplinary Journal, 4*, 107–144.

Olson, R., Wise, B., Connors, F., Rack, J., & Fulker, D. (1989). Specific deficits in component reading and spelling skills: Genetic and environmental influences. *Journal of Learning Disabilities, 22*, 339–349.

Paulesu, E., Frith, U., Snowling, M., Gallagher, A., Morton, J., Frakowiak, R. *et al.* (1996). Is developmental dyslexia a disconnection syndrome? Evidence from PET scanning. *Brain, 119*, 143–157.

Plaut, D., McClelland, J., Seidenberg, M., & Patterson, K. (1996). Understanding normal and impaired reading: Computational principles in quasi-regular domains. *Psychological Review, 103*, 56–115.

Perfetti, C., Beck, I., Bell, L., & Hughes, C. (1987). Phonemic knowledge and learning to read are reciprocal: A longitudinal study of first grade children. *Merrill–Palmer Quarterly, 33*, 283–319.

Rack, J., Hulme, C., Snowling, M., & Wightman, J. (1994). The role of phonology in young children's learning of sight words: The direct mapping hypothesis. *Journal of Experimental Child Psychology, 57*, 42–71.

Rack, J., Snowling, M., & Olson, R. (1992). The nonword reading deficit in developmental dyslexia: A review. *Reading Research Quarterly, 27*, 29–53.

Rohl, M., & Pratt, C. (1995). Phonological awareness, verbal working memory and acquisition of literacy. *Reading and Writing: An Interdisciplinary Journal, 7*, 327–360.

Raz, I. S., & Bryant, P. E. (1990). Social background, phonological awareness and children's reading. *British Journal of Developmental Psychology, 8*, 209–225.

Savage, R. (1997). Do children need concurrent prompts in order to use lexical analogies in reading? *Journal of Child Psychology and Psychiatry, 38*, 235–246.

Sawyer, D. (1987). *Test of awareness of language segments, TALS*. Austin, TX: Pro-Ed.

Scarborough, H. (1990). Very early language deficits in dyslexic children. *Child Development, 61*, 1728–1743.

Scarborough, H. (1991). Antecedents to reading disability: Preschool language development and literacy experiences of children from dyslexic families. *Reading and Writing, 3*, 219–233.

Scarborough, H., & Dobrich, W. (1994). On the efficacy of reading to preschoolers. *Developmental Review, 14*, 245–302.

Schneider, W., & Naslund, J. C. (1993). The impact of metalinguistic competencies and memory capacity on reading and spelling in elementary schools: Results of the Munich longitudinal study on the genesis of individual competencies (LOGIC). *European Journal of Psychology of Education, 8*, 273–287.

Schonell, F., & Goodacre, E. (1971). *The psychology and teaching of reading* (5th ed.). Edinburgh & London: Oliver & Boyd.

Seidenberg, M., & McClelland, J. (1989). A distributed developmental model of word recognition and naming. *Psychological Review, 96*, 523–568.

Senechal, M., LeFevre, J., Thomas, E., & Daley, K. (1998). Differential effects of home literacy experiences on the development of oral and written language. *Reading Research Quarterly, 33*, 96–116.

Seymour, P. H. K., & Duncan, L. (1997). Small versus large unit theories of reading acquisition. *Dyslexia, 3*, 125–134.

Shankweiler, D., Crain, S., Brady, S., & Macaruso, P. (1992). Identifying the causes of reading disability. In P. B. Gough, L. C. Ehri, & R. Treiman (Eds.), *Reading acquisition* (pp. 275–305). Hillsdale, NJ: Erlbaum Associates.

Snowling, M. (1998). Reading development and its difficulties. *Educational and Child Psychology, 15*, 44–58.

Snowling, M. (2000). *Dyslexia*. Oxford: Blackwell Publishers Ltd.

Snowling, M., Chiat, S., & Hulme, C. (1991). Words, nonwords and phonological processes: Some comments on Gathercole, Ellis, Emslie and Baddeley. *Applied Psycholinguistics, 12*, 369–373.

Snowling, M., & Hulme, C. (1994). The development of phonological skills. *Transactions of the Royal Society, B346*, 21–28.
Snowling, M., Stothard, S., & MacLean, J. (1996). *The graded nonword reading test*. Bury St Edmunds, UK: Thames Valley Test.
Spector, J. E. (1995). Phonemic awareness training: Application of principles of direct instruction. *Reading and Writing Quarterly: Overcoming Learning Difficulties, 11*, 37–51.
Stanovich, K. E., Cunningham, A. E., & Cramer, B. B. (1984). Assessing phonological processes in kindergarten children: Issues of task comparability. *Journal of Experimental Child Psychology, 38*, 175–190.
Stuart, M. (1990). Processing strategies in a phoneme deletion task. *Quarterly Journal of Experimental Psychology, 42*, 305–327.
Stuart, M., & Masterson, J. (1992). Patterns of reading and spelling in 10-year-old children related to pre-reading phonological abilities. *Journal of Experimental Child Psychology, 54*, 168–187.
Tabachnick, B., & Fidell, L. (1989). *Using multivariate statistics*. New York: Harper Collins.
Torgesen, J., & Bryant, B. (1994). *Test of phonological awareness, TOPA*. Austin, TX: Pro-Ed.
Torgesen, J. K., Wagner, R. K., Rashotte, C. A., Rose, E., Lindamood, P., Conway, T. *et al.* (1999). Preventing reading failure in young children with phonological processing disabilities: Group and individual responses to instruction. *Journal of Educational Psychology, 91*, 579–593.
Treiman, R., Weatherston, S., & Berch, D. (1994). The role of letter names in children's learning of phoneme–grapheme relationships. *Applied Psycholinguistics, 15*, 97–122.
Tunmer, W. E. (1989). The role of language-related factors in reading disability. In D. Shankweiler, & I. Y. Liberman (Eds.), *Phonology and reading disability: Solving the reading puzzle, IARLDM* (pp. 91–131). Ann Arbor: University of Michigan Press.
Tunmer, W. E., & Chapman, J. W. (1998). Language prediction skill, phonological recoding, and beginning reading. In C. Hulme & R. Joshi (Eds.), *Reading and spelling: Development and disorders* (pp. 33–67). Mahwah, NJ: Erlbaum & Associates.
Tunmer, W. E., & Hoover, W. A. (1992). Cognitive and linguistic factors in learning to read. In P. B. Gough, L. C. Ehri, & R. Treiman (Eds.), *Reading acquisition* (pp. 175–214). Hillsdale, NJ: Erlbaum & Associates.
Vellutino, F. (1979). *Dyslexia: Research and theory*. Cambridge, MA: MIT Press.
Vellutino, F., & Scanlon, D. (1991). The pre-eminence of phonologically based skills in learning to read. In S. Brady, & D. Shankweiler (Eds.), *Phonological processes in literacy: A tribute to Isabelle Y. Liberman*. Hillsdale, NJ: Erlbaum & Associates.
Wagner, R. K., & Torgesen, J. K. (1987). The nature of phonological processing and its causal role in the acquisition of reading skills. *Psychological Bulletin, 101*, 192–212.
Wagner, R.K, Torgesen, J. K., & Rashotte, C. A. (1994). The development or reading-related phonological processing abilities: New evidence of bi-directional causality from a latent variable longitudinal study. *Developmental Psychology, 30*, 73–87.
Wagner, R. K., Torgesen, J. K., Rashotte, C. A., Hecht, S. A., Barker, T. A., Burgess, S. R. *et al.* (1997). Changing relations between phonological processing abilities and word-level reading as children develop from beginning to skilled readers: A 5-year longitudinal study. *Developmental Psychology, 33*, 468–479.
Wagner, R. K., Torgesen, J. K., & Rashotte, C. A. (1999). *Comprehensive test of phonological processing: CTOPP*. Austin, TX: Pro-Ed.
Walley, A. (1993). The role of vocabulary development in children's spoken word recognition and segmentation ability. *Developmental Review, 13*, 286–350.
Wechsler, D. (1967). *Wechsler preschool and primary scale of intelligence—WPPSI*. London: Psychological Corporation, Harcourt Brace.
Wechsler, D. (1992). *Wechsler intelligence scale for children III UK*. London: The Psychological Corporation, Harcourt Brace.
Wechsler, D. (1993). *Wechsler objective reading dimensions, WORD*. London: Psychological Corporation, Harcourt Brace.

White, K. R. (1982). The relation between socioeconomic status and academic achievement. *Psychological Bulletin, 91*, 461–481.

Whitehurst, G., & Lonigan, C. (1998). Child development and emergent literacy. *Child Development, 69*, 848–872.

Willows, D., & Ryan, E. (1986). The development of grammatical sensitivity and its relationship to early reading achievement. *Reading Research Quarterly, 21*, 253–266.

Wimmer, H., Landerl, K., & Schneider, W. (1994). The role of rhyme awareness in learning to read a regular orthography. *British Journal of Developmental Psychology, 112*, 469–484.

Wise, B., & Olson, R. (1995). Computer-based phonological awareness and reading instruction. *Annals of Dyslexia, 45*, 99–122.

Wolf, M. (1997). A provisional, integrative account of phonological and naming-speed deficits in dyslexia: Implications for diagnosis and intervention. In B. Blachman (Ed.), *Foundations of reading acquisition in dyslexia* (pp. 67–92). Hillsdale, NJ: Erlbaum & Associates.

Wolf, M., & O'Brien, B. (2001). On issues of time, fluency and intervention. In A. Fawcett (Ed.), *Dyslexia: Theory and good practice* (pp. 124–40). London, UK: Whurr.

Yopp, H. (1988). The validity and reliability of phonemic awareness tests. *Reading Research Quarterly, 23*, 159–177.

6

Verbal Memory in the Learning of Literacy

Susan J. Pickering

It is widely believed that verbal memory, and in particular verbal short-term memory, plays an important role in our ability to acquire literacy skills. Moreover, difficulties with verbal short-term memory are often evident in children and adults with literacy difficulties, such as developmental dyslexia (e.g., Jorm, 1983). Turner (1997) suggests that immediate verbal memory deficits have been seen by many as the "hallmark of dyslexia" (p. 86), while Rack (1994) states that "one of the most reliable and often-quoted associated characteristics of developmental dyslexia is an inefficiency in short-term memory" (p. 9).

It is no surprise, then, that a great deal of research has been carried out in order to understand the nature of the relationship between verbal short-term memory and literacy. Although the literature on this topic is vast, very few clear findings have emerged and the extent to which short-term memory is critical for successful literacy acquisition has yet to be established.

In this chapter, we will examine the current status of our understanding of the role of short-term memory in literacy and literacy disorders. However, before we begin to consider this issue directly, it is necessary to understand how our view of short-term memory has changed over the last 30 years. Models and theories of memory inevitably guide research by specifying the memory processes to be investigated and the methods by which such investigations can be carried out.

Susan J. Pickering, University of Bristol, Graduate School of Education, Bristol BS8 IJA, UK.

The Study of Dyslexia, edited by Turner and Rack.
Kluwer Academic Publishers, New York, 2004.

In the last 30 years, two major theories of short-term memory have dominated research in this field. We will consider these models in the following section.

TWO MODELS OF MEMORY: THE MODAL MODEL AND THE WORKING MEMORY MODEL

A popular conceptualization of memory during the 1960s and 1970s was the "modal model" of memory (e.g., Atkinson & Shiffrin, 1968). This model of memory was particularly influential because it specified a separation between short-term memory and long-term memory. A schematic representation of the modal model is shown in Figure 1.

According to the model, information from the world enters our memory system via sensory registers that feed into a limited capacity short-term store. The short-term store is capable of maintaining information long enough for it to enter a long-term store by a process of rehearsal. Rehearsal involves the repeated re-presentation of a memory list by an individual, often in the form of saying the list over and over to themselves (either aloud or silently). The more often a list is rehearsed, the greater the likelihood of it being remembered (Rundus, 1971). Although the modal model did acknowledge that input from the world can be represented in memory in a range of ways including visual, auditory, and semantic (based on the meaning of memory items), the majority of research carried out with this model tended to focus on verbal coding in short-term memory. Experimental procedures for assessing short-term memory were largely verbal and, therefore, significant findings from this research were limited to the verbal domain.

As is often the case with models of psychological processes, the modal model was found to have a number of limitations that brought into question its validity as an accurate account of the way that human memory operated. One of these limitations centered on the assertion that short-term memory was served by a unitary store. A series of studies by Baddeley and colleagues (Baddeley & Hitch, 1974; Baddeley, Lewis, Eldridge, & Thomson, 1984) found that when a test of short-term memory, a digit span task, was combined with another activity requiring the use of short-term memory (such as the comprehension of prose passages), performance on these two activities was not completely impaired (as would be the case if there were only one short-term memory store). This finding suggested that short-term memory was composed of more than one component and led to the

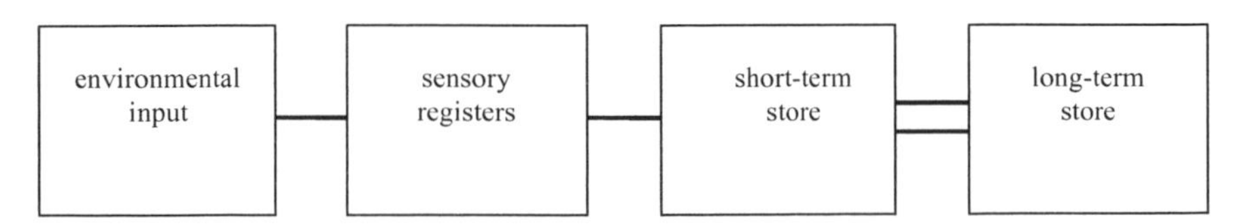

Figure 1. A schematic representation of the modal model of memory.

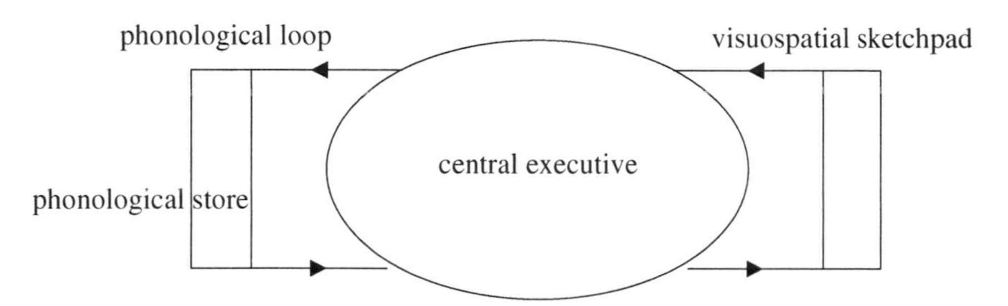

Figure 2. The working memory model.

development of a second account of short-term memory, the working memory model.

On the basis of the above findings, Baddeley and Hitch (1974; and later Baddeley, 1986) suggested that human temporary memory operated as a *working memory* in which a controlling attentional system (the *central executive*) supervises and coordinates a number of subsidiary slave systems. Two slave systems were initially proposed: the *phonological loop* and the *visuospatial* sketchpad, responsible for dealing with information in verbal and nonverbal form, respectively. Figure 2 presents a typical representation of the working memory model.

Since its initial proposal, a vast amount of research has been devoted to understanding more about how working memory operates. Information from a range of sources provides converging evidence for a number of phenomena that appear to reflect core characteristics of our temporary memory processes. Much information has been gathered from healthy adults and children (see Gathercole, 1999, for a recent review), while other research has focused on neurological patients and other clinical populations who exhibit significant deficits in working memory (see Baddeley, 1998). Particularly important are individuals whose impairment is found to be located in only one of the working memory components. Such cases provide strong evidence to support the existence of a multicomponent working memory system. In addition to the great number of experimental and neuropsychological studies of working memory, sophisticated brain imaging techniques have also been used to study working memory. Evidence from these studies also supports a multicomponent working memory by implicating different brain areas in the activity of the phonological loop, visuospatial sketchpad, and central executive (e.g., Smith & Jonides, 1999). Most importantly perhaps, while a vast range of studies have provided evidence to support this account of memory, very little evidence exists to challenge it.

THE STRUCTURE AND FUNCTION OF WORKING MEMORY

Our understanding of working memory is continually increasing. However, our knowledge of the three specified components of the system is not equal—much more is known about the phonological loop than the other two components. One important factor in this is the availability of reliable techniques for measuring the activity of the three components of working memory. While phonological loop tasks have been relatively easy to develop, sampling the visuospatial sketchpad and central executive has proved more difficult. This situation is beginning to change; in the last decade, much more attention has been given to these two components of working memory, with a corresponding increase in our knowledge about how they operate.

The Phonological Loop

A wealth of evidence to suggest that verbal coding was important in short-term memory was already available (e.g., Conrad & Hull, 1964) before the working memory model came into existence. However, in proposing the phonological loop, Baddeley and Hitch took our understanding of verbal short-term memory one step further, by specifying its precise structure. It was suggested that the phonological loop was composed of two subcomponents: a phonological store (capable of holding information for a limited period) and a phonological rehearsal system (capable of refreshing the contents of the phonological store and consequently lengthening the duration of the memory trace).

Evidence to support the structure of the phonological loop has been obtained from a number of experimental studies. One group of studies has focused on a phenomenon called the *phonological similarity effect* (Baddeley, 1966; Conrad & Hull, 1964). If participants are given a memory list composed of letters and asked to recall the list in order, they will find a list like "P G T D C" much more difficult than "R H X W Y." It has been suggested that the former list is harder because the phonological store holds information in a speech-based code and therefore items with similar codes (e.g., P and G) will be harder to discriminate from one another. Another demonstration of the importance of speech-based codes in the phonological loop comes from studies of *unattended speech* (e.g., Colle & Welsh, 1976; Salame & Baddeley, 1987). When participants are asked to recall memory lists while being presented with background noise composed of speech, performance is significantly impaired. This is thought to occur because the unattended speech gains obligatory access to the phonological loop. Importantly, it appears that speech rather than noise in general is critical for eliciting such an effect (Salame & Baddeley, 1989).

The *word length effect* is yet another phenomenon that has played an important role in mapping the characteristics of the phonological loop. In general, participants can remember significantly more short words (such as cat, bag) than long words (such as aluminum, constitutional) in a list recall task. Moreover, it seems that it is the spoken duration of a word rather than the number of syllables that determines how well it will be recalled (Baddeley, 1997). The word length effect clearly illustrates the *limited capacity* of the phonological loop. Research by Baddeley, Thomson, and Buchanan (1975) indicated that a person's memory span actually represents the number of memory items that can be uttered in about 2 s. Young children's memory span is significantly lower than that of adults and increases over childhood. One explanation for this developmental increase is a change in the rate of articulation with age (Hulme, Thomson, Muir, & Lawrence, 1984; Nicolson, 1981). Speed of articulation, therefore, seems to be critical in determining how many items we can maintain in working memory. It is possible that our rate of articulation determines how quickly the phonological rehearsal mechanism of the phonological loop can re-present memory items in order to refresh the steadily decaying contents of the phonological store (Roodenrys *et al.*, 1993).

Information can enter the phonological loop either by being heard or by being read. It has been suggested that the phonological rehearsal mechanism of the phonological loop features heavily in the conversion of visual information. Evidence to support this comes from research using a technique referred to as *articulatory suppression*. Articulatory suppression involves asking participants to continually utter a word or phrase (e.g., "the, the, the" or "teddy bear, teddy bear"), while carrying out a memory task. In a similar manner to that for the unattended speech effect described earlier, performance on the memory task is significantly disrupted. In the case of articulatory suppression, it is because the participant's own articulations gain obligatory access to the phonological rehearsal mechanism of the phonological loop. Such articulations, therefore, interfere with rehearsal of verbally presented information and the phonological re-coding and rehearsal of visually presented items.

The choice of materials to include in a test of phonological loop function is important in determining to what extent other parts of the cognitive system might influence performance. If memory items are words, numbers, or other familiar elements, performance on a memory task can be supported by our stored knowledge in long-term memory, and in some cases verbal items can be processed using visual and semantic strategies in addition to phonological processes. One method of minimizing the role of non-phonological factors in a phonological loop task is to create nonsense words for use as memory items. The Children's Test of Nonword Repetition (Gathercole & Baddeley, 1996) is a standardized test of phonological loop function that involves the presentation of 40 single nonsense words such as "doppelate" and "blonterstaping." When children hear nonsense words such as these, they have very little other than the phonological loop to rely on while trying to recall each word. Performance on tests involving nonsense words is usually much worse than tests involving real words, a phenomenon referred to as the *lexicality effect* (e.g., Hulme, Maughan, & Brown, 1991).

A number of studies have tried to locate the mechanisms responsible for the lexicality effect. Two important sources of influence have been found. The first is our stored knowledge of words that exists in long-term memory. It has been suggested that when words in the phonological store decay, they can still be recovered even if partially decayed—by a process known as redintegration (Hulme *et al.*, 1991; Schweickert, 1993). That is, a partially decayed memory item can be compared to similar items in long-term memory with a high chance of selecting the right word to output. This process cannot be applied to nonsense words, as these words do not exist in our long-term store. However, we do appear to have knowledge of the typical sound patterns of our language (phonotactics), and this may be called upon in the reconstruction of partially decayed memory traces (Gathercole, Pickering, Hall, & Peaker, 2001).

It is clear from the above review that the structures and functions of the phonological loop of working memory are already well specified. Some consideration has also been given to what role a structure such as this has in human everyday functioning. In a recent review of this part of working memory, Baddeley, Gathercole, and Papagno (1998) make the suggestion that the primary role of the phonological loop is as a language learning device, that is, it acts to store unfamiliar sound patterns while more permanent memory records are being created. Other domains in which the phonological loop is hypothesized to be important include language comprehension and reading.

The Visuospatial Sketchpad

Less is known about the structure and function of the visuospatial sketchpad than its phonological counterpart. Research activity concerning this component of working memory has not been as great as that for the phonological loop and one possible explanation is the relative lack of tasks available for measuring "pure" visuospatial working memory. When presented with information in visual form, it is highly likely that we will recode it into phonological form—that is, we name it or describe it to ourselves and store the information in the phonological loop as well as, or instead of, the visuospatial sketchpad. Hitch and colleagues (e.g., Hitch, Halliday, Dodd, & Littler, 1989; Hitch, Halliday, Schaafstal, & Schraagen, 1988) have demonstrated this phenomenon very clearly in a number of studies. They have shown that while very young children do tend to remember visually presented information using a visual code, from the age of 8 years, children tend to remember visually presented items using a phonological strategy. Tests involving familiar images are therefore unlikely to form good measures of "pure" visuospatial working memory. Only items that cannot be verbally recoded can be included in such tests.

In recent years, two tests have emerged as relatively good measures of visuospatial sketchpad function. One task, the Corsi blocks task (e.g., Milner, 1971) involves the recall of sequences tapped out on a series of blocks attached to the board. In contrast, the Visual Patterns Test (Della Sala, Gray, Baddeley, & Wilson, 1997) requires participants to view abstract patterns made up of black and white squares in a matrix and then to recall the location of the black squares. Both tasks have played an important role in furthering our understanding of the visuospatial sketchpad. Studies that have utilized both tasks have provided evidence to challenge earlier notions that the visuospatial sketchpad is a unitary component (e.g., Della Sala, Gray, Baddeley, Allamano, & Wilson, 1999; Logie &

Pearson, 1997). It is now suggested that visuospatial working memory is composed of two dissociable components which Logie (1995) has referred to as the "visual cache" and the "inner scribe." Evidence from neuropsychological case studies also lends weight to the view that visuospatial working memory can be fractionated. It is, however, still very early days in our understanding of this part of working memory and a great deal of debate exists regarding the structures and processes that are carried out by the visuospatial sketchpad. Pickering (2001) provides a recent review of this component of working memory and the evidence regarding its fractionation.

Relatively little is known about the role of the visuospatial sketchpad in everyday life. Children and adults with visuospatial difficulties have been found to exhibit some problems with visuospatial learning and the retention of routes, however (e.g., Hanley, Young, & Pearson, 1991; Nichelli & Venneri, 1995; Temple, 1992).

The Central Executive

The central executive is perhaps the least well specified of the three working memory components. In a way similar to the visuospatial sketchpad, the design of tasks to measure central executive function has proved particularly difficult, partly due to the suggestion that this part of working memory plays a number of quite different roles (Baddeley, 1996). Baddeley and Logie (1999) have suggested, for example, that the central executive is involved in the control and regulation of the whole working memory system and plays various executive functions such as coordinating the slave systems, focusing and switching attention, and activating representations in long-term memory.

The central executive is also believed to be involved in carrying out tasks that involve maintaining information in memory while simultaneously manipulating that information. A good example of this type of task is mental arithmetic, whereby digits are held in memory while an operation such as addition or subtraction is carried out on those digits. A number of tasks have been designed to sample this aspect of working memory function. Collectively known as complex span tasks, these tests involve the requirement to hold information while processing other information. In the listening or reading span tasks (e.g., Daneman & Carpenter, 1980), participants hear or read sentences and have to respond to the sentences by either judging their veracity or providing a missing word. After hearing or reading a number of sentences, they are then required to recall the last word of each sentence in the order that they were encountered. Another complex span task, counting span (Case, Kurland, & Goldberg, 1982) involves counting arrays of dots and recalling successive dot tallies. These tasks are extremely taxing and adults often find them difficult to carry out.

Interestingly, one task that has been used extensively to measure verbal memory may actually represent a better test of central executive function. Backward digit span, a task that requires participants to recall a sequence of digits in reverse order has recently been recognized by many as more than a simple phonological loop task (e.g., Turner, 1997). As the heard list of digits is maintained in memory, participants are simultaneously engaged in the reversal of the list. In many ways, therefore, this task resembles the complex span tasks described above.

Baddeley and Logie (1999) have acknowledged that in many ways the central executive acts as a "useful ragbag" (p. 39) containing all of the working memory phenomena that have not been accounted for by other components of the system. However, the authors also stress that the process of specifying a range of possible central executive functions increases the extent to which this part of working memory becomes empirically tractable. This, in fact, is what has been happening over recent years. As different central executive processes are hypothesized, tasks can be designed to measure such processes. The increase in understanding of central executive function represents an extremely important development in immediate memory research. For the first time we are able to move forward from the basic concept of a visual and verbal short-term memory (as embodied by early conceptualizations of memory such as the modal model), to consider a dynamic system that is capable of dealing flexibly with information in a range of domains.

Moreover, the contribution of central executive function to our everyday activities is likely to be considerable. This type of working memory activity has been found to be associated with a range of different activities including language and reading comprehension (e.g., Daneman & Carpenter, 1980; Dixon, LeFevre, & Twilley, 1998; Engle, Nations, & Cantor, 1990; Turner & Engle, 1989; Yuill, Oakhill, & Parkin, 1989), reading achievement (Siegel, 1994; Swanson, 1994), arithmetic development during childhood (Bull, Johnson, & Roy, 1999; Siegel & Linder, 1984; Siegel & Ryan, 1989), the conceptual aspect of vocabulary acquisition (Daneman & Green, 1986), college entrance and achievement scores (Daneman & Carpenter, 1983; Jurden, 1995), and occupational success (Kyllonen & Christal, 1990). In our own research (Gathercole & Pickering, 2000, 2001), we have found that children who underperform in school or who have been identified as having special educational needs score particularly poorly on tests of central executive function. Overall, it is hard to envisage a complex activity that would not draw on the resources of a memory component like the central executive.

WORKING MEMORY AND LITERACY: A REVIEW OF THE AVAILABLE EVIDENCE

So how, then, do the three components of working memory feature in literacy acquisition? Two major approaches have been taken to investigate this issue. One involves the investigation of specific working memory processes in individuals with literacy difficulties. This approach accounts for most of the research in this area. The second approach is to study children as they move from being nonreaders to skilled readers and to measure the extent to which working memory plays a role in this. Such longitudinal studies are difficult to carry out but provide important information about direction of causality. Ultimately we wish to know whether working memory difficulties are the causes of literacy difficulties (rather than just another manifestation of a different cause).

The Role of the Phonological Loop in Literacy and Literacy Disorders

Any review of the literature on working memory and literacy immediately reveals one important fact: almost all studies in this area concern verbal memory (or phonological loop function). There are a number of reasons why this might be the case. One important factor is the influence of the modal model on early research in this area. As verbal short-term memory was more intensively investigated using this model, it is likely to have generated processes and techniques which were adopted by researchers interested in literacy. Also, research in this field became quite intensive at the time when dyslexia and literacy difficulties were beginning to be viewed as a language-based disorder, with many of the early memory studies being carried out by researchers in the highly influential Haskins Laboratories (e.g., Mann, Liberman, & Shankweiler, 1980; Shankweiler, Liberman, Mark, Fowler, & Fischer, 1979) responsible for the redefinition of dyslexia as a phonological problem. It is no surprise, therefore, that the nature of verbal memory in literacy was of greatest interest to researchers in the 1970s and for some years afterwards.

One of the first verbal memory phenomena to feature in such investigations was phonological similarity. Across a range of studies that contrasted children with literacy difficulties with control participants, it was found that the poor reading children were not as affected by the phonological similarity of the memory items as the controls (Mann *et al.*, 1980; Mark, Shankweiler, Liberman, & Fowler, 1977; Shankweiler *et al.*, 1979). That is, although poor readers performed worse than controls overall, the difference between their scores in the phonologically similar and phonologically different conditions was not as great as that for the control participants. It was subsequently concluded that poor readers experience difficulties in accessing phonic representations during the memory task (e.g., Torgesen & Houck, 1980).

In subsequent studies of this kind, the picture has become less clear, however. While some studies have replicated this finding (e.g., Siegel & Linder, 1984), others have not (e.g., Hall, Wilson, Humphries, Tinzmann, & Bowyer, 1983; Johnson, 1982; Johnson, Rugg, & Scott, 1987). One interpretation of this pattern of findings is that a lack of phonological similarity effect is only found in younger poor readers, perhaps because they find the overall memory task very difficult (Johnson, 1993). When task difficulty is increased to the point that even good readers find it demanding, phonological similarity effects disappear in this group too (Holligan & Johnson, 1988). In summary, although poor readers perform poorly on tests of immediate verbal memory, they do seem to be able to use a phonological code to at least some extent (Brady, Mann, & Schmidt, 1987; Olson, Davidson, Kliegl, & Davies, 1984), leading to the presence of phonological similarity effects in studies in which task difficulty is controlled appropriately. Evidence to suggest problems with phonological codes has also been obtained using word length effects (Macaruso, Locke, Smith, & Powers, 1996).

An important factor in the ability to retain information in verbal short-term memory is the rehearsal of memory items. It has been suggested that children with literacy problems show less efficient rehearsal processes than unimpaired controls. In order for rehearsal to benefit memory performance, it has to occur in a cumulative fashion. That is, as each item in the memory list is heard, we need to remember it in addition to any items that have come before it. When very young children carry out a digit span task, they can often be heard repeating each digit to themselves as they hear it. What they fail to do, however, is to string the digits together and remember them as a whole list. By the time the list has been completed, the child has no idea of what the first items were as these were not recently rehearsed.

A study of memory processes in dyslexic boys carried out by Spring and Capps in 1974 found that this group were much less likely to carry out cumulative rehearsal than controls. Correspondingly, this group was much worse at recalling all but the most recently presented items in the memory lists. Across all participants, it was found that the use of a cumulative rehearsal strategy was related to the speed of naming, suggesting that this is an important factor in memory task performance. Other studies have provided support for the idea that phonological rehearsal is deficient in poor readers (e.g., Bauer, 1977; Macaruso *et al.*, 1996). Macaruso and colleagues (1996) have suggested that poor readers experience a breakdown in cumulative phonological rehearsal, particularly in the middle of a memory list. This may occur because of a switch in coding strategy away from phonological codes.

This view ties in with evidence to suggest that poor readers are more likely to use non-phonological representational codes to remember information in a phonological memory task (Byrne & Shea, 1979; Hicks, 1980; Rack, 1985). Byrne and Shea (1979), for example, found that poor readers were more likely to use a semantic (meaning-based) code for maintaining information in a phonological memory task. In a study by Rack (1985) where items in a memory list had visual (rather than phonological) similarity, it was found that poor readers were significantly more affected by this manipulation than controls, suggesting a greater use of visual codes for retaining information in working memory (see also Palmer, 2000).

In our review of phonological loop function earlier, it was noted that speed of articulation appeared to play a critical role in the amount of information we can hold in phonological short-term memory. A study of short-term memory in good, average, and poor readers by McDougall, Hulme, Ellis, and Monk (1994) found that observed differences in memory performance could be accounted for by differences in articulation rate between the groups. Macaruso *et al.* (1996) have suggested that poor readers may have problems in phonological memory tasks because of a combination of less precise phonological codes (which may be more susceptible to confusion during rehearsal) and reduced speed of articulation (which limits the amount of information that can be rehearsed).

When poor readers have been asked to carry out tasks that involve repeating nonsense words, their scores have been found to be significantly lower than those of controls (Snowling, 1981; Snowling, Goulandris, Bowlby, & Howell, 1986). A task such as nonsense word repetition has been suggested to be a good index of the role of the phonological loop as a language learning device (Baddeley *et al.*, 1998) as it requires the precise laying down of unfamiliar sound patterns in short-term memory, without assistance from other types of representations in memory. A recent study by Roodenrys and Stokes (2001) found that poor readers appear to be affected by the lexicality of verbal memory items in much the same way as controls. This suggests that, in contrast to the observed deficits in short-term memory, long-term memory is working normally to support immediate memory for familiar information.

Longitudinal studies of the relationship of verbal short-term memory to literacy acquisition provide another line of evidence of the importance of this type of cognitive activity. A number of such studies have found that memory performance before or at school entry is predictive of later reading performance (e.g., Jorm, Share, MacLean, & Matthews, 1984; Mann & Liberman, 1984). However, the picture again becomes more complex when we consider a study by Ellis (1991), which shows the opposite pattern: reading performance predicts later memory scores. At least two possible explanations for this finding exist. One is that reading assists in the development of phonological codes for use in memory (Frick, 1984), while another is that both reading and verbal short-term memory are determined by a third factor (e.g., phonological awareness). This second point will be discussed in greater detail later in this chapter.

Overall, it appears that when poor readers are asked to carry out memory tasks involving phonological material, they perform worse than would be expected. This is true whether the task involves list recall (as described above), list learning (Douglas & Benezra, 1990; Felton, Wood, Brown, & Campbell, 1987; Kinsbourne, Rufo, Gamzu, Palmer, & Berliner, 1991; McGee, Williams, Moffitt, & Anderson, 1989; Michaels, Lazar, & Risucci, 1997; Rudel & Helfgott, 1984), story recall (Felton *et al.*, 1987; O'Neill & Douglas, 1991) or paired-associate learning (Helfgott, Rudel, & Kairam, 1986). Whether the cause of such problems lies in the nature of the phonological codes, the use of rehearsal or both, is unclear at this time.

Visuospatial Sketchpad Function and Literacy

Until the 1970s, dyslexia was viewed as a predominantly visual disorder (e.g., Orton, 1937) and correlations between visual short-term memory and reading had been reported by a number of researchers (e.g., Carroll, 1973). However, the idea of visuospatial short-term memory problems in poor readers was strongly refuted by Vellutino and colleagues (Vellutino, Pruzek, Steger, & Meshoulam, 1973; Vellutino, Steger, DeSetto, & Phillips, 1975) who found that poor readers did not differ from controls in their immediate memory for unfamiliar visual items. What did seem important in tasks of visuospatial memory, however, was the extent to which stimuli could be phonologically recoded. In a study by Swanson (1978), poor readers and controls were given nonsense shapes to remember. Half of the participants were taught names for each nonsense shape while the other half were not. Several interesting findings emerged from this study. First, the performance of control participants was much better than that for the poor readers in the named condition. Second, the good and poor readers performed at equivalent levels in the unnamed condition. Finally, poor readers performed marginally (although non-significantly) better in the unnamed condition than in the named condition. Taken together, these findings point to unimpaired visual short-term memory function in poor readers. It is only when visual items can be named that problems appear to occur, a finding that concurs with the phonological short-term memory problems described above.

Other studies have provided evidence to support this view. For example, Liberman, Mann, Shankweiler, and Werfelman (1982) presented good and poor readers with two kinds of visuospatial information that could not easily be phonologically recoded: unfamiliar faces and abstract line drawings. The two groups of participants did not differ in their recognition of the stimuli. In another study, Katz, Shankweiler, and Liberman (1981) presented sequences of common objects and doodle drawings in a memory task. Good and poor readers did not differ in their performance with the doodle drawings but the good readers significantly outperformed the poor readers with the (nameable) objects. Studies by Hulme (1981) and Vellutino (1979) reveal similar results. Research involving visuospatial sequential memory as measured by the Corsi blocks task (e.g., Gould & Glencross, 1990) has also failed to find differences between good and poor readers.

One question that is often asked with respect to the nature of memory in poor readers is "what about the problems encountered by visual dyslexics?" Is it the case, for example, that the smaller group of individuals viewed as having a more predominantly visual form of reading disability show corresponding problems with visuospatial memory? Howes, Bigler, Lawson, and Burlingame (1999) carried out a study to investigate this issue by administering the Test of Memory and Learning (TOMAL, Reynolds & Bigler, 1994) to children classified as either auditory or visual dyslexics using Boder's (1973) typology. They found that all participants, regardless of classification, showed deficits in verbal memory compared to controls. It was the case, however, that some children manifested visual memory deficits in addition to their verbal problems. It seems, therefore, that while visuospatial memory problems would not be found in the majority of poor readers, there may be a smaller number of individuals who would not be found to perform as Vellutino and others have found. In addition, there may be language-specific aspects to the failure to find visuospatial short-term memory deficits in poor readers. A study by Meyler and Breznitz (1998) found that visual short-term memory was related to reading when the language in question was Hebrew. The reason for this finding is not clear; however, the need to study working memory function in poor readers whose language is not English is emphasized by such results.

The Role of the Central Executive in Literacy Performance

Over recent years, a number of studies have attempted to understand how central executive function influences our abilities in literacy. One investigator, in particular, has contributed significantly to our understanding of this issue. Across a range of studies, Swanson and colleagues have presented good and poor readers with tasks designed to tap central executive function (e.g., Swanson, 1992, 1993a, 1993b, 1994, 1999; Swanson & Alexander, 1997; Swanson & Ashbaker, 2000; Swanson, Ashbaker, & Lee, 1996).

In one important study, Swanson (1994) examined how central executive function related to literacy in children and adults. In this study, participants were given measures of verbal and visuospatial short-term memory and measures of central executive function in order to contrast the relative importance of the first two types of process (storage of information only) with the third type of process (storage and processing of information). Swanson found that central executive processes made the greater contribution to reading recognition. Additionally, it was found that while the short-term memory processes related to performance in reading comprehension for poor readers, it made no such contribution to the reading comprehension of controls. Two very critical points, therefore, arise from this study: memory deficits in poor readers go beyond short-term memory (storage only), and different memory processes may be used by individuals with and without reading problems when carrying out reading-related tasks.

Subsequent studies have attempted to clarify the nature of the observed deficit in central executive function found in poor readers. For example, Swanson and Ashbaker (2000) examined the role that articulation speed played in the relationship between executive processing and reading. They found that the short-term memory and central executive performance of poor readers was worse than that for controls (both chronological age and reading age matched) even after the contribution of articulation rate had been removed from the picture. Moreover, central executive performance was found to predict reading independently of the contribution of both short-term memory and articulation rate (see also Cohen & Heath, 1990; Henry, 1994). These findings suggest that poor readers have a deficit in a central executive system that is independent of any deficits that they may have in the phonological loop. In other words, perhaps two parts of the working memory system are causing a problem for poor readers rather than just one.

Other researchers have attempted to relate the issue of central executive function to reading. For example, Siegel (1994) found that executive processes were related to reading in individuals ranging from 6 to 49 years of age. Isaki and Plante (1997) also found that adults with and without reading problems scored differently on tasks of central executive and short-term memory function. Similar findings are reported in a study by de Jong (1998). These findings have led de Jong to suggest that poor readers' deficits in central executive function reflect a general lack of capacity for simultaneous processing and storage of information rather than a processing deficiency or specific problem with verbal short-term memory. However, de Jong does suggest that the fundamental problem of poor readers may be their poor ability to store phonological information in short-term memory—a problem that manifests more severely when such codes are dynamically manipulated, as in the case of a central executive task. In other words, as noted above, poor readers will show problems in phonological loop function, but much more severe problems with central executive function. The failure to efficiently store and manipulate information in working memory is likely to have a huge impact on literacy due to the many occasions where complex activities (such as storing the phonemes of a word long enough to allow blending, storing the words in a sentence long enough to achieve meaning, remembering the phonological and visual form of a word while writing it down) are essential to the use of written language.

Palmer (2000) presents further evidence of central executive deficits in a group of dyslexic teenagers. These participants showed signs of perseveration errors on a task commonly used to measure an aspect of central executive function related to the ability to switch attention and inhibit responses (The Wisconsin Card Sorting Task). In Palmer's view, the central executive is important in the development of literacy because it facilitates the ability to inhibit the use of visual processing strategies in favor of phonological ones during the reading process. Her sample of dyslexic teenagers appeared to be using visual coding strategies at an age at which this approach is no longer an efficient route to reading success. Hence, another way in which the central executive may be important in literacy is in the development of phonological strategies for reading in place of the early visual strategies that appear to characterize the approach to reading taken by very young children.

Some progress in assessing the status of central executive function in poor readers has clearly been made. Indeed, it may have been the case that central executive problems were actually manifesting themselves in poor readers for many years, but our understanding of the central executive did not allow us to realize it. Administration of the digit recall subtest that features in a number of standardized attainment batteries (e.g., the Wechsler Intelligence Scale for Children, Wechsler, 1996) involves asking children to recall digits in both forward and reverse order. While many children are found to score at average levels in the forward condition, their performance in the backward condition is often strikingly bad. If we view the backward condition of this task as a test of central executive function, and the forward condition as a test of phonological loop function, this pattern of findings begins to make more sense.

We are still a long way from fully understanding the nature of central executive function in individuals with literacy difficulties; however, as research in this areas moves forward, more techniques for assessing different types of central executive function are being developed. If we continue to apply these tasks to the issue of literacy attainment, we provide the basis for understanding how the entire working memory system functions in reading disability. This is an extremely important goal for both theoretical and applied reasons. From a theoretical point of view, it is essential to our understanding of literacy and literacy disorders that we attempt to map out the cognitive processes that underlie reading and its associated activities.

From an applied point of view, it is critical that we understand not just what aspects of memory function cause problems for poor readers, but those aspects of memory function that are unimpaired. It is only by assessing the relative strengths and weaknesses of working memory in poor readers that we can understand the specific needs of each individual and provide appropriate remedial approaches that teach to the areas of strength to overcome the areas of weakness. This is a well-established philosophy in the reading disorders world; however, the first step to developing such an approach is to assess working memory comprehensively across all three components of the system. Until recently, no tool has been available for this purpose; however, over the last few years, Sue Gathercole and I have been developing a memory test for children to serve this very purpose. The Working Memory Test Battery for Children (WMTB-C, Pickering & Gathercole, 2001) is a standardized test of working memory function for use with children between the ages of 5 and 15 years. It includes measures of phonological loop, visuospatial sketchpad, and central executive function and provides the user with the ability to profile memory performance across the three components of working memory, hence identifying areas of strength and weakness. Importantly from the point of view of those interested in literacy, this test has been used in a study of working memory in children with dyslexia (Pickering & Gathercole, 2004).

RELATIONSHIPS BETWEEN WORKING MEMORY COMPONENTS IN INDIVIDUALS WITH LITERACY DISORDERS

Pickering and Gathercole (2004) administered a battery of working memory tests to 15 children with dyslexia aged between 7 and 14 years. The performance of each child was compared against two matched non-dyslexic children, one of the same chronological age and one with the same reading age. All children were given the prototype version of the WMTB-C. This battery included five phonological loop tasks (plus the Children's Test of Nonword Repetition), four tests of visuospatial sketchpad function, and three test of central executive function. These tests are described in Table 1.

The performance of the dyslexic group varied across the different working memory tests. In the case of the phonological loop, it was found that the

Table 1. Tests included in the prototype version of the WMTB-C

Test	Component	Description
Digit Recall	Phonological loop	Sequences of digits are presented in spoken format. The child is required to recall each list immediately, in the correct order.
Word List Matching	Phonological loop	Sequences of one syllable words are presented twice in spoken format. The child is asked to detect whether changes in serial order occur in the second presentation of the sequence.
Nonword List Matching	Phonological loop	As for Word List Matching except that stimuli are one syllable nonsense words.
Word List Recall	Phonological loop	Sequences of one syllable words are presented in spoken format. The child is required to recall each list immediately, in the correct order.
Nonword List Recall	Phonological loop	As for Word List Recall except that stimuli are one syllable nonsense words.
Matrices Static	Visuospatial sketchpad	Abstract patterns composed of black and white squares in a matrix are presented. The child is required to recall the location of the black squares.
Matrices Dynamic	Visuospatial sketchpad	Each black square in the matrices pattern is presented one at a time in a special sequence. The child is required to recall the sequence.
Mazes Static	Visuospatial sketchpad	Routes through two-dimensional mazes are presented as line drawings. The child is required to draw the route into an empty maze.
Mazes Dynamic	Visuospatial sketchpad	Routes through two-dimensional mazes are traced with the experimenter's finger. The child is required to draw the route into an empty maze.
Listening Recall	Central executive	A series of sentences are presented for which the child is required to respond either "true" or "false." Following this, the child is required to recall the final word from each sentence, in the order that they were heard.
Counting Recall	Central executive	The child is presented with a series of cards bearing dots. The child counts each card of dots and then recalls the dot totals, in the order that they were encountered.
Backward Digit Recall	Central executive	Sequences of digits are presented in spoken format. The child is required to recall each list immediately, in reverse order.

dyslexics' scores were lower than those of the chronological age controls but slightly better than those of the reading age controls. On the Children's Test of Nonword Repetition (Gathercole & Baddeley, 1996), however, the dyslexic group scored marginally lower than even the reading age control group, underlining their notable difficulties with immediate repetition of unfamiliar phonological forms.

In contrast to this, visuospatial sketchpad performance was highly similar to that for the chronological age controls on all but one of the tests given. The dyslexic group appeared to be performing at age appropriate levels on the two Mazes tasks and the Matrices Dynamic task. They did, however, perform noticeably poorly on the static version of the Matrices task. The reasons for this finding are unclear; however, previous research with tasks of this kind has found some evidence to suggest that phonological recoding of these stimuli might be possible (Miles, Morgan, Milne, & Morris, 1996). If this is the case, dyslexic children may not engage in phonological recoding, or may not benefit from phonological recoding to the same extent as the non-dyslexic controls. Further assessment is needed in order to fully understand why this finding occurred.

Performance on the three tasks of central executive function proved the most interesting of all the results from this study. On two of the three tasks (Counting Recall and Backward Digit Recall), the performance of the dyslexic group was very poor, and noticeably lower than that for the reading age controls. This finding reinforces the points made earlier in our discussion of central executive deficits in dyslexia. Performance on these tasks does not appear to reflect the reading levels of the children, but seems to index something much more fundamental to their dyslexic deficits. However, one surprising finding was the very good performance of the dyslexic group on the Listening Recall task. Their performance on this test was a little better than that for the same age controls, a finding that would not be predicted by the presence of a central executive deficit in dyslexia. However, there may be an explanation for this finding that does not compromise the view that central executive deficits are found in dyslexia. This is that while this task proved very difficult for all participants, leading to low scores for all groups, the dyslexic group was able to use alternative strategies to those used by the controls in order to carry out the task. That is, while the controls may have been treating this task as primarily a phonological one (i.e., retaining the memory items in phonological form), the dyslexic group may have utilized semantic and/or visual approaches in order to carry it out (as suggested by the findings of Byrne & Shea, 1979; and Rack, 1985). Additional research is required to establish whether this explanation is correct.

Overall, therefore, the findings of a study which samples not just one, but all three components of working memory provides us with the following picture. Children with dyslexia do seem to have problems with phonological loop function; however, their performance in this area is not dissimilar to that for their reading age controls, suggesting that this aspect of memory is in line with the level of reading ability attained. Visuospatial sketchpad performance, on the other hand, appears unimpaired, unless the tasks used to sample it allow for the possibility of phonological recoding of memory items. Finally, central executive performance does seem to be highly impaired in dyslexia, perhaps to a greater degree than that suggested by a child's reading level. Yet even here there may be scope for improved performance if the kind of strategies *with which* dyslexic children may already be *familiar with can be* utilized.

There is one caveat in this account of working memory in literacy. The tasks used to measure central executive function all have one feature in common, that is, they are all highly biased toward *phonological coding* of memory items. This means that when we assess central executive function, we are assessing the storage and manipulation of that kind of information that has repeatedly been shown to cause problems for dyslexic individuals. In order to show definitively that a central executive deficit exists, we need to present participants with tasks that do not require phonological processing. Such tasks are extremely difficult to design, as they require the use of memory items that cannot be phonologically recoded. Another approach to this would be to study central executive task performance while asking participants to carry out articulatory suppression to block the use of the phonological loop in the central executive task. This too may be difficult, as there would be a need to make sure that the articulatory suppression task was equally demanding for both dyslexic and control participants. Without this kind of evidence, it appears that the common component of all of the memory tasks on which dyslexics are found to perform poorly is the use of a phonological code. This point has led some to suggest that working memory *per se* is not a critical factor in literacy and literacy disorders. Instead, it has been argued that phonological awareness skills are the fundamental cognitive function upon which literacy (and working memory) skills depend. The evidence for and against this view is outlined in the next section.

SPECIFIC WORKING MEMORY IMPAIRMENT OR BROADER LINGUISTIC DEFICIT?

Hulme and Roodenrys (1995) have argued that the verbal short-term memory problems found in poor readers are an index of other phonological deficits rather than a causal factor in the reading disability. Evidence in favor of this view is provided by McDougall and colleagues' (1994) study in which it was found that phonological awareness predicted reading skill even when IQ and short-term memory had been accounted for. Short-term memory is therefore suggested to be related to literacy via a third factor that is related to both literacy and short-term memory. This third factor is phonological awareness, a cognitive skill with which poor readers are usually found to have difficulty (Hulme & Snowling, 1991).

Other studies on this issue have provided evidence of a different picture, however. Hansen and Bowey (1994), for example, found that although phonological awareness and verbal short-term memory shared a lot of common variance, each type of cognitive activity appeared to tap different reading-related skills. Gathercole, Baddeley, and colleagues have reported similar findings (Baddeley & Gathercole, 1992; Gathercole, Willis, & Baddeley, 1991). In particular, they found that phonological loop function, as indexed by nonword repetition, was significantly related to reading performance about 1 year after the start of reading instruction, although this effect appeared to decline after this point. The authors suggest that this finding is explained in terms of the acquisition of letter–sound correspondences at this stage in reading acquisition, a process that is dependent upon the capacity of the phonological loop. Baddeley and Gathercole (1992) also note that the capacity to store phonological information and the capacity to analyze speech sounds, although not the same thing, are both required for successful reading acquisition. The findings of an investigation by Rohl and Pratt (1995) further suggest that, as children learn to read, verbal short-term memory and phonological awareness become increasingly differentiated from one another.

When phonological awareness and central executive function are examined with respect to their relationship with literacy, it appears that working memory does account for unique variance in reading performance (Cormier & Dea, 1997; Leather & Henry, 1994). Although central executive function, phonological loop function, and phonological awareness were found to be related, Leather and Henry (1994) provided data to show that central executive function was still able to predict reading performance when phonological awareness and phonological loop function had been accounted for.

Thus, the issue of whether literacy is dependent upon a specific working memory deficit or a more general linguistic deficit is not easily resolved. Tasks of phonological awareness such as the rhyme oddity task designed by Bradley and Bryant (1983) make significant demands on short-term memory (Kirtley, Bryant, MacLean, & Bradley, 1989; Wagner & Torgesen, 1987). In this task, children are presented with four words, one of which differs in either the onset or the rime. The child is asked to identify the odd one out. Likewise, many phonological loop tasks, such as nonword repetition, involve phonological awareness. Indeed Snowling and colleagues (1986) included this task in their research as a measure of speech segmentation processes, rather than as a short-term memory task. To complicate matters further, Oakhill and Kyle (2000) have found that central executive function is related to performance on the rhyme oddity task, indicating that this task may involve storage and processing of task items. All of this evidence underlines the difficulty of designing any task that can operate as a "pure" measure of phonological awareness, phonological loop, or central executive function. Until such tasks are available, the possibility of testing the two accounts of the nature of poor readers' difficulties is not open to us, and we must be forced to conclude that a range of different cognitive processes are important in the task of learning to read.

WORKING MEMORY AND LITERACY: IMPLICATIONS FOR PRACTICE

Given that very few clear conclusions can be drawn from the huge body of research on the issue of short-term memory and literacy, can any useful implications for practice be proposed? I think there can, and moreover, I believe that there is still a lot that we can learn about the role of immediate memory in reading, especially if we ground our research in the well-established and well-supported models of cognition that are currently available to us.

From the existing evidence on working memory and literacy, it appears that two main areas of weakness are present in at least some individuals with literacy problems: phonological loop processes and central executive function. These areas are likely to cause problems for individuals as they attempt to learn how to read and spell. However, the presence of an unimpaired visuospatial short-term memory makes a range of approaches to language learning possible. In many cases these approaches are already being used with great success. Multisensory techniques that present information to a range of modalities help to avoid unnecessary focus on the less efficient parts of the cognitive system.

But how do we know what the strengths and weaknesses of the individual in question are? To answer this question, we need to look toward the process of building up a detailed cognitive profile of the individual. Assessment of working memory provides a very valuable way of doing this as working memory can be thought of as one of the most fundamental of all cognitive activities, and one upon which many other cognitive activities depend. The WMTB-C (Pickering & Gathercole, 2001) is a tool that allows the detailed specification of performance in this critical area.

On the basis of assessment of working memory function, it is possible to provide training in certain important memory-related skills. For example, for some individuals it might be possible to provide instruction and support in the development of cumulative rehearsal strategies for remembering sequences of information. For other individuals it may be possible to give assistance in the development of simultaneous processing and storage activity by initially breaking up these two task components and practicing them separately. If training is not feasible, external support for working memory difficulties is also an option. For example, use of visual rather than verbal approaches to tasks where possible. Where central executive difficulties are present, it may be useful to alter the way in which the individual is asked to process information. For example, it might be possible to reduce memory load by making the information to be processed available to the individual (e.g., in an arithmetic task), therefore restricting the cognitive processing required to achieve a solution. Instructions can be given one at a time instead of several at once. The possibilities are endless if we know the cognitive status of the individual in question.

SUMMARY

For a great many years, we have believed that short-term memory problems are a cardinal feature of disorders of literacy; however, experimental evidence to support this view has been more difficult to obtain. By making research in this area engage with a highly successful account of immediate memory (the working memory model), the scope of the field has been increased significantly. Working memory does appear to be a problem for poor readers and the specific reasons for this may become clearer over time. Until then, we can assess working memory in each individual of interest and provide solutions tailored to the specific problems that they experience.

REFERENCES

Atkinson, R. C., & Shiffrin, R. M. (1968). Human memory: A proposed system and its control processes. In K. W. Spence (Ed.), *The psychology of learning and motivation: Advances in research and theory* (Vol. 2, pp. 89–195). New York: Academic Press.

Baddeley, A. D. (1966). Short-term memory for word sequences as a function of acoustic, semantic and formal similarity. *Quarterly Journal of Experimental Psychology, 18*, 362–265.

Baddeley, A. D. (1986). *Working memory*. Oxford: OUP.

Baddeley, A. D. (1996). Exploring the central executive. *Quarterly Journal of Experimental Psychology, 49A*, 5–28.

Baddeley, A. D. (1997). *Human memory: Theory and practice*. Hove, UK: Erlbaum.

Baddeley, A. D. (1998). Recent developments in working memory. *Current Opinion in Neurobiology, 8*, 234–238.

Baddeley, A., & Gathercole, S. (1992). Learning to read: The role of the phonological loop. In J. Alegria, D. Holender, J. Juncade Morais, & M. Radeau (Eds.), *Analytic approaches to human cognition* (pp. 153–167). Oxford, England: North Holland.

Baddeley, A., Gathercole, S., & Papagno, C. (1998). The phonological loop as a language learning device. *Psychological Review, 105*, 158–173.

Baddeley, A. D., & Hitch, G. (1974). Working memory. In G. A. Bower (Ed.), *Recent advances in learning and motivation* (Vol. 8, pp. 47–90). New York: Academic Press.

Baddeley, A. D., Lewis, V., Eldridge, M., & Thomson, N. (1984). Attention and retrieval from long-term memory. *Journal of Experimental Psychology: General, 113*, 518–540.

Baddeley, A., & Logie, R. (1999). Working memory: The multiple component model. In A. Miyake, & P. Shah (Eds.), *Models of working memory* (pp. 28–61). New York: Cambridge University Press.

Baddeley, A. D., Thomson, N., & Buchanan, M. (1975). Word length and the structure of short-term memory. *Journal of Verbal Learning and Verbal Behaviour, 14*, 575–589.

Bauer, R. H. (1977). Memory processes in children with learning disabilities: Evidence for deficient rehearsal. *Journal of Experimental Child Psychology, 24*, 415–430.

Boder, E. (1973). Developmental dyslexia: A diagnostic screening procedure based on three characteristic patterns of reading and spelling. In B. Bateman (Ed.), *Learning disorders* (pp. 293 321). Seattle, WA: Special Child.

Bradley, L., & Bryant, P. E. (1983). Categorising sounds and learning to read—A causal connection. *Nature, 31*, 419–421.

Brady, S., Mann, V., & Schmidt, R. (1987). Errors in short-term memory for good and poor readers. *Memory and Cognition, 15*, 444–453.

Bull, R., Johnson, R. S., & Roy, J. A. (1999). Exploring the roles of the visuo-spatial sketchpad and central executive in children's arithmetical skills: Views from cognition and developmental neuropsychology. *Developmental Neuropsychology, 15*, 421–442.

Byrne, B., & Shea, P. (1979). Semantic and phonetic memory codes in beginning readers? *Memory and Cognition 7*, 333–338.

Carroll, J. (1973). Assessment of short-term visual memory and its educational implications. *Perceptual and Motor Skills, 37*, 383–388.

Case, R. D., Kurland, M., & Goldberg, J. (1982). Operational efficiency and the growth of short-term memory span. *Journal of Experimental Child Psychology, 33*, 386–404.

Cohen, R. L., & Heath, M. (1990). The development of serial short-term memory and the articulatory loop hypothesis. *Intelligence, 14*, 151–171.

Colle, H. A., & Welsh, A. (1976). Acoustic masking in primary memory. *Journal of Verbal Learning and Verbal Behaviour, 8*, 240–247.

Conrad, R., & Hull, A. J. (1964). Information, acoustic confusion and memory span. *British Journal of Psychology, 55*, 429–432.

Cormier, P., & Dea, S. (1997). Distinctive patterns of relationship of phonological awareness and working memory with reading development. *Reading and Writing, 9*, 193–206.

Daneman, M., & Carpenter, P. A. (1980). Individual differences in working memory and reading. *Journal of Verbal Learning and Verbal Behaviour, 19*, 450–466.

Daneman, M., & Carpenter, P. A. (1983). Individual differences in integrating information within and between sentences. *Journal of Experimental Psychology: Learning, Memory, and Cognition, 9*, 561–584.

Daneman, M., & Green, I. (1986). Individual differences in comprehending and producing words in context. *Journal of Memory and Language, 25*, 1–18.

de Jong, P. F. (1998). Working memory deficits of reading disabled children. *Journal of Experimental Child Psychology, 70*, 75–96.

Della Sala, S., Gray, C., Baddeley, A., & Wilson, L. (1997). *Visual patterns test.* Bury St Edmunds, UK: Thames Valley Test.

Della Sala, S., Gray, C., Baddeley, A., Allamano, N., & Wilson, L. (1999). Pattern span: A tool for unwelding visuo-spatial memory. *Neuropsychologia, 37*, 1189–1199.

Dixon, P., LeFevre, J., & Twilley, L. C. (1998). Word knowledge and working memory as predictors of reading skill. *Journal of Educational Psychology, 80*, 465–472.

Douglas, V. I., & Benezra, E. (1990). Supraspan verbal memory in attention deficit disorder with hyperactivity, normal, and reading-disabled boys. *Journal of Abnormal Child Psychology, 18*, 617–638.

Ellis, N. C. (1991). Spelling and sound in learning to read. In M. J. Snowling, & M. Thompson (Eds.), *Dyslexia: Integrating theory and practice* (pp. 80–94). London: Whurr.

Engle, R. W., Nations, J. K., & Cantor, J. (1990). Is "working memory capacity" just another name for word knowledge? *Journal of Educational Psychology, 82*, 799–804.

Felton, R. H., Wood, F. B., Brown, I. S., & Campbell, S. K. (1987). Specific verbal memory and naming deficits in attention deficit disorder and reading disability. *Brain and Language, 31*, 171–184.

Frick, R. W. (1984). Using both an auditory and a visual short-term store to increase digit span. *Memory and Cognition, 12*, 507–514.

Gathercole, S. E. (1999). Cognitive approaches to the development of short-term memory. *Trends in Cognitive Science, 3*, 410–418.

Gathercole, S. E., & Baddeley, A. D. (1996). *The children's test of nonword repetition.* London: The Psychological Corporation.

Gathercole, S. E., Willis, C., & Baddeley, A. D. (1991). Differentiating phonological memory and awareness of rhyme: Reading and vocabulary development in children. *British Journal of Psychology, 82*, 387–406.

Gathercole, S. E., & Pickering, S. J. (2000). Working memory deficits in children with low achievements in the national curriculum at seven years of age. *British Journal of Educational Psychology, 70*, 177–194.

Gathercole, S. E., & Pickering, S. J. (2001). Working memory deficits in children with special educational needs. *British Journal of Special Education, 28*, 89–97.

Gathercole, S. E., Pickering, S. J., Hall, M., & Peaker, S. M. (2001). Dissociable lexical and phonological influences on serial recall and serial recognition. *Quarterly Journal of Experimental Psychology, 54A*, 1–30.

Gould, J. H., & Glencross, D. J. (1990). Do children with a specific reading disability have a general serial-ordering deficit? *Neuropsychologia, 28*, 271–278.

Hall, J. W., Wilson, K. P., Humphries, M. S., Tinzmann, M. B., & Bowyer, P. M. (1983). Phonemic-similarity effects in good vs. poor readers. *Memory and Cognition, 11*, 520–527.

Hanley, J. R., Young, A. W., & Pearson, N. A. (1991). Impairment of the visuo-spatial sketch pad. *Quarterly Journal of Experimental Psychology, 43A*, 101–125.

Hansen, J., & Bowey, J. A. (1994). Phonological analysis skills, verbal working memory, and reading ability in second-grade children. *Child Development, 65*, 938–950.

Helfgott, E., Rudel, R. G., & Kairam, R. (1986). The effect of piracetam on short and long term verbal retrieval in dyslexic boys. *International Journal of Psychophysiology, 4*, 53–61.

Hicks, C. (1980). The ITPA visual sequential memory task: An alternative interpretation and the implications for good and poor readers. *British Journal of Educational Psychology, 50*, 16–25.

Hitch, G. J., Halliday, M. S., Dodd, A., & Littler, J. E. (1989). Development of rehearsal in short-term memory: Differences between pictorial and spoken stimuli. *British Journal of Developmental Psychology, 7*, 347–362.

Hitch, G. J., Halliday, M. S., Schaafstal, A. M., & Schraagen, J. M. C. (1988). Visual working memory in children. *Memory and Cognition, 16*, 120–132.

Holligan, C., & Johnson, R. S. (1988). The use of phonological information by good and poor readers in memory and reading tasks. *Memory and Cognition, 16*, 522–532.

Howes, N. L., Bigler, E. D., Lawson, J. S., & Burlingame, G. M. (1999). Reading disability subtypes and the test of memory and learning. *Archives of Clinical Neuropsychology, 14*, 317–339.

Hulme, C. (1981). *Reading retardation and multi-sensory teaching.* London: Routledge & Kegan Paul.

Hulme, C., Maughan, S., & Brown, G. D. A. (1991). Memory for familiar and unfamiliar words: Evidence for a longer term memory contribution to short-term memory span. *Journal of Memory and Language, 30*, 685–701.

Hulme, C., & Roodenrys, S. (1995). Practitioner review: Verbal working memory development and its disorders. *Journal of Child Psychology and Psychiatry, 36*, 373–398.

Hulme, C., & Snowling, M. (1991). Phonological deficits in dyslexia: A "sound" reappraisal of the verbal deficit hypothesis? In N. Singh, & I. Beale (Eds.), *Progress in learning disabilities*. Berlin: Springer-Verlag.

Hulme, C., Thomson, N., Muir, C., & Lawrence, A. (1984). Speech rate and the development of short-term memory span. *Journal of Experimental Child Psychology, 38*, 241–253.

Isaki, E., & Plante, E. (1997). Short-term and working memory differences in language/learning disabled and normal adults. *Journal of Communication Disorders, 30*, 427–437.

Johnson, R. S. (1982). Phonological coding in dyslexic readers. *British Journal of Psychology, 73*, 455–460.

Johnson, R. S. (1993). The role of memory in learning to read, write and spell: A review of recent research. In G. M. Davies, & R. H. Logie (Eds.), *Memory in everyday life* (pp. 59–77). Elsevier.

Johnson, R. S., Rugg, M. D., & Scott, T. (1987). Phonological similarity effects, memory span and developmental reading disorders: The nature of the relationship. *British Journal of Psychology, 78*, 205–211.

Jorm, A. F. (1983). Specific reading retardation and working memory: A review. *British Journal of Psychology, 74*, 311–342.

Jorm, A. F., Share, D. L., MacLean, R., & Matthews, R. (1984). Phonological confusablity in short-term memory for sentences as a predictor of reading ability. *British Journal of Psychology, 75*, 393–400.

Jurden, F. H. (1995). Individual differences in working memory and complex cognition. *Journal of Educational Psychology, 87*, 93–102.

Katz, R. B., Shankweiler, D., & Liberman, I. Y. (1981). Memory for item order and phonetic recoding in the beginning reader. *Journal of Experimental Child Psychology, 32*, 474–484.

Kinsbourne, M., Rufo, D. T., Gamzu, E., Palmer, R. L., & Berliner, A. K. (1991). Neuropsychological deficits in adults with dyslexia. *Developmental Medicine and Child Neurology, 33*, 763–775.

Kirtley, C., Bryant, P., MacLean, M., & Bradley, L. (1989). Rhyme, rime, and the onset of reading. *Journal of Experimental Child Psychology, 48*, 224–245.

Kyllonen, P. C., & Christal, R. E. (1990). Reasoning ability (is little more than) working memory capacity. *Intelligence, 14*, 389–433.

Leather, C. V., & Henry, L. A. (1994). Working memory span and phonological awareness tasks as predictors of early reading ability. *Journal of Experimental Child Psychology, 58*, 88–111.

Liberman, I. Y., Mann, V. A., Shankweiler, D., & Werfelman, M. (1982). Children's memory for recurring linguistic and non-linguistic material in relation to reading ability. *Cortex, 18*, 367–375.

Logie, R. H. (1995). *Visuo-spatial working memory*. Hove, UK: Lawrence Erlbaum Associates.

Logie, R. H., & Pearson, D. G. (1997). The inner eye and the inner scribe of visuo-spatial working memory: Evidence from developmental fractionation. *European Journal of Cognitive Psychology, 9*, 241–257.

Macaruso, P., Locke, J. L., Smith, S. T., & Powers, S. (1996). Short-term memory and phonological coding in developmental dyslexia. *Journal of Neurolinguistics, 9*, 135–146.

Mann, V., & Liberman, I. Y. (1984). Phonological awareness and verbal short-term memory: Can they presage early reading success? *Journal of Learning Disabilities, 17*, 592–599.

Mann, V. A., Liberman, I. Y., & Shankweiler, D. (1980). Children's memory for sentences and word strings in relation to reading ability. *Memory and Cognition, 8*, 329–335.

Mark, L. S., Shankweiler, D., Liberman, I. Y., & Fowler, C. A. (1977). Phonetic recoding and reading difficulty in beginning readers. *Memory and Cognition, 5*, 623–629.

McDougall, S., Hulme, C., Ellis, A., & Monk, A. (1994). Learning to read: The role of short-term memory and phonological skills. *Journal of Experimental Child Psychology, 58*, 112–133.

McGee, R., Williams, S., Moffitt, T., & Anderson, J. (1989). A comparison of 13-year-old boys with attention deficit or reading disorder on neuropsychological measures. *Journal of Abnormal Child Psychology, 17*, 37–53.

Meyler, A., & Breznitz, Z. (1998). Developmental associations between verbal and visual short-term memory and the acquisition of decoding skill. *Reading and Writing, 10*, 519–540.

Michaels, C. A., Lazar, J. W., & Risucci, D. A. (1997). A neuropsychological approach to the assessment of adults with learning disabilities in vocational rehabilitation. *Journal of Learning Disabilities, 30*, 544–551.

Miles, C., Morgan, M. J., Milne, A. B., & Morris, E. D. M. (1996). Developmental and individual differences in visual memory span. *Current Psychology, 15*, 53–67.

Milner, B. (1971). Interhemispheric differences in the localisation of psychological processes in man. *British Medical Bulletin, 27*, 272–277.

Nichelli, P., & Venneri, A. (1995). Right hemisphere developmental learning disability: A case study. *Neurocase, 1*, 173–177.

Nicolson, R. (1981). The relationship between memory span and processing speed. In M. Friedman, J. P. Das, & N. O'Connor (Eds.), *Intelligence and learning* (pp. 179–184). New York: Plenum Press.

Oakhill, J., & Kyle, F. (2000). The relation between phonological awareness and working memory. *Journal of Experimental Child Psychology, 75*, 152–164.

Olson, R. K., Davidson, B. J., Kliegl, R., & Davies, S. E. (1984). Development if phonetic memory in disabled and normal readers. *Journal of Experimental Child Psychology, 37*, 187–206.

O'Neill, M. E., & Douglas, V. I. (1991). Study strategies and story recall in attention deficit disorder and reading disability. *Journal of Abnormal Child Psychology, 19*, 671–692.

Orton, S. (1937). *Reading, writing, and speech problems in children*. New York: W.W. Norton.

Palmer, S. E. (2000). The retention of a visual encoding strategy in dyslexic teenagers. *Journal of Reading Research, 23*, 28–40.

Pickering, S. J. (2001). Cognitive approaches to the fractionation of visuo-spatial working memory. *Cortex, 37*, 457–473.

Pickering, S. J., & Gathercole, S. E. (2001). *The working memory test battery for children*. London: Psychological Corporation.

Pickering, S. J., & Gathercole, S. E. (2004). *Working memory deficits in dyslexia: Are they located in the phonological loop, visuo-spatial sketchpad or central executive?* Manuscript under revision.

Rack, J. P. (1994). Dyslexia: The phonological deficit hypothesis. In A. Fawcett, & R. Nicolson (Eds.), *Dyslexia in children* (pp. 5–38). London: Harvester Wheatsheaf.

Rack, J. (1985). Orthographic and phonetic coding in normal and dyslexic readers. *British Journal of Psychology, 76*, 325–340.

Reynolds, C. R., & Bigler, E. D. (1994). *Test of memory and learning*. Austin, TX: Pro-Ed.

Rohl, M., & Pratt, C. (1995). Phonological awareness, verbal working memory and the acquisition of literacy. *Reading and Writing, 7*, 327–360.

Roodenrys, S., Hulme, C., & Brown, G. (1993). The development of short-term memory span: Separable effects of speech rate and long-term memory. *Journal of Experimental Child Psychology, 56*, 431–442.

Roodenrys, S., & Stokes, J. (2001). Serial recall and nonword repetition in reading disabled children. *Reading and Writing, 14*, 379–394.

Rudel, R. G., & Helfgott, E. (1984). Effects of piracetam on verbal memory of dyslexic boys. *Journal of the American Academy of Child Psychiatry, 23*, 695–699.

Rundus, D. (1971). Analysis of rehearsal processes in free recall. *Journal of Experimental Psychology, 89*, 63–77.

Salame, P., & Baddeley, A. D. (1987). Noise, unattended speech and short-term memory. *Ergonomics, 30*, 1185–1193.

Salame, P., & Baddeley, A. D. (1989). Effects of background music on phonological short-term memory. *Quarterly Journal of Experimental Psychology, 41A*, 107–122.

Schweickert, R. (1993). A multinomial processing tree for degradation and redintegration in immediate recall. *Memory and Cognition, 21*, 168–175.

Shankweiler, D., Liberman, I. Y., Mark, L. S., Fowler, C. A., & Fischer, F. W. (1979). The speech code and learning to read. *Journal of Experimental Psychology: Human Learning and Memory, 5*, 531–545.

Siegel, L. S. (1994). Working memory and reading: A life-span perspective. *International Journal of Behavioural Development, 17*, 109–124.

Siegel, L. S., & Linder, B. A. (1984). Short-term memory processes in children with reading and arithmetic learning disabilities. *Developmental Psychology, 20*, 200–207.

Siegel, L. S., & Ryan, E. B. (1989). The development of working memory in normally achieving and subtypes of learning disabled children. *Child Development, 60*, 973–980.

Smith, E. E., & Jonides, J. (1999). Storage and executive processes in the frontal lobes. *Science, 283*, 1657–1661.

Snowling, M. J. (1981). Phonemic deficits in developmental dyslexia. *Psychological Research, 43*, 219–234.

Snowling, M. J., Goulandris, N., Bowlby, M., & Howell, P. (1986). Segmentation and speech perception in relation to reading skill: A developmental analysis. *Journal of Experimental Child Psychology, 41*, 489–507.

Spring, C., & Capps, C. (1974). Encoding speed, rehearsal, and probed recall of dyslexic boys. *Journal of Educational Psychology, 66*, 780–786.

Swanson, H. L. (1978). Verbal encoding effects on the visual short-term memory of learning disabled and normal readers. *Journal of Educational Psychology, 70*, 539–544.

Swanson, H. L. (1992). Generality and modifiability of working memory among skilled and less skilled readers. *Journal of Educational Psychology, 84*, 473–488.

Swanson, H. L. (1993a). Working memory in learning disability subgroups. *Journal of Experimental Child Psychology, 56*, 87–114.

Swanson, H. L. (1993b). Executive processes in learning disabled readers. *Intelligence, 17*, 117–149.

Swanson, H. L. (1994). Short-term memory and working memory: Do both contribute to our understanding of academic achievement in children and adults with learning disabilities? *Journal of Learning Disabilities, 27*, 34–50.

Swanson, H. L. (1999). Reading comprehension and working memory in learning-disabled readers: Is the phonological loop more important than the executive system? *Journal of Experimental Child Psychology, 72*, 1–31.

Swanson, H. L., & Alexander, J. E. (1997). Cognitive processes as predictors of word recognition and reading comprehension in learning-disabled and skilled readers: Revisiting the specificity hypothesis. *Journal of Educational Psychology, 89*, 128–158.

Swanson, H. L., Ashbaker, M. H., & Lee, C. (1996). Learning-disabled readers' working memory as a function of processing demands. *Journal of Experimental Child Psychology, 61*, 242–275.

Swanson, H. L., & Ashbaker, M. H. (2000). Working memory, short-term memory, speech rate, word recognition and reading comprehension in learning disabled readers: Does the executive system have a role? *Intelligence, 28*, 1–30.

Temple, C. M. (1992). Developmental memory impairment: Faces and patterns. In R. Campbell (Ed.), *Mental lives* (pp. 198–215). Oxford: Blackwell.

Torgesen, J. K., & Houck, D. G. (1980). Processing deficiencies of learning-disabled children who perform poorly on the digit span test. *Journal of Educational Psychology, 72*, 141–160.

Turner, M. (1997). *Psychological assessment of dyslexia*. London: Whurr.

Turner, M. L., & Engle, R. W. (1989). Is working memory capacity task dependent? *Journal of Memory and Language, 28*, 127–154.

Yuill, N., Oakhill, J., & Parkin, A. (1989). Working memory, comprehension ability and the resolution of text anomoly. *British Journal of Psychology, 80*, 351–361.

Vellutino, F., Pruzek, R., Steger, J., & Meshoulam, U. (1973). Immediate visual recall in poor and normal readers as a function of orthographic-linguistic familiarity. *Cortex, 9*, 370–386.

Vellutino, F., Steger, J., DeSetto, L., & Phillips, F. (1975). Immediate and delayed recognition of visual stimuli in poor and normal readers. *Journal of Experimental Child Psychology, 19*, 223–232.

Vellutino, F. R. (1979). *Dyslexia: Theory and research*. Cambridge, MA: MIT Press.

Wagner, R. K., & Torgesen, J. K. (1987). The nature of phonological processing and its causal role in the acquisition of reading skills. *Psychological Bulletin, 101*, 192–212.

Wechsler, D. (1996). *Wechsler intelligence scale for children—Revised*. New York: Psychological Corporation.

7

Accelerating Word Reading, Spelling, and Comprehension Skills with Synthetic Phonics

Rhona S. Johnston and Joyce Watson

The earliest writing systems used pictures, but although this approach was effective in representing easily picturable objects, it did not represent speech in a linear fashion and it was not good at expressing complex ideas. Logographic writing systems such as Chinese are much more effective, underpinned by the idea that one character represents a word, and that characters can be decoded into spoken words in sequence to form sentences. The same principle underlies alphabetic writing systems, but here the individual sounds of the spoken word are represented by letters. Originally, the sounds represented were at the syllable level, which is quite salient in speech. In the Phoenician script, syllables were represented by their first consonants. However, in a language with a lot of syllables, a large number of symbols would have to be learnt. A major development was the creation of the Greek alphabet, whereby each consonant and vowel was represented. With this approach, new words can easily be represented in writing, although the writer has the challenge of breaking the sounds of words into artificial segments, that is, phonemes. This of course is the system used in English spelling.

Early teaching methods capitalized on the alphabetic nature of the English spelling system. The alphabetic spelling method was used in Britain from early

Rhona S. Johnston, Department of Psychology, University of Hull, Hull HU6 7RX, UK.
Joyce Watson, School of Psychology, University of St Andrews, St Andrews KY16 9JU, UK.

The Study of Dyslexia, edited by Turner and Rack.
Kluwer Academic Publishers, New York, 2004.

times until well into the 19th and 20th centuries (Moyle, 1968; Thompson, 1996), but it was not a phonic system. When early reading instruction was in Latin in the Middle Ages, letter names were taught first, the pronunciation of the letters being a good to guide than the pronunciation of the words. The medieval "primer" was a basic prayer book with the alphabet displayed at the beginning (Clanchy, 1984).

The phonic method started to develop in the 19th century (Morris, 1984). In this approach, the sounds rather than the names of the letters of the alphabet are taught, and children learn the correspondences between letters and groups of letters, and their pronunciations (Adams, 1990). In analytic phonics, the predominant method in the United Kingdom, letter sounds are taught after reading has already begun, children initially learning to read some words by sight, often in the context of meaningful text. However, the analytic phonics component of the reading program is generally taught in a separate lesson devoted to word study (Watson, 1998). In order to teach the letter sounds, whole words sharing a common initial letter sound are presented to children, for example, "milk," "man," "mother" (Harris & Smith, 1976). Attention is drawn to the /m/ sound heard at the beginning of the words. When all of the letter sounds have been taught in this way, attention is then drawn to letters at the ends of words, then in the middle, in consonant–vowel–consonant (CVC) words. Therefore, children learn about letter sounds in the context of whole words. At this stage, which can be at the end of the first year at school, children may also be taught to sound and blend CVC words, for example, /c/ /a/ /t/ → cat, but this is not a feature of all analytic phonics schemes, although it used to be. After mastering CVC words, children are taught about vowel and consonant digraphs and shown word families of similarly spelt words, for example, "cake," "bake," "make," "lake"; "coat," "boat," "float," etc. These spelling patterns used to be learnt by rote, with children chanting the words in unison in class, although this approach is not used now. Phonic readers also used to be widely available, some of which used very stilted text to reinforce phonic spelling patterns.

The analytic phonics method fell foul of the growing move toward child-centered education, which sought to introduce a greater emphasis on meaning and purpose in educational activities. Piaget, in a philosophical tradition stemming back to Kant and Rousseau, theorized that children were active learners, who constructed knowledge for themselves. Piaget did not specifically address learning to read, but his work encouraged teachers to tailor the teaching of reading and writing to the individual child's learning rate. At its extreme, all structured lessons were abandoned, as at the William Tyndale School in London, where it was believed that child-centered education implies standing back from direct teaching in order to avoid interfering with natural growth (Blenkin & Kelly, 1987). However, many children failed to learn to read and write at this school.

Analytic phonics fell into disfavor because it was often implemented in a rote manner, and because it was usually carried out without reference to the reading of meaningful text. As part of the emphasis on children learning for themselves and carrying out meaningful activities, the whole language approach to reading developed. It was felt that it was of paramount importance that children read meaningful material; it was thought that they could learn for themselves the relationship between letters and sounds. Unfamiliar words were to be identified by using context, rather than the "bottom up" approach of looking at individual words and applying phonic knowledge to decode the words. Added to this was the view that as some words in the English language are irregularly spelt, the phonic approach is ineffective and leads to inaccurate pronunciation, the word "yacht" being an extreme example of a word not amenable to being read by a phonic approach.

Although the whole language approach was very influential in the primary curriculum throughout the United Kingdom, the phonic method was kept as a core component in most Scottish schools, whereas in England many schools dropped this element of teaching altogether. When concerns were raised about attainment levels in reading in England, the lack of phonics tuition was widely considered to be an element in the poor attainment figures.

Starting in the 1980s, Johnston and Thompson (1989), Johnston, Thompson, Fletcher, and Holligan (1995), and Johnston, Connelly, and Watson (1995) looked at the effects of phonics teaching on how children read. Children in Scotland, learning by a program that included an analytic phonics component, were compared with children in New Zealand learning by a book experience approach. The latter method is a type of whole language approach where the emphasis is on children predicting reading responses from story and sentence contexts. One approach used is shared reading. Using a Big Book with large print that the whole class can see, the teacher starts a discussion of the general story topic without revealing the actual story line. The children are encouraged to predict what the text will be about from the title of the story. The teacher then reads the story out, pointing to the words, and pausing to discuss predictions about events and words. Another element in the New Zealand method is guided reading, where a child or group of children follow a story with their own copy of the text, the teacher actively working with the children in predicting events and words in the story. Children are encouraged to read silently in these sessions, followed by the teacher asking comprehension questions to gauge how well they have understood the text. Letter names are taught but not letter sounds; if a child cannot guess an unfamiliar word from context, they are encouraged to look at the first letter of the word (New Zealand Department of Education, 1985). In the early stages of learning to write, a language experience approach is used. The child draws a picture and then dictates a story about it to the teacher. She then writes down the story, and the child copies it.

Our studies involved matching children taught by the book experience and analytic phonics methods on word recognition skills using the Burt Word Reading Test. However, it was found that the New Zealand norms required fewer correct responses to gain the same reading age as on the Scottish norms. This showed that word recognition attainment levels were lower in New Zealand at each age level. We therefore matched the groups on the number of items correct, and were then able to compare nonword reading ability and comprehension skills. The Scottish children were found to be ahead in both nonword reading (Johnston & Thompson, 1989), and comprehension (Connelly, Johnston, & Thompson, 2001). The superiority of nonword reading was a predictable outcome of the phonics teaching method, in which children are taught about the importance of letter sound correspondences in individual words. Children learning by the book experience method are taught to decode words in context, so new words presented in isolation are likely to pose more of a problem for them. However, given that this method emphasizes reading text for meaning, it had been expected that they would be at an advantage in reading comprehension, but in fact they were significantly behind.

In England, in order to improve the teaching of literacy in schools, the National Literacy Strategy was started in 1998. A Literacy Hour was established in which teachers are encouraged to use a book experience method similar to the one used in New Zealand, and also an analytic phonics approach. The latter was made more explicit by the subsequent publication of Progression in Phonics (DFES, 2001), for children in reception and years 1 and 2. As a first step, children are taught to discriminate general environmental sounds, such as vehicle noises, birds singing, water being poured/splashed, and the like. A considerable amount of phonological awareness training is also advocated at the beginning of the program, that is, training to hear rhymes and phonemes in spoken words. This has been introduced as it has been shown that measures of preschool phonological awareness ability correlate with later reading skill (Bradley & Bryant, 1983; Lundberg, Olofsson, & Wall, 1980; Share, Jorm, Maclean, & Matthews, 1984; Stanovich, Cunningham, & Cramer, 1984; Stuart & Coltheart, 1988). Prereaders have limited ability to detect phonemes in spoken words, this being a skill that develops largely through learning to read in an alphabetic language (e.g., Morais, Bertelson, Cary, & Alegria, 1986; Morais, Cary, Alegria, & Bertelson, 1979). One skill that prereaders are rather better at, however, is awareness of rhyme (Goswami & Bryant, 1990). It has been proposed that children will make better progress in learning to read if their phonological awareness skills, especially rhyme, are well-developed before starting a reading program (Fraser, 1997; Goswami, 1999; Maclean, Bryant, & Bradley, 1987). However, of the many studies carried out to establish the effectiveness of phonological awareness training without print in enhancing literacy skills, by the early 1990s, only three had produced positive results (Cunningham, 1990; Lie, 1991; Lundberg, Frost, & Petersen, 1988). The majority of the phonological awareness studies that found an enhancement of reading skills used letters as part of the training program (e.g., Ball & Blachman, 1991; Bradley & Bryant, 1983; Byrne & Fielding-Barnsley, 1989, 1991; Fox & Routh, 1984; Hatcher, Hulme, & Ellis, 1994; Williams, 1980). In a recent statistical meta-analysis of 52 studies, Ehri *et al.* (2001) have found that when all of the studies are combined together, phonological, awareness training on its own does enhance reading skill. However, they conclude that phonological awareness ability should not be trained in isolation as it is more effective when taught together with letter sound correspondences and when applied to reading and writing.

In Goswami's (1994a) view, reading tuition should build on the phonological awareness skills children have at the start of schooling. She proposes that those who are weak in this area should be given rhyme and alliteration training before starting reading instruction, as these are skills that develop before phoneme awareness. Children can then be shown how to read by making analogies between known sight words and unfamiliar words. She advocates that the major focus should be on the rhymes in spoken and written words, as children can make analogies with these large units. Letter sound correspondences should only be taught at the beginning of words in the onset position as young children generally only have awareness of phonemes in this position (Goswami, 1995). In fact, Goswami (1994b) has argued that children should not learn by a method where they are taught initially about the spelling–sound correspondences for all the phonemes in spoken words, as this poses unnecessary difficulties for them. The National Literacy Strategy has been heavily influenced by this view, and proposes that at the start of reading tuition, children receive a substantial amount of phonological awareness training without exposure to print, and that when letter sounds are first taught, they are shown in the initial position of words (DFES, 2001).

We have carried out a number of studies to examine the effects of different types of teaching programs on children's progress in learning to read. Watson (1998) carried out a study of 228 children learning to read in Scotland, where an analytic phonics scheme was a core component of the reading program. The children started to learn to read by sight, but also had phonics lessons where they learnt about letter sounds at the beginning of words. This latter process was completed around March of the first year at school. When tested at this stage, the children were reading 5 months below chronological age on the British Ability Scales (BAS) Word Reading Test (Elliott, Murray, & Pearson, 1977). The children were then taught about CVC words, for example, "cat," "sun," "pen," with attention being drawn to letters in all position of words. Near the end of the summer term, around 2 months after the previous test phase, the children were reading only 1 month below chronological age. Toward the end of the third year at school, the girls were reading words 6.6 months above chronological age and were age appropriate in spelling. However, there was a much poorer outcome for boys, who were reading words 3 months above chronological age, but were 4 months behind for their age in spelling (Schonell & Schonell, 1952). When comprehension was measured at the end of the year, the girls were reading text appropriately for chronological age, but the boys were 5 months behind. At this point, nearly 10% of the children were reading 12 or more months behind chronological age, 9.4% of the girls and 10.4% of the boys.

However, in carrying out this study, Joyce Watson noticed that one class was making better progress than the others. The pace of analytic phonics teaching was accelerated in this class; the children were learning about letters in all positions of CVC words several months earlier than the other classes, and were taught to sound and blend letters to pronounce unfamiliar words. The gains these children made compared to the other classes were still apparent at the end of the third year at school. This led us to look at synthetic phonics, which is a very accelerated form of phonics that does not begin by establishing an initial sight vocabulary. With this approach, before children are introduced to books, they are taught letter sounds. After the first few of these have been taught, they are shown how these sounds can be blended together to build up words (Feitelson, 1988). For example, when taught the letter sounds /t/ /p/ /a/, and /s/, the children can build up the words "tap," "pat," "pats," "taps," "a tap," and the like. The children are not told the pronunciation of the new word by the teacher either before it is constructed with magnetic letters or indeed afterwards; the children sound each letter in turn and then synthesize the sounds together in order to generate the pronunciation of the word. Thus, the children construct the pronunciation for themselves. Most of the letter sound correspondences, including the consonant and vowel digraphs, can be taught in the space of a few months at the start of their first year at school. This means that the children can read many of the unfamiliar words they meet in text for themselves, without the assistance of the teacher. By contrast, in analytic phonics, whole words are presented and pronounced by the teacher, and the children's attention is only subsequently drawn to the information given by letter sound correspondences. Typically in Scotland with the analytic phonics approach, it would not be until the third term of the first year at school that children would be made aware of the importance of letter sound correspondences in all positions of words, whereas in synthetic phonics this is done at the start of the year. The full analytic phonics scheme is usually not completed until the end of the third year at school.

STUDY 1

We have carried out a number of experimental studies comparing different types of phonics teaching (see Johnston & Watson [in press] and Watson & Johnston, 1998, for further details). In Study 1, we looked at initial readers who had just started school at an average age of 5. They were being taught by the classroom analytic phonics program advocated by their local education authority, learning one letter sound a week at the beginning of words. We extracted the children from their classrooms for extra tuition in addition to their normal reading programs. They were taught in groups of 4–5, starting 6 weeks after entering school. The children in the three experimental groups were drawn equally from four classes, thereby controlling for minor differences in the implementation of the analytic phonics method used by the class teachers. The intervention lasted for 10 weeks, there being two 15-min training lessons a week, 19 sessions in all. The same print vocabulary was used for all three programs. One group was told how these words were pronounced, but the children were not taught letter sounds in the training program, and their attention was not drawn to letter sounds within the words (sight word + analytic phonics control group). A second group learnt letter sounds at the rate of 2 a week, and the children's attention was drawn to letter sounds in the initial position of words (accelerated analytic phonics control group). The third group learnt letter sounds at the rate of 2 a week, and the children had their attention drawn to letter sounds in all positions of words (analytic + synthetic phonics group).

Ninety-two children were studied, forty-six boys and forty-six girls. There were 29 children in the sight word + analytic phonics control group, 33 in the accelerated analytic phonics control group, and 30 children in the analytic + synthetic phonics group. The participants were matched into three groups on chronological age, sex, vocabulary knowledge (British Picture Vocabulary Scale, Dunn & Dunn, 1982), letter knowledge, emergent reading (Clay Ready to Read Test, Clay, 1979), phoneme segmentation (Yopp–Singer Test, Yopp, 1988), and rhyme generation ability. No differences were found between the groups on these measures. These tasks were administered again at the first posttest, straight after the end of the program, with the addition of the BAS Word Reading Test (Elliott *et al.*, 1977). At the second posttest, 9 months after the end of the intervention, the Schonell Spelling test (Schonell & Schonell, 1952) and a nonword naming task were additionally administered.

At the first posttest, at the end of the program, the analytic + synthetic phonics group performed better than the sight word + analytic phonics and the accelerated analytic phonics control groups on the BAS Word Reading Test, the Clay Ready to Read Test, and the test of letter knowledge; the other two groups did not differ from each other on these measures. On the test of phoneme awareness (Yopp, 1988), the difference between groups was not quite significant, and on the test of rhyme production ability, there was clearly no difference between the groups. See Table 1 for means and standard deviations.

The second posttest measures were taken at the start of the second year at school, 9 months after completion of the intervention program. By this time all of the children had learnt about letter sounds in all positions of words, in their normal analytic phonics classroom programs. The analytic + synthetic phonics group performed better than the sight word + analytic phonics and accelerated analytic phonics control groups on the BAS Word Reading Test, the Clay Ready to Read Test, the Schonell Spelling Test, and the test of letter knowledge; the other two groups did not differ from each other on these tests. At this stage, we tested nonword reading, and found that the analytic + synthetic phonics group read more nonwords correctly than the sight word + analytic phonics and accelerated analytic phonics control groups, there being no difference between the latter two groups. On the measures of phoneme awareness (Yopp, 1988) and of rhyme production ability, the analytic + synthetic phonics group performed better than the sight word + analytic phonics and accelerated analytic phonics, the latter two groups not differing from each other. See Table 1 for means and standard deviations.

It was concluded from this study that even when controlling for exposure to new print items, children additionally taught by a synthetic phonics approach made better progress in reading and spelling than children taught solely by an analytic phonics approach. This was the case even when letter sounds were taught at the same accelerated pace as in the synthetic phonics program, and when there

Table 1. Mean chronological age, reading age (British Ability Scales), emergent reading (Clay Ready to Read Test), letter sound knowledge, phoneme segmentation (Yopp–Singer Test), rhyme skills, and nonword reading (standard deviations in brackets), first and second posttest, Study 1

Research group	Age	Reading age	Spelling age	Emergent reading	Letter knowledge	Phonemic segmentation	Rhyme skills	Nonwords
First posttest								
Sight word + analytic phonics controls, $n = 29$	5.2 (0.3)	5.0 (0.5)	—	8.0 (18.3)	30.4 (24.0)	9.7 (21.6)	47.1 (38.1)	—
Accelerated analytic phonics controls, $n = 33$	5.2 (0.3)	5.0 (0.3)	—	10.3 (16.0)	37.1 (26.8)	9.0 (19.7)	36.4 (37.7)	—
Analytic + synthetic phonics, $n = 30$	5.3 (0.3)	5.4 (0.3)	—	25.6 (17.7)	51.8 (21.7)	21.2 (26.5)	33.6 (35.9)	—
Second posttest								
Sight word + analytic phonics controls, $n = 29$	6.0 (0.3)	5.6 (0.9)	5.6 (0.8)	24.8 (24.4)	68.1 (22.6)	26.8 (36.2)	41.4 (44.1)	14.6 (26.8)
Accelerated analytic phonics controls, $n = 33$	6.0 (0.3)	5.5 (0.8)	5.4 (0.7)	27.3 (30.6)	68.1 (24.9)	25.8 (36.7)	32.5 (38.3)	12.1 (24.6)
Analytic + synthetic phonics, $n = 33$	6.0 (0.3)	6.3 (1.3)	6.3 (0.8)	49.9 (28.8)	82.0 (20.1)	69.3 (36.1)	61.6 (44.9)	54.6 (40.7)

was the same exposure to new print. These gains were long lasting and could be found at the start of the second year at school, yet the intervention had lasted for only 9.5 hr. Furthermore, the initial advantage in reading for the synthetic phonics group had doubled between the first and second posttests without further intervention. It seems very likely that children who are taught the technique of sounding and blending early on are able to continue to expand their reading vocabulary by using the approach to decode the unknown words they encounter when reading text.

STUDY 2

In Study 1, the intervention was carried out on a group basis, with small groups of 4–5 children. This raised the question of whether the synthetic phonics approach would be effective in the classroom situation, delivered by class teachers. It had also been found in Study 1 that synthetic phonics was very effective at increasing children's phonemic awareness skills. A further issue was, therefore, whether analytic phonics teaching would be found to be as effective in developing reading and spelling skills as synthetic phonics if there was an additional phonological awareness training program. Altogether 304 children were studied in 13 classes in Clackmannanshire in Scotland. Our interventions began shortly after the children started school at an average age of 5. Four classes were taught about the relationship between letters and sounds using an analytic phonics approach (analytic phonics only group). Another four classes carried out a program where in addition to analytic phonics teaching, children were taught how to segment and blend spoken words at the level of both rhymes and phonemes, without the aid of print or letters (analytic phonics + phonological awareness group). In the third program, five classes of children were taught by a synthetic phonics approach (synthetic phonics group).

It was predicted that the synthetic phonics program would be the most effective in developing reading, spelling, and phonemic awareness showing that children can use a grapheme–phoneme conversion level approach right at the start of reading instruction. It was further predicted that the teaching of explicit phoneme and rhyme awareness in the absence of print, in addition to an analytic phonics program, would confer no advantages in terms of reading and spelling attainment over a control group taught wholly by the analytic phonics method. The programs lasted for 16 weeks, the children receiving their interventions via scripted whole class programs which lasted for 20 min a day. Reading scheme books were introduced at around 6 weeks after the intervention started.

The same pretest and posttest measures as for Study 1 were administered. In addition, at the posttest, a separate analysis was made of ability to read irregular words on the BAS Word Reading Test, and a test of analogy reading skills was made. At pretest, the children in the three groups were found to be matched on all tasks except for knowledge of letter sounds; the analytic phonics only group knew more letter sounds than the other two groups.

At the posttest directly after the intervention (see Table 2 for means and standard deviations), it was found that the synthetic phonics group had a significantly higher mean reading age on the BAS Word Reading Test than the other two groups, who did not differ from each other. The groups also differed on the Clay Ready to Read Test, which is a more sensitive test of emergent reading. The synthetic phonics group performed better than the other two groups, but the group taught only by the analytic phonics method performed better than the group that was taught analytic phonics + phonological awareness. On nonword reading, the groups again differed, the synthetic phonics group reading nonwords better than the other two groups, who did not differ from each other. Spelling ability was also best in the synthetic phonics group, the other two groups not differing from each other. Knowledge of letter sounds was also differentially affected by the training schemes, the synthetic phonics group being significantly ahead of the other two groups, although at the pretest they had been behind the analytic phonics group. On the phonemic awareness task (Yopp, 1988), the synthetic phonics group was significantly ahead of both of the other groups; it was also found that the analytic phonics + phonological awareness group was significantly better at this task than the analytic phonics only group. Thus, the analytic phonics group that got additional phonological awareness training did improve its performance on a phonemic awareness task, but did not improve its literacy skills beyond the level of the analytic phonics only group, and was in fact inferior on the test of emergent reading. On rhyme generation ability, the synthetic phonics group outperformed the analytic phonics only group, but not the analytic phonics + phonological awareness group; the two analytic phonics groups did not differ. For irregular word reading, it was found that the synthetic phonics taught children read these items better than the other two groups, who did not differ. See Table 3 for means and standard deviations.

An examination was also made at the posttest at the end of the intervention of ability to read words by analogy. See Table 3 for means and standard deviations. The children were asked to read a list of 40 words. They then read 5 clue words that would assist them in reading the 40 words by analogy on second showing, that is, prior exposure to "ring" should facilitate the pronunciation of "sing." These clue words were then removed, and the 40 words shown again. The items were taken from Muter, Snowling, and Taylor (1994). The gain in reading skill after exposure to the clue words was assessed. It was found that the synthetic phonics children were the only group to show an increase in reading skill between pre- and posttest, showing that they alone could benefit from exposure to the clue words. This indicates a qualitative difference in the approach to reading taken by the synthetic phonics taught children compared with the analytic phonics taught children: the former were able to read by analogy while the latter were not.

After the posttest at the end of the intervention, the two analytic phonics groups carried out the synthetic phonics program, completing it by the end of their first year at school. Posttests are still being carried out on a yearly basis; we report

Table 2. Mean chronological age, reading age (British Ability Scales Word Reading Test), spelling age (Schonell Spelling Test), emergent reading (Clay Ready to Read Test), letter sound knowledge, phoneme segmentation (Yopp–Singer Test), rhyme skills, and nonword reading (standard deviations in brackets), first posttest, Study 2

Research group	Age	Reading age	Spelling age	Emergent reading	Letter knowledge	Phonemic segmentation	Rhyme skills	Nonword
First posttest								
Analytic phonics controls, $n = 104$	5.4 (0.3)	5.4 (0.6)	5.2 (0.4)	37.8 (24.0)	58.1 (24.7)	17.2 (27.4)	26.4 (36.6)	8.8 (22.4)
Analytic phonics + phonological awareness, $n = 75$	5.4 (0.3)	5.4 (0.7)	5.3 (0.5)	23.9 (25.6)	59.9 (24.8)	34.7 (44.6)	36.4 (36.4)	15.8 (29.3)
Synthetic phonics, $n = 113$	5.5 (0.3)	6.04 (0.8)	6.0 (0.7)	53.4 (30.1)	90.1 (14.5)	64.8 (37.9)	46.5 (29.1)	53.3 (41.2)

here the performance of the children on reading and spelling tests at the end of their third year at school, 2 years after the intervention ended. See Table 4 for means and standard deviations. We report on the 190 children for whom we have a complete set of data over the years, 100 boys and 90 girls. In all previous analyses, we had found no differences between boys and girls in reading and spelling ability. However, we now found that the boys read better than the girls, but that the two groups spelt equally well. Although the boys showed an 8 months advantage in reading over the girls, this did not indicate poor performance by the girls. The girls were reading a significant 15 months ahead of chronological age, whereas the boys were reading a significant 23 months ahead of chronological age. In terms of spelling, the boys and girls were both a significant 10 months ahead of chronological age. Only 1 child (a boy) was reading 12 or more months behind chronological age. When scores 3 or more years above chronological age were examined, there were found to be 30 boys in this category and only 12 girls. There were also gains in reading comprehension on the Primary Reading Test (France, 1981). See Table 4 for means and standard deviations. The boys were reading 5 months above chronological age, whereas with the analytic phonics method, boys had been 5 months behind (Watson, 1998). The girls' comprehension of text

Table 3. Mean percentage correct on Analogy Reading Task and Irregular Word Task at the end of training program (first posttest), Study 2

Research group	Analogy task		Irregular word task	
	Pretest scores	Clue word reading scores	Posttest scores	Irregular words
Analytic phonics controls, $n = 104$	2.9 (12.0)	6.3 (18.3)	2.6 (9.3)	21.4 (19.5)
Analytic phonics + phonological awareness, $n = 75$	4.9 (15.8)	11.4 (27.3)	5.5 (16.2)	15.3 (23.1)
Synthetic phonics, $n = 113$	16.9 (25.7)	30.5 (32.9)	22.7 (23.7)	30.2 (25.4)

Table 4. Comparison of performance of synthetic phonics taught boys and girls in Study 2, at the end of the third year at school; chronological age, word reading age, reading comprehension, age, and spelling age (standard deviations in brackets)

Participants	Chronological age	Reading age (word recognition)	Reading age (comprehension)	Spelling age
Girls, $n = 90$	7.7 (0.3)	9.0 (1.7)	8.0 (1.1)	8.5 (1.0)
Boys, $N = 100$	7.8 (0.3)	9.7 (2.0)	8.2 (1.1)	8.6 (1.1)

was 4 months ahead of chronological age. However, some children are not included in these comparisons, because of incomplete data over the years. When examining all the children available for just the Primary 3 Reading Test, it was found that 2% were reading 12 or more months below chronological age, 3% of the boys (4 out of 132) and 0.8% of the girls (1 out of 120), the maximum deficit being 33 months. At the upper end of the scale, 34% of boys and 18% of girls were reading 36 or more months above chronological age.

DISCUSSION

We have found in two studies that children can be taught to read English by the synthetic phonics approach, using the grapheme to phoneme conversion ("small unit") level right from the start of reading tuition. In Study 1, children were taught by the normal classroom analytic phonics method, but also had additional training outside the classroom. When both speed of letter learning and exposure to new print vocabulary were controlled for, the children additionally taught by the synthetic phonics approach read and spelt better than the children taught only by the analytic phonics method. In the longitudinal study (Study 2), at the first posttest, the children taught phonics by the synthetic method were found to read and spell better than the analytic phonics taught groups, again despite exposure to the same new print vocabulary. Only the synthetic phonics group was able to read words by analogy; these children also showed better reading of words with irregular spellings than the other groups. At the end of the third year at school, the boys read significantly better than the girls (spelling and reading comprehension being equivalent); this is in contrast to an analytic phonics program in which boys were behind girls in all areas (Watson, 1998). Synthetic phonics teaching was also found to produce a significant acceleration of phoneme awareness skills in both studies. In Study 2, these gains were greater in the synthetic phonics group than in the analytic phonics group that had received extra phonological awareness training without exposure to print. Furthermore, phonological awareness training without print conferred no advantages in literacy attainment, confirming Hatcher *et al.*'s (1994) findings with poor readers.

It is worth focusing in more detail on the finding that the synthetic phonics approach was found to be particularly effective for boys. In Study 2, by the end of the third year at school, the boys had word reading 8 months ahead of the girls; both groups were reading well above chronological age, boys were 23 months ahead and girls were 15 months ahead. Spelling was 10 months ahead of chronological age for both groups. The boys were comprehending text 5 months above chronological age, performing slightly ahead of the girls on this task. Given that boys had been found to have comprehension skills 5 months below chronological age in a longitudinal analytic phonics study, and to be spelling 4 months below age level (Watson, 1998), these are very significant improvements over standard teaching methods. It is not generally the case that boys read as well as or better than girls; in an international comparison of reading comprehension, boys read less well than girls in 19 out of 26 countries (Elley, 1992).

As the boys had an advantage in reading but not spelling, we propose that it is the process of synthesizing sounds for pronunciation when reading that is particularly beneficial for boys; they showed no advantage compared with girls in spelling, which was taught by analyzing spoken words into phonemes. Naglieri and Rojahn (2001) have recently shown that boys are less good than girls at attending to or planning how to tackle a cognitive task, and they propose that boys may need to be taught to focus their attention and to be more strategic. Synthetic phonics may encourage children to focus their attention on the information given by letter sounds in words, and the sequential blending of the sounds is a strategy they can apply to many of the unknown words they meet in text. There may be a greater need for boys to have their attention focused in this way, and they may become more assiduous in applying this strategy than girls.

At the end of the third year at school, only 2% of the children were a year or more behind chronological age in reading, compared to nearly 10% of children taught by the analytic phonics method (Watson, 1998). Slightly more boys than girls had reading scores at the lower end of the distribution, although this was counterbalanced by the number of boys reading more than 3 years above chronological age. Although we expect the proportion of underachievers to rise as the children get older, our research suggests that fewer children taught by the synthetic phonics method will need dyslexia assessments, and fewer will be categorized as having special needs. Children with reading problems seem to be particularly prone to developing a form of word reading that is not well underpinned by phonological information (Ehri, 1992), so it may be that focusing on the sequence of letter sounds in words is of particular benefit to them. The sight word element in the analytic phonics method may reinforce a rather holistic approach to word recognition, which may disadvantage dyslexic children.

Why do the gains in reading continue to accumulate? A gain of 7 months over chronological age at the end of the program became a gain of 19 months just over 2 years later. If children are taught by a synthetic phonics method from the beginning, sounding and blending unfamiliar words at the start of reading tuition, they will have a self-teaching mechanism that they can apply for themselves when they meet unknown words in text. Indeed synthetic phonics is a method that enables children to construct knowledge for themselves, rather than depending on the help of their teacher when faced with an unfamiliar word. However, children may need to be encouraged to apply their sounding and blending skills when reading text. A recent report (Ofsted, 2002) has found that even when sounding and blending has been taught within the analytic phonics method advocated by the National Literacy Strategy, teachers often did not encourage children to apply this technique when reading text.

There is some confusion over what constitutes a synthetic phonics method. The approach advocated by the National Literacy Strategy is an analytic one, as a sight vocabulary is established prior to children learning phonics. That is, the children learn to recognize words and then how to analyze them phonically. Analytic phonics has traditionally included sounding and blending, so the use of this procedure does not mean that a synthetic phonics method is being used. In a synthetic phonics approach, children must learn to sound and blend right at the start of reading tuition, finding for themselves the pronunciation of new words. Children are taught a small group of letter sounds before they have any exposure to sight word learning, and they only hear the pronunciation of new words when they have sounded and blended the letters for themselves. These differences may appear to be subtle, but they have a significant impact on children's reading attainment and the development of their phonemic awareness skills. Historically, the development of the alphabetic writing system was a major achievement; it required the insight that the spoken word can be broken into small subunits and that these sounds can be represented by letters. We have found that children progress better when learning to read by a method that shows them this alphabetical principle right at the start of reading tuition. Many children can work this out for themselves, but a substantial proportion do not.

CONCLUSIONS

We conclude that it is more effective to teach synthetic than analytic phonics to children at the start of schooling, and its advantages are not just due to the accelerated letter sound learning typical of this approach. Synthetic phonics develops phonemic awareness skills, and gives rise to word reading, reading comprehension, and spelling skills significantly above chronological age. Furthermore, it is a method of teaching reading that is particularly beneficial for boys.

ACKNOWLEDGMENTS

We would like to thank Clackmannanshire Council and the Scottish Executive for their support of Study 2. The opinions expressed here do not necessarily reflect their views.

REFERENCES

Adams, M. J. (1990). *Beginning to read: Thinking and learning about print*. Cambridge, MA: MIT Press.

Ball, E., & Blachman, B. (1991). Does phoneme awareness training in kindergarten make a difference in early word recognition and developmental spelling? *Reading Research Quarterly, 26*, 46–66.

Blenkin, G. M., & Kelly, A. V. (1987). *Early childhood education. A developmental curriculum*. London: Paul Chapman.

Bradley, L., & Bryant, P. E. (1983). Categorizing sounds and learning to read—a causal connection. *Nature, 301*, 419–421.

Byrne, B., & Fielding-Barnsley, R. (1989). Phonemic awareness and letter knowledge in the child's acquisition of the alphabetic principle. *Journal of Educational Psychology, 81*, 313–321.
Byrne, B., & Fielding-Barnsley, R. (1991). Evaluation of a program to teach phonemic awareness to young children. *Journal of Educational Psychology, 83*, 451–455.
Clanchy, M. T. (1984). Learning to read in the Middle Ages and the role of mothers. In G. Brooks, & A. K. Pugh (Eds.), *Studies in the history of reading*. Reading, MA: Centre for the Teaching of Reading.
Clay, M. M. (1979). *The early detection of reading difficulties*. London: Heinemann.
Connelly, V., Johnston, R. S., & Thompson, G. B. (2001). The effects of phonics instruction on the reading comprehension of beginning readers. *Reading and Writing, 14*, 423–457.
Cunningham, A. E. (1990). Explicit versus implicit instruction in phoneme awareness. *Journal of Experimental Child Psychology, 50*, 429–444.
Department for Education and Skills. (2001). *The National Literacy Strategy: Progression in Phonics (PIPS)—Materials for whole class teaching*. London: Author.
Dunn, L. M., & Dunn, L. M. (1982). *British picture vocabulary scale*. Windsor: NFER-Nelson.
Ehri, L. C. (1992). Reconceptualising the development of sight word reading and its relationship to recoding. In P. B. Gough, L. C. Ehri, R. Treiman (Eds.), *Reading Acquisition* (pp. 107–143). Hillsdale, NJ: Lawrence Erlbaum Associates.
Ehri, L. C., Nunes, S. R., Willows, D. M., Schuster, B. V., Yaghoub-Zadeh, & Shanahan, T. (2001). Phonemic awareness instruction helps children learn to read: Evidence from the National Reading Panel's meta-analysis. *Reading Research Quarterly, 36*, 250–287.
Elley, W. B. (1992). *How in the world do students read?* Oxford: Elsevier Science.
Elliott, C. D., Murray, D. J., & Pearson, L. S. (1977). *The British Ability Scales*. Windsor: NFER-Nelson.
Feitelson, D. (1988). *Facts and fads in beginning reading. A cross-language perspective*. Norwood, NJ: Ablex.
Fox, B., & Routh, D. K. (1984). Phonemic analysis and synthesis as word attack skills: Revisited. *Journal of Educational Psychology, 76*, 1059–1064.
France, N. (1981). *Primary reading test*. Windsor: NFER-Nelson.
Fraser, H. (1997). *Early intervention: A literature review*. Edinburgh: Moray House Institute of Education.
Goswami, U. (1994a). The role of analogies in reading development. *Support for Learning, 9*, 22–26.
Goswami, U. (1994b). Phonological skills, analogies and reading development. *Reading, 28*, 32–37.
Goswami, U. (1995). Phonological development and reading by analogy: What is analogy and what is not? *Journal of Research in Reading, 18*, 139–145.
Goswami, U. (1999). Causal connections in beginning reading: The importance of rhyme. *Journal of Research in Reading, 22*, 217–240.
Goswami, U. C., & Bryant, P. E. (1990). *Phonological skills and learning to read*. Hove: Lawrence Erlbaum Associates.
Harris, L. A., & Smith, C. B. (1976). *Reading instruction: Diagnostic teaching in the classroom* (2nd ed.). London: Holt, Rinehart & Winston.
Hatcher, P. J., Hulme, C., & Ellis, A. W. (1994). Ameliorating early reading failure by integrating the teaching of reading and phonological skills: The phonological linkage hypothesis. *Child Development, 65*, 41–57.
Johnston, R. S., Connelly, V., & Watson, J. (1995). The effects of phonics teaching on reading development. In P. Owen, & P. Pumfrey (Eds.), *Children learning to read: International concerns* (Vol. 1). London: Falmer Press.
Johnston, R. S., & Thompson, G. B. (1989). Is dependence on phonological information in children's reading a product of instructional approach? *Journal of Experimental Child Psychology, 48*, 131–145.
Johnston, R. S., Thompson, G. B., Fletcher, C., & Holligan, C. (1995). The functions of phonology in the acquisition of reading: Lexical and sentence processing. *Memory and Cognition, 23*, 749–766.

Johnston, R. S., & Watson, J. (in press). Accelerating the development of reading, spelling and phonemic awareness skills in initial readers, *Reading and Writing*.

Lie, A. (1991). Effects of a training programme for stimulating skills in word analysis in first grade. *Reading Research Quarterly, 26*, 234–249.

Lundberg, I., Frost, J., & Petersen, O.-P. (1988). Effects of an intensive programme for stimulating phonological awareness in preschool children. *Reading Research Quarterly, 23*, 263–284.

Lundberg, I., Olofsson, A., & Wall, S. (1980). Reading and spelling skills in the first school years predicted from phonemic awareness skills in kindergarten. *Scandinavian Journal of Psychology, 21*, 159–173.

Maclean, M., Bryant, P., & Bradley, L. (1987). Rhymes, nursery rhymes and reading in early childhood. *Merrill–Palmer Quarterly Journal of Developmental Psychology, 33*, 255–281.

Morais, J., Bertelson, P., Cary, L., & Alegria, J. (1986). Literacy training and speech segmentation. *Cognition, 24*, 45–64.

Morais, J., Cary, L., Alegria, J., & Bertelson, P. (1979). Does awareness of speech as a sequence of phones arise spontaneously? *Cognition, 7*, 323–331.

Morris, J. (1984). Phonics: From an unsophisticated past to a linguistics-informed future. In G. Brooks, & A. K. Pugh (Eds.), *Studies in the history of reading*. Reading, MA: Centre for the Teaching of Reading.

Moyle, D. (1968). The teaching of reading. London: Ward Lock Educational.

Muter, V., Snowling, M., & Taylor, S. (1994). Orthographic analogies and phonological awareness: Their role and significance in early reading development. *Journal of Child Psychology and Child Psychiatry, 35*, 293–310.

Naglieri, J. A., & Rojahn, J. (2001). Gender differences in Planning, Attention, Simultaneous, and Successive (PASS) cognitive processes and achievement. *Journal of Educational Psychology, 93*, 430–437.

New Zealand Department of Education. (1985). *Reading in junior classes (with guidelines to the revised ready to read series)*. Wellington, New Zealand: Government Printer.

Ofsted (2002). The National Literacy Strategy: The first four years 1998–2002. London: Office for Standards in Education.

Schonell, F. J., & Schonell, F. E. (1952). *Diagnostic and attainment testing* (2nd ed.). Edinburgh: Oliver & Boyd.

Share, D. L., Jorm, A. F., Maclean, R., & Matthews, R. (1984). Sources of individual differences in reading acquisition. *Journal of Educational Psychology, 76*, 466–477.

Stanovich, K. E., Cunningham, A. E., & Cramer, B. B. (1984). Assessing phonological awareness in kindergarten children: Issues of task comparability. *Journal of Experimental Child Psychology, 38*, 175–190.

Stuart, M., & Coltheart, M. (1988). Does reading develop in a sequence of stages? *Cognition, 30*, 139–181.

Thompson, G. B. (1996). The teaching of reading. *Encyclopedia of language and education* (Vol. 2). The Netherlands: Kluwer Academic.

Watson, J. (1998). *An investigation of the effects of phonics teaching on children's progress in reading and spelling*. PhD thesis, University of St Andrews.

Watson, J. E., & Johnston, R. S. (1998). Accelerating reading attainment: The effectiveness of synthetic phonics. *Interchange, 57*, SOEID.

Williams, J. P. (1980). Teaching decoding with an emphasis on phoneme analysis and phoneme blending. *Journal of Educational Psychology, 72*, 1–15.

Yopp, H. K. (1988). The validity and reliability of phonemic awareness tests. *Reading Research Quarterly, 23*, 159–177.

8

Review of Research Evidence on Effective Intervention

John Rack

This chapter contains a review of recent findings from studies which have looked at the effectiveness of intervention programs for children, who are identified as having particular problems in learning to read—those often referred to as dyslexic or reading disabled. There is a larger body of evidence concerned with the effectiveness of the most effective general methods of teaching literacy skills to young children which will not be covered here in any depth. However, that evidence has recently been summarized in a report from the National Reading Panel in the United States. The report concluded:

> To summarise the results of the meta-analysis, the Panel examined 96 cases, each comparing a treatment group that received Phonological Awareness training, to a control group that received an alternative form of instruction or no special instruction; ... PA training improved children's ability to read and spell in both the short and long term (Langenburg *et al.*, 2000, pp. 2–28).

This evidence is consistent with a further strand of research, some of which will be reviewed here, which has established that phonological skills have a causal role in the development of reading skills in young children. Thus, following a meta-analysis of 36 research studies, Bus and van Ijzendoorn concluded: "The training studies settle the issue of the causal role of phonological awareness in learning to read: Phonological training reliably enhances phonological and reading skills.

John Rack, The Dyslexia Institute, The Henry Wellcome Building for Psychology, The University of York, Heslington, York YO10 5DD, UK.

The Study of Dyslexia, edited by Turner and Rack.
Kluwer Academic Publishers, New York, 2004.

About 500 studies with null results in the file drawers of disappointed researchers would be needed to turn the current results to non-significance" (p. 411).

There is also a large body of evidence that shows that dyslexic individuals tend to have specific difficulties in phonological processing. This is seen in reading skills (Rack, Snowling, & Olson, 1992) and in phonological language skills (Olson *et al.*, 1990), and these difficulties in phonological skills have been found to extend into adulthood (Bruck, 1992). This convergence of evidence of different kinds strongly supports a theory of dyslexia that has, at its heart, deficits in phonological processing. In a review of much of this evidence, Rack (1994) referred to this as the Phonological Deficit Hypothesis. Stanovich (1988) articulated a similar view in his "phonological core—variable difference" hypothesis and Snowling (1987) developed a very similar argument emphasizing the importance of compensatory processes.

Before reviewing a number of key intervention studies, it is worth pausing to consider what predictions might be made regarding the development of reading skills amongst children who experience reading difficulties. A strong version of the phonological deficit theory (reading depends on phonology) predicts that reading will improve, though slowly, to the extent that phonological difficulties can be remediated. A weaker version of the theory predicts that reading may develop, but in a qualitatively different fashion, to the extent that skills and strategies other than phonological ones can be fostered through teaching (Snowling, 1987; Stanovich, 1988). Both versions of the theory might predict differential outcome, depending on the "suitability" of the teaching for the individual's pattern of difficulties. Thus, we need to consider the following practical questions: can teaching methods be adapted to develop phonological skills in the face of phonological difficulties, and how can alternative skills and strategies be encouraged to compensate for phonological difficulties? As will be seen, most work has addressed the former question; relatively little has been done to look at effective ways of fostering compensatory strategies.

ESTABLISHING CAUSALITY

Associations between different skills and abilities are frequently observed in psychology and education, and those associations are of many and varying kinds. For example, it has been suggested that there are more people with balance problems amongst those with reading difficulties; others have suggested that poor readers have poorer visual skills. But these associations tell us nothing about causality—what causes what? For example, research on eye movements in the late 1980s established that dyslexic children's patterns of eye movements were no different from children without reading difficulties *provided* they were given passages to read appropriate to their ability. If the dyslexic children were given too difficult text, then, perhaps unsurprisingly, their eyes tended to "go all over the place." But, quite clearly, this was a consequence of their reading difficulties rather than its cause. Trickier still is the problem of the "third factor" (or *tertium quid*): some underlying difficulty at a more fundamental level may cause, for example, both reading problems and (for some) balance problems. Thus, these two types of problem would tend to occur together, without there being any direct connection between balance and reading. If this were so, an attempt to improve reading by training balance would be very unlikely to produce any significant specific benefit. An intervention study allows us to disentangle some of the problems of establishing causal relationships between different skills. One group receives an intervention that alters its skills, and then the effects of that alteration are assessed on a second skill. In our balance example, we would improve the balance skills of one group and then assess improvements in reading, in comparison to a control group whose balance skills had not been improved.

Thus, intervention studies are particularly important in developing theories about causal factors in reading ability and reading disability. However, these studies are very difficult to carry out well, as a recent review by Troia (1999) showed. One particular problem is in demonstrating a specific effect, over and above the "Placebo" or "Hawthorne Effect" through which improvements may come about simply because the participants are receiving attention of some kind. The usual solution to this is to allow more than one group to receive an intervention. Another useful check is to give both groups a task in which, on the theory being tested, a group difference would not be expected. If the intervention group is better than the comparison group in everything, then this suggests a rather nonspecific effect; if the difference is just on the targeted skills, then this is much more likely to be a specific effect.

THE RELATIONSHIP BETWEEN PHONOLOGY AND READING—EARLY LANDMARKS IN OXFORD AND SCANDINAVIA

This question of causality was addressed in the seminal intervention study of Bradley and Bryant (1985). They selected a sample of 65 children in the Oxford area who had previously been found to have poor phonological skills as assessed using a sound categorization task (Bradley & Bryant, 1983). Pupils were allocated to four different groups: (a) Sound categorization training; (b) Sound categorization training supported by concrete materials (plastic letters); (c) Semantic categorization training; (d) No treatment control. The semantic categorization group was included as a "placebo" group—they were receiving individual attention but, according to the theory being tested, this would not be expected to have an effect on reading. Or rather, it was predicted that the sound categorization training would produce a specific effect over and above any general effect of receiving attention. The subjects were initially matched on the basis of age, sex, sound categorization ability, and IQ, and the training sessions were given weekly over a period of 2 years. The children were trained individually and progressed through the program at their own rates. After 2 years, the effects of training were evaluated by standardized tests of reading and spelling. Sound categorization training was found to have a beneficial effect on later reading and spelling, but it was only significantly better than the semantic categorization control when plastic letters were also used as part of the training. The group that received sound categorization supplemented by concrete materials (Group 2) was 9 months ahead of the semantic categorization group (Group 3) in reading and 17 months ahead in spelling. The group that had sound categorization without the plastic letters (Group 1) was 4 months ahead of the control group in both reading and spelling. Groups 1, 2, and 3 did not differ in the progress made in maths indicating a specific role for the sound categorization training.

In an important study conducted in Denmark, where formal reading typically starts rather later than in the United Kingdom and United States, Lundberg, Frost, and Peterson (1988) sought to alter the phonological skills of a group of 235 kindergarten children before they had begun to learn to read and then to assess the impact of that training on emerging reading skills. The training took place during preschool where no formal reading instruction was given and included a range of exercises. A posttest measure confirmed that the training was effective; the experimental group performed better than a comparison group of 155 children on a range of measures of phonological skill. General language comprehension and letter knowledge increased equally for both groups over the training period showing that the effects of training were specific to the phonological domain. It is important to show specific effects of an intervention to rule out the possibility that their gains have arisen simply because they received attention and the control group did not. The effects of the phonological awareness training were investigated by measuring reading and spelling some 7 months into the children's first year in school and again in the middle of their second year. The experimental group did significantly better than the control group on spelling in Grade 1 and on both reading and spelling in Grade 2. Interestingly, the group differences were significantly larger in Grade 2 indicating that the benefits of phonological skills training may take some time to become established and may be seen first in spelling. Recall, that Bradley and Bryant (1985) also found greater benefits for spelling.

One potential problem with the Lundberg *et al.* study is that the control group received no special attention during the intervention period. However, the control group was found to perform better than the phonological intervention group on outcome measures of arithmetic strongly suggesting that the benefit of the phonological awareness training was a specific and not a general effect. It has been suggested by some that the phonological training in the Lundberg *et al.* (1988) study did not produce substantial benefits. However, to suggest this is to confuse two different purposes of intervention studies. The Lundberg *et al.* study was designed to establish whether there was a causal link between phonology and learning to read, by altering phonological skills at "step 1" and then observing the effects on reading development at "step 2." Although their results showed a positive effect of beginning the process of learning to read with enhanced phonological skills, it does not follow that this would be the optimal way of teaching in practice. A different kind of intervention study is needed to address this question.

LINKING READING AND PHONOLOGY AND MODIFIED READING RECOVERY—SIGNPOSTS IN CUMBRIA AND NEW ZEALAND

In another important study, Hatcher, Hulme, and Ellis (1994) explored the possibility that phonological training linked to reading experience might be more effective than phonological skills training given alone or reading instruction given alone. This hypothesis—which they termed the phonological linkage hypothesis—was based on Bradley and Bryant's finding that the significant gains in reading come from training in sound categorization with the use of plastic letters. Hatcher *et al.* conducted a county-wide screening program in Cumbria to identify 7-year-old children who were experiencing difficulties learning to read who were then assigned to four matched groups. The groups received either "no training," "reading alone," "phonological skills training alone," and "reading with phonology." The reading program was based on the Reading Recovery program of Clay (1985) which includes extensive practice in reading from books, with progression to the next level of difficulty only when a consistent level of accuracy (94%) has been achieved. The "phonology-alone" group followed a program of activities including rhyme detection, identification of sounds within words, segmenting, substituting, and deleting sounds from words. The work done by this group involved only listening and speaking; no visual support was given. The Reading plus Phonology (R + P) group had roughly half the amount of time on each of the R and P programs plus specific linkage activities that included letter–sound associations, sound categorization supported by plastic letters, and phonologically based writing tasks.

The Phonology Group (P) spent the most time on phonological activities and, perhaps unsurprisingly, they made the most progress on measures of phonological skill. By the same logic, it might have been expected that the "reading alone" group, who received the greatest exposure to printed materials, would progress most in reading; but this was not the case. The greatest gains in reading and spelling were made by the R + P group. Thus, the greatest gains come from a program in which phonological activities and reading experience are linked explicitly.

Very similar results were obtained by Iversen and Tunmer (1993) who compared the Reading Recovery program (Clay, 1985) with a modified version that included explicit training in phonological recoding. They selected a young "at-risk sample" whose average age was 6 years 2 months. These children had been designated by their schools as needing extra reading support at the end of their kindergarten year. One group of children received the standard reading recovery program and a second group received, in addition, explicit training in phonological skills. This training included activities similar to those used by Bradley and Bryant (1985) plus an emphasis on "phonograms" or "rime" units.

These are the word-family patterns which receive a more consistent pronunciation (e.g., -ight, -all) than do single letters. The success of the program was assessed in terms of the number of sessions required to bring the children up to a level whereby they were no longer judged as needing special educational measures. Both the reading recovery group and the modified reading recovery group did better than a third control group at the end of the programs; thus, both could be said to be effective. However, the modified reading recovery group reached the same end point in significantly fewer sessions (42 on average, compared to 57). Iversen and Tunmer point out that the standard reading recovery program was thus 37% less efficient than the program that also included explicit training in phonological awareness.

THE STRUCTURE OF EFFECTIVE TUITION PROGRAMS—DISCUSSION

The evidence from Hatcher *et al.*'s work is that improving poor readers' phonological skills does not, on its own, bring about substantial improvement in reading skills. Rather, development of phonological skills linked to reading experience and practice is required. The inclusion of phonological skills training is important, since reading experience and practice, on its own, do not produce significant gains. Iversen and Tunmer's work underlines this by showing that a reading-experience-based program is significantly less effective than a program that has both reading experience and phonological skills training. Hatcher *et al.*'s work, building on Bradley and Bryant's work, suggests that children with poor phonology need the linkage between spoken sounds and written letters to be made explicit. Moreover, it suggests that the teaching of word-level and sub-word-level decoding strategies must be explicitly linked to reading of connected text. In the R + P group, the teachers would highlight and consolidate taught rules during structured reading practice. This is an important practical point: It is not simply a matter of teaching phonic rules and then practicing them in isolation; rather, the practice provides opportunities to reinforce the rules and to demonstrate their use explicitly in context.

So what is happening in an effective intervention such as Sound Linkage? It seems most likely that the "linked teaching" is enabling development of reading skills despite poor underlying phonological difficulties. Thus, children will be helped to acquire reading skills beyond the level that would be expected on the basis of their phonological skills. Hatcher *et al.*'s data also provides evidence against the strong version of the phonological deficit theory: those whose phonology improved the most (the P-alone) group were not the ones who improved the most in reading. (It is possible that this is, to some extent, a "test artifact." On the face of it, the "phonology only group" [who did not have the linked reading activities] improved more on measures of phonology than the "reading only "or the "R + P" groups. It may be, however, that these children were taught how to do better on the phonology tests. Whatever interpretation one prefers, the finding serves to emphasize that reading and performance on phonological skills tests are not always tightly linked.)

COMPUTER READING WITH SPEECH FEEDBACK—THE COLORADO INTERVENTION STUDIES

Olson and Wise and their colleagues in Boulder approached the question of how best to teach poor readers from a rather different perspective, more in line with the predictions from the "strong version" of the phonological deficit hypothesis. They hypothesized that disabled readers might develop reading skills more effectively if they were taught about links between letter-patterns and sounds at the level of syllables or other larger orthographic units. To test this prediction, they used a computer reading system that provided feedback on words targeted by a mouse-click. Different levels of feedback were explored. For example, at the whole-word level, the word would be highlighted and the spoken word form provided; at the onset–rime level, the word would be segmented by syllable and within the syllable by onset and rime (e.g., "s-uit/c-ase"). In initial short-term training studies (described in Olson, Wise, Conners, & Rack, 1990), it was found that feedback at the level of individual letters and letter–sounds was unhelpful as the memory and blending demands were too great. There was also a suggestion of an advantage for segmented feedback at the level of the "onset and rime" (e.g., G-AR/D-EN), but in a later study (Wise, 1987) whole-word, syllable segmentation, and onset–rime segmentation were found to be similar. Indeed, for the younger and less able readers, onset–rime segmentation for polysyllabic words proved unhelpful. In a later, longer term intervention, subjects received either whole-word, syllable, onset–rime, or a "combined" feedback condition which involved both syllable and onset–rime feedback. Overall, segmented feedback was more effective than whole-word feedback, although whole-word training was no less effective in producing gains on speeded reading measures. Olson *et al.* (1990) speculated that the optimal form of feedback might vary for individuals according to their underlying phonological abilities. In line with this prediction, Olson and Wise (1992) found that children with low levels of phonological awareness gained less than children with high levels of phonological awareness from the segmented feedback training. Extending this argument, Wise and Olson (1995) predicted that the computer training would be more effective if it were given alongside, or following, training in phonological skills. They therefore compared two groups of children who both received computer training. One group received a version of the Lindamood and Lindamood (1975) "Auditory Discrimination in Depth" (ADD) program. The second group received small-group training on comprehension strategies. Both groups were found to make more progress than a third "control" group who received normal classroom instruction. As predicted, the children in the ADD group gained more in phoneme awareness and phonemic decoding than the group trained in comprehension strategies. This advantage was evident, though to a reduced extent, on standardized measures of word recognition.

The Colorado group developed their methods in a second, more extended intervention study (Olson, Wise, Ring, & Johnson, 1997; Wise, Ring, & Olson, 2000). Noting that the students in their earlier studies had sometimes had difficulty working without supervision on the computer reading activities, they used a model of delivery in which one teacher worked with a group of four students each of whom had their own computer. Two hundred children aged between 7 and 12 took part; they were selected as being in the lowest 10% on a measure of word reading, using local norms (because the study took place in a relatively affluent area, the local norms were higher than the national norms). The students received training for 50 half-hour sessions over a period of 4 months; they were posttested at the end of the intervention and a year later. Students were placed in one of two intervention groups or a "waiting-list control" group who received intervention at a later time. Both groups were given computer reading experience in which they read graded stories on the screen with the opportunity to request speech feedback on unknown words. The computer reading included checks on comprehension. The "phonological analysis" (PA) group was also given activities and computer exercises learning about phonemes through their association with articulatory motor movements (Lindamood & Lindamood, 1975, 1998). The "accurate reading in context" group (ARC) had a greater emphasis on accurate reading of stories, on and off the computer, and was given explicit training in comprehension strategies along the lines described by Palincsar and Brown (1984).

Both the ARC and PA programs were found to be effective in comparison to the "regular" classroom reading instruction, and it should be emphasized that the "regular classrooms" were in "good schools" in relatively affluent areas. (The rates of growth per hour compare favorably to studies that used one-to-one tuition with a specialist tutor [reviewed by Torgesen *et al.*, 2001].) As might be expected, the PA group showed an advantage on the post-training tests on measures of phoneme awareness and phonological decoding (nonword reading) and this advantage was maintained on the 1-year follow-up tests. Thus, like Hatcher, Hulme, and Ellis, the group that received most training in phonological awareness made the greatest gains on measures of that skill. However, the benefit of improved phonological skills did not translate to advantages on other measures of word identification and comprehension or on measures of spelling. The exception to this was for the younger children in the study who did show advantages in word reading, but only on untimed tests. The ARC group, on the whole, showed advantages on timed reading tests. Thus, the more phonologically explicit training program was effective in producing benefits in terms of phonological skills, but not in terms of more general reading skills (and at some expense in terms of speed).

There are a number of possible explanations for the failure to find differences between the two intervention groups. The finding that younger children benefit more is consistent with the findings of earlier studies (Bradley & Bryant, 1985; Hatcher *et al.*, 1994; Lundberg *et al.*, 1988) in which the participants were typically quite young. Thus, one important implication of these studies is that it is better to deliver phonological skills training during the early stages of learning to read. A further possibility is that the 25-hr training used by Olson and Wise was not long enough to establish differences in methods. Both the interventions had elements that would be predicted to be successful and it would thus be possible that advantages of one method over the other would only emerge over time.

INTENSIVE SPECIALIST INTERVENTION—TORGESEN *ET AL.* (2001)

Torgesen *et al.* (2001) addressed the issues of duration and intensity of intervention in an intervention study that involved an average of almost 68 hr of instruction over a 2-month period. Sixty pupils between the ages of 8 and 12 took part in the study, selected to be in the lower 2% of the population on national norms. The two intervention groups were both taught letter–sound decoding rules, but the emphasis of each program was different. The ADD Group was similar to the Colorado Group's PA group, being also based on the Lindamood program. The other condition was called Embedded Phonics which involved an emphasis on letter–sound decoding rules, but in the context of text reading. This program has some similarities to the Hatcher Sound Linkage Intervention Programmes in which rules are applied and reinforced in context.

The table below shows the standard scores on a measure of word identification for the two intervention groups immediately before and after the programs and at two follow-up points. The average for standard scores is 100 with 70 reflecting performance two standard deviations below the mean, which is equivalent to performance in the lower 2% of the population. There are different conventions for expressing the average range, but a score of 90 is well accepted as being at the low end of the average range. In the table below, the percentage of children scoring below 90 at three of the times of measurement is shown in brackets. It can be seen that participants in the Torgesen *et al.* study made substantial progress over the 2 months of the intervention and, more importantly, these gains were maintained at the 1-year and 2-year follow-ups.

	Pretest	Posttest	1-year follow-up	2-year follow-up
Word attack				
ADD	68.5 (100)	96.4 (16)	90.7	91.8 (31)
EP	70.1 (100)	90.3 (54)	87.0	89.9 (46)
Word identification				
ADD	68.9 (100)	82.4 (72)	82.7	87.0 (61)
EP	66.4 (100)	80.5 (83)	78.2	83.9 (67)
Passage comprehension				
ADD	83.0 (65)	91.0 (40)	92.8	94.7 (15)
EP	82.2 (75)	92.0 (46)	91.5	96.9 (21)

The ADD group was found to make greater gains on measures of phonological decoding, during the intervention period, consistent with the results from the Colorado Group. This is seen most strikingly on the posttest scores for word attack where only 16% of the group scored below 90, compared to 54% for the EP group. However, as the table shows, this advantage for the ADD group was not maintained at the follow-up tests. Moreover, this short-term advantage in terms of phonological decoding did not translate to more general benefits in terms of other reading skills. Posttest and follow-up scores on measures of single word identification and passage comprehension were remarkably similar for the two groups.

Torgesen *et al.* suggest that a useful measurement of progress is the standard score gain divided by the number of hours intervention. For Word Identification, the gain per hour was 0.20 for the ADD group and 0.21 for the EP group. Putting this another way, 10 hr of intervention would produce a gain of 2 points; 50 hr would produce a gain of 10 points and so on. The figure of about 0.2 compares very closely with the results of other studies. For example, the gain ratios for word recognition were 0.22 from the Wise *et al.* (1999) study, 0.21 from a study by Truch (1994), and 0.23 from a study by Alexander *et al.* (1991). The gain ratios for phonological decoding in the Torgesen *et al.* study were 0.41 for the ADD program and 0.30 for EP, again comparing well with 0.31 from Wise *et al.* (1999) and 0.34 from Alexander *et al.* (1991). Gains for passage comprehension were the least impressive in the Torgesen study at 0.12 of the ADD group and 0.15 for the EP group and the same was true for the Wise *et al.* study where a gain of 0.14 was obtained. While this might be thought of as rather disappointing, given that the ultimate goal of reading is comprehension, the participants in these studies were typically starting out with better comprehension levels. At the end of the intervention, as the above table shows, comprehension remained the least of the children's difficulties.

Torgesen *et al.*'s results are clearly impressive in demonstrating that quite substantial gains can be achieved in a short period with intensive specialist tuition. However, the participants had remaining difficulties on measures of reading speed and on measures of spelling. Standard Scores are shown in the table below, along with results for arithmetic, which was included as a test for specific versus general effects.

	Pretest	Posttest	1 year	2 year
Reading rate (Gray)				
ADD	71.3	75.4	75.0	72.7
EP	71.5	75.1	72.1	70.7
Spelling (Kaufman)				
ADD	75.6	77.5	76.7	76.2
EP	74.4	80.0	74.0	75.3
Calculation (Woodcock)				
ADD	93.1	95.0	87.7	89.6
EP	89.0	86.9	86.5	89.4

The above table shows that overall standard scores for spelling and for reading rate were not substantially different at the 2-year follow-up when compared to the pretest. Two points need emphasizing however. In relation to rate, we should note that at the end of the study, the children were reading more difficult passages and they were reading more quickly; it is just that they have not caught up with their peers. Likewise for spelling, the pupils have got better at spelling—they have improved, but not fast enough to catch up to their peers.

A potential problem with the Torgesen *et al.* study is that there is no control group so we cannot know what might have happened to the pupils had they been left in their regular classrooms. However, data were available from about 16 months before the pretests were given, albeit using slightly different tests. Thus, it was possible to compare the gains made during the intervention with the gains that were being made beforehand which, in all likelihood, would have continued had the intervention not been given. During the period before the intervention, pupils in both groups made very little progress in standard score terms, the means for both groups remaining at about 78 on a composite reading measure of Passage Comprehension, Word Identification, and Word Attack. Again, it should be stressed that pupils were progressing before the intervention, as would be hoped given that they were receiving special educational programs. However these programs were, as Torgesen (2001) expressed it, "stabilising" their reading problems rather than "remediating" them.

PROMOTING GENERALIZATION—THE TORONTO STUDIES

Lovett *et al.* (2000) conducted a large-scale intervention study to look at the effective components of remedial teaching and, in particular, to address the problem of generalization from what has been taught into other situations. One of the ideas behind this study was that children with reading difficulties might need explicit instruction in learning when and how to apply decoding strategies. They compared the effectiveness of four teaching programs. One of the groups had 70 hr of a Phonic/Code-based Direct Instruction program and a second group had 70 hr of a Metacognitive/Strategy-based program. The third and fourth groups had a combination, involving 35 hr of both of these programs, either in the order Code-based followed by Strategy-based or vice versa. A final control group received 35 hr of Classroom Strategy Instruction followed by a 30-hr, Direct Instruction Maths program. The Code-based program was based on the Direct Instruction approach of Engelman and Carnine (1982). In this, pupils were taught sound segmentation and blending skills using a structured, cumulative program with the emphasis on letter-to-sound rules and with additional fluency exercises. The Strategy group were taught key "sight words" and then encouraged to use these to decode new words using a range of "analogy" strategies. These were (a) a "compare/contrast" analogy strategy, (b) looking for a known part of the word, (c) attempting variable vowel pronunciations, and (d) prefix/suffix removal.

The control group covered age-appropriate study and organizational skills (e.g., for the younger ones telling the time; for the older ones, research/problem-solving strategies), and a structured maths program which involved contact with a teacher for the same amount of time as was involved in the reading interventions.

The pupils in the study were around $9\frac{1}{2}$ years of age on average, ranging from 6 years 9 months to 13 years 9 months at the start. Eighty-five children took part, with approximately twenty-seven in each of the five groups. They were described as severely impaired in reading; average standard scores on a variety of measures of reading were typically below 70, that is, more than two standard deviations below the mean. The mean Verbal and Performance IQs (WISC-3) were 90 (range 60–122) and 97 (range 64–130), respectively. In addition to standardized measures, pupils were given a range of tests to assess progress on the materials taught and a range of tests to measure transfer of learning to new materials.

As might be expected, the four reading intervention groups outperformed the control group on tests of "trained content" but, more importantly, this advantage extended to both "transfer words" and "transfer nonwords." There were also interesting differences between the different intervention groups. Those who had the combined programs made more progress (on taught content and on transfer materials) than the groups who had the single programs. This advantage for the combined programs was not due to the amount of teaching; those in the combined programs had half as much of both programs. On measures of speed on tests of taught content, those who had just the Code-based approach did better than those who had just the Strategy-based program. Furthermore, when comparing the two combined programs, it was found that the Code-based followed by Strategy-based was better than the other way round. There were no differences between the two combination groups or between the two single-program groups on the transfer words and nonwords, although there was some evidence that the Code-based-only group outperformed the Strategy-only group on nonword reading until about 52 hr of treatment.

Results were also reported for the raw scores on the Standardized measures (but, rather surprisingly, the posttest scores are not reported in standard score terms). Lovett *et al.* point out that standardized tests tend to have steep item-difficulty gradients and are not, therefore, particularly sensitive for comparing programs. Nevertheless, an overall treatment effect was found on measures of word identification and on measures of passage comprehension, but not on spelling. Importantly, the control group that received maths instruction was found to outperform the reading intervention groups on posttests of arithmetic abilities.

The results of the Lovett *et al.* study provide further evidence that an emphasis on phonic decoding alone does not produce the optimum benefits. In general, a combination of phonic decoding and explicit training in the application of decoding strategies was more effective than either of those two approaches alone. There was some evidence that an emphasis on decoding skills followed by the strategy-based instruction was better than the other way round. The evidence of Hatcher *et al.* (1994) would lead us to predict that even more benefit would have been obtained had the two elements been combined throughout the program.

INDIVIDUAL DIFFERENCES—DISCUSSION

The evidence discussed up to now has been mainly concerned with differences between teaching programs of various kinds. Most of the studies have also considered the question of individual differences between the children taking part in the studies. While there may be no substantial differences between methods overall, it is quite possible that one method might be more effective for children having certain characteristics. A first step, therefore, is to look at the relationship between learning outcomes and different measures of ability assessed before the program begins. There are two basic methods for doing this. One is to calculate the slope of the growth curve in reading during the course of the intervention and then see what initial measures correlate with that. The other method is to use a prediction (regression) model that involves finding out which measures of initial ability account for differences (variance) in reading outcome. A problem with the prediction model is that predictor measures are often related; so, for example, the children with the poorest phonological skills may also be the worst readers. The usual procedure in building a prediction model is to account for differences in reading outcome first in terms of preexisting differences and then see whether other measures can account for additional differences.

Iversen and Tunmer used a variation of the regression approach that is called path analysis. Interestingly, they found very strong relationships between initial phonological skill and reading outcome, even after allowing for initial differences in reading. This is probably because the young children in their study were "at the floor" of the reading test and so this was not a particularly sensitive measure. Iversen and Tunmer concluded:

> Overall the results of the path analysis suggest that phonological awareness is primarily responsible for the development of phonological recoding ability. That phonological recoding ability is in turn primarily responsible for the development of context free word recognition ability, and that context free word recognition ability is in turn primarily responsible for the ability to read connected text (Iversen & Tunmer, 1993, p. 123).

As already noted, Wise *et al.* (1999) found that children with poor phonological skills responded least well to their computer-based reading program. It was this finding that led them to suggest that more intensive phonological awareness training needed to be provided alongside the computer reading activities which might then enable a better response for all, regardless of initial ability.

TEACHER ADAPTATION

Research studies are designed to compare different methods of teaching and, at some cost to good relationships on the research project, the researchers will place some restrictions on the teachers and will normally collect information about the amount of time spent on different activities. Several researchers have commented that skilled teachers will often find a way of responding to children's most pressing needs within the constraints of the program. Or, in other words, a good teacher will be doing different things with different children even though that activity might be listed as the same thing on a research protocol. For example, Iversen and Tunmer were struck by the finding that both the standard and their modified reading recovery programs proved effective—at discontinuation and at the end of the year both groups were performing at average reading levels. They noted that such positive results had not typically been obtained in evaluations of reading recovery and speculated that this may be because the teachers in other studies were typically nonspecialists. In their study, the programs were delivered by highly skilled specialists with masters certificates in reading. They suggest that the adaptable nature of the reading recovery program may have enabled these skilled specialists to respond to the individual needs of the poor readers. This effect might provide some explanation as to why the Florida and the Colorado studies have typically failed to find different outcomes for different teaching programs.

THE IMPORTANCE OF PARENTAL SUPPORT

It is widely recognized that parents acting as educators of their children at home have positive effects on learning (Topping, 1986; Wolfendale & Topping, 1996). Paired reading is an approach to the remediation of literacy difficulties that is commonly used by parents at home and has been shown to be effective (see Topping, 1995, for a review of research into its effectiveness). In particular, paired reading has been reported to have a positive effect on children's reading comprehension (Topping & Lindsay, 1992). However, Brooks, Flanagan, Henkhuzens, and Hutchison (1998) found that interventions such as paired reading were only effective if they were used to support the children by trained staff or parents.

Cupolillo *et al.* evaluated the paired reading procedure with children who were repeating the first year of primary school in the Brazilian school system (ages are not given). In order to progress through the Brazilian school system, children must show sufficient academic progress each year. If they do not perform well enough, they are required to repeat a year (which can happen indefinitely), and as a result, many drop out of school altogether. Cupolillo *et al.*'s research was aimed at remediating literacy difficulties using paired reading as an alternative or supplementary (to that of the teacher) strategy. They chose paired reading as it was felt it would empower families as well as individual children and thus offer a longer term effect of increasing literacy in the home. Parents who agreed to participate were expected to attend training sessions each week for 6 weeks and carry out daily 10–15-min sessions of reading at home.

The parents were supported throughout by professional staff, who could give advice and encouragement when needed, as well as monitor progress. Assessments of fluency, comprehension, reading habits, and confidence in reading were carried out at the beginning and end of the 6-week training period. Progress was also evaluated through seeking the opinions of the children, the children's parents and teachers, and the researchers. Good progress in all areas assessed was noted, with 73% of the children subsequently succeeding in securing approval to progress to the second year of primary school.

INTEGRATION OF APPROACHES

The evidence from recent "objective" evaluations of intervention to alleviate literacy difficulties thus shows an encouraging consistency with the approaches that have been advocated for some time by practitioners. Recently, Brooks *et al.* (1998) reviewed evaluations of intervention available for children categorized as slow readers in the United Kingdom. They found that focusing on phonological skills was effective in increasing children's reading ability and attainment, but only if it was embedded within a broad approach or framework. In other words, phonological skills were not just taught in isolation but together with other training strategies to improve reading such as paired reading, or work on reading comprehension.

In an earlier study by Lovett *et al.* (1989), three different types of intervention for alleviating literacy difficulties caused by dyslexia were compared. Lovett *et al.* monitored literacy progress in sample groups receiving a training program in word recognition and decoding skills, oral and written language skills, or classroom survival skills. They found that children in the groups receiving word recognition and decoding skills training, and oral and written language training made significantly better progress in literacy than their counterparts receiving training in classroom survival skills. Greater generalization of the skills learnt was observed in the children who received training in word recognition and decoding skills than those who received the training in oral and written language skills. However, the children who received training in word recognition and decoding did not improve in their knowledge of letter–sound correspondence.

THE DYSLEXIA INSTITUTE "SPELLIT" STUDY

The Dyslexia Institute has just completed a longitudinal intervention study (Rack & Hatcher, 2002), which focused on the question of individual differences in response to three different learning programs. These were:

1. A Home Support program (HSP) consisting of activities and exercises to be done at home for around 15 min per day, for 5 days per week over a 30-week period.
2. Structured, multisensory teaching using the Dyslexia Institute's approach, twice weekly over a 24-week period in sessions each lasting 1 hr.

3. A Combined program involving 1 hr per week of structured teaching for 24 weeks and HS activities in 15-min sessions, three times per week over a 30-week period.

We took the decision not to seek only participants who were thought to be "dyslexic." Rather, it was decided to seek referrals from schools of those children who were experiencing difficulties with literacy learning. In this way, it should be possible to make statements about the effectiveness of the programs that would readily generalize—we wanted to include the whole range of those with literacy difficulties in order to find out which factors or individual characteristics most influenced outcome. We obtained data on almost 450 7-year-old pupils, assessed about 350, and selected 240 to take part in the study. Two hundred and fifteen pupils, who had been assessed at the beginning and end of the intervention period, were in our final sample of pupils. On average, the pupils were reading and spelling in the lowest 10% on standardized tests—at around the 6-year age equivalent level compared to their chronological age of $7\frac{1}{2}$. The sample covered a wide range on measures of general intellectual ability with the mean for the whole sample almost exactly 100, the expected population average.

One of the key findings from the SPELLIT research was that different children responded in different ways to the different learning programs. The overall results therefore give only a partial picture. For various reasons, it was only possible to make direct comparisons between HS (only), Teaching (only), and Waiting groups. Data from pupils in the Combined (Teaching plus HS) group are given, but these need to be looked at separately.

The teaching took place over a 9-month period. During this time, the average reading age for those in the teaching group went from 5 years 9 months to 6 years 8 months. This is reflected in a gain of just over 2 standard score points, on average, showing that there is some "catching up" to age-group norms. The Waiting group did make progress—from 5 years 10 months, on average, to 6 years 4 months. This reflects progress at a slower rate than would be expected and thus they were falling further behind their age-peers. Their standard scores were, on average, 2 points lower than at the start of the program. The HS (only) group progressed at a rate somewhere in between. Their reading increased on average, from 5 years 10 months to 6 years 7 months, and their standard score increased by about 1 point, on average.

The gains made by pupils in the Teaching and HS groups may sound like small gains, but they show that children were, on average, keeping pace with their peers at a time when their peers are moving ahead quite fast. This is underlined by the scores for the Waiting group who, although they are making gains, are falling further behind their peers. Thus, the "benefit" of the 2 hr per week teaching, in relation to the control group is 4 standard score points, and for the HSP group, it is about $1\frac{1}{2}$ points.

The results for the Combined program were mixed. One group of children, all from the same school, who received this program, did not do particularly well.

Unfortunately, we did not have children from this school in any of the other programs and so we could not draw comparisons. It was clear, from other evidence, that the disappointing results in this case were, at least in part, due to a "school" or "area" effect. In other areas, where pupils had an opportunity to take part in either the HS (only) or Combined programs, we found clear benefits of the teaching element. Thus, we concluded that the Combined program can be effective, but much depends on the success in linking the Teaching and HS elements. At least in some cases, parents appeared not to put as much effort into the HSP element when teaching was also taking place.

DIFFERENTIAL RESPONSE TO THE TEACHING PROGRAM AND HSP

For comparison with previous studies, which have tended to select participants with poor phonological skills and poor reading, we divided our sample into subgroups. The resulting subgroups were small and the findings must therefore be regarded as preliminary. We found that those pupils with more severe reading difficulties and more severe phonological difficulties responded particularly well to the Teaching program. Children with these characteristics did poorly in the HSP and Waiting groups. The difference in reading outcome between Teaching and Waiting groups was almost 10 standard score points. On the other hand, the pupils with relatively good reading and phonological skills did not respond particularly well to the teaching program—they did no better than those in the Waiting group. Children with these characteristics who had the HSP responded much better. This is a very important finding. It is not the case that one program is always best; one program is better for pupils with one set of characteristics and another program is better for children with a different set of characteristics.

The children with weakest reading and weakest phonological skills were receiving most support in school, indicating that schools were generally identifying and supporting those with the severest difficulties. However, those in the Teaching group were receiving no more support in school than those in the Waiting group. This shows that the positive benefit of teaching was additional to any benefit of school support. An implication of these results is that the kind of support provided in school was not, on its own, effective for children with this pattern of most severe difficulties.

It was perhaps unsurprising that the children with relatively better reading and relatively better phonological skills responded best to the HSP. However, it was surprising that children with these characteristics who had the HSP did better than similar children who had received 2 hr per week teaching. Both groups made more progress than children with similar characteristics in the Waiting group who were receiving support in school. The children—with the milder difficulties—might have been expected to respond quite well to school-based support.

The SPELLIT results suggest that those children with the more severe difficulties with reading and phonological awareness are those who need the fine-grained structured language teaching used in the Dyslexia Institute and other programs. However, those children who had less severe reading and phonological difficulties seemed to benefit more from aspects of the HSP. Our tentative interpretation is that the HSP was providing more opportunities to practice decoding skills and other strategies in context and thus promoting greater reading experience (giving more "exposure to print"). These results suggest that we need to broaden our ideas about what makes particular kinds of interventions effective, to ask instead, what kinds of interventions are effective for which children and at what stages of literacy learning.

SUMMARY

We have known for a long time that some children do not respond to the normal classroom and home-based teaching of early literacy skills. There is considerable evidence that the majority of these "dyslexic" children have poor phonological skills; they lack awareness of the phonological units within spoken words and of the correspondence between these and printed letters. Particularly when phonological difficulties are more severe, dyslexic children fail to benefit from experiential learning and tend to fall further and further behind their peers unless they are given explicit, structured teaching of decoding skills. The research studies reviewed here allow us to draw a number of conclusions about the effectiveness of teaching of this kind.

The evidence from Hatcher *et al.*'s work is that improving poor readers' phonological skills does not, on its own, bring about substantial improvement in reading skills. Rather, development of phonological skills linked to reading experience and practice is required. The inclusion of phonological skills training is important, since reading experience and practice, on its own, do not produce significant gains. Iversen and Tunmer's work underlines this by showing that a reading-experience-based program is significantly less effective than a program that has both reading experience and phonological skills training. Hatcher *et al.*'s work, building on Bradley and Bryant's work, suggests that children with poor phonology need the linkage between spoken sounds and written letters to be made explicit. Moreover, it suggests that the teaching of word-level and sub-word-level decoding strategies must be explicitly linked to reading of connected text. In the R + P group, the teachers would highlight and consolidate taught rules during structured reading practice. This is an important practical point: It is not simply a matter of teaching phonic rules and then practicing them in isolation; rather, the practice provides opportunities to reinforce the rules and to demonstrate their use explicitly in context. This conclusion is reinforced by the findings of the study conducted by Lovett *et al.* They showed that a combination of direct instruction in decoding rules, with the teaching of strategies to learn when and how to apply those rules, was optimally effective.

The work of Olson and Wise, and of Torgesen and colleagues, is consistent in showing that the same kinds of results can be obtained using two rather different methods. Both research groups used a program that emphasized decoding rules with a highly structured progression and an emphasis on awareness of sound patterns via articulatory gestures. These were "the PA" in the Colorado studies and the ADD in the Florida study. Both groups also used an approach that emphasized the application of word decoding skills in context—ARC in Colorado and Embedded Phonics in Florida. It should be stressed that the programs were delivered by experienced specialist teachers and that the differences were more in the "mix" of the activities rather than one program having all of one type of activity and the contrasting program having completely different activities. Nevertheless, the finding that, in both Colorado and Florida, there were no appreciable differences in learning outcomes between the two types of programs is surprising.

The findings of the Dyslexia Institute's "SPELLIT" study offer a partial explanation for these somewhat surprising null results. In the SPELLIT study, the sample was deliberately heterogeneous—the children were poor at reading but their other abilities such as phonological skills and general conceptual abilities were free to vary. The suggestion from those results was that the differences in program outcomes were more likely to be found for subgroups. If there are some children who benefit more from one program than the other, that effect would not show up if the sample contained a second group of children who show the opposite pattern.

The most striking aspect of the Florida studies is the rapid gains in standard scores that can be made with very intensive intervention. As has been pointed out, these gains "per hour" are not necessarily better than has been achieved by some other studies over a longer period, but the effect for the individual is dramatic. Importantly, these rapid gains seem also to be sustainable. The argument is similar to that for early intervention—progress can be made at a later time, but it is clearly much better to effect the "catch-up" at an early time and so avoid a constant feeling of being behind despite making progress.

Finally, let us consider where these findings leave the phonological deficit hypothesis. It seems most likely that the "linked teaching"—phonological decoding rules with opportunities for practice and generalization through reading in context—is enabling development of reading skills despite poor underlying phonological difficulties. Thus, children will be helped to acquire reading skills beyond the level that would be expected on the basis of their phonological skills. We therefore have to reject a strong version of the phonological deficit theory which predicts that those whose phonological skills improve the most will improve the most in reading. It seems instead that there is a certain threshold of phonological awareness that needs to be reached in order to support the use of reading skills in context. Interestingly, it seems that further development of reading does not happen automatically once that threshold is reached, rather, direct teaching of strategies and provision of structured opportunities for practice is needed. To practitioners it may sound obvious, but the evidence reviewed here seems consistent with the idea that both approaches are important for effective teaching but that the balance will shift as skills develop. In the early stages, the balance is in favor of decoding skills and in the later stages of application of skills in context; but at all stages there should be some of each approach. Having concluded that teaching phonological skills is necessary but not sufficient for the development of reading skills in dyslexic children, future research is needed to understand more about the additional factors that play a role, in particular, to understand more about what can be achieved through teaching to address those factors.

REFERENCES

Alexander, A., Anderson, H., Heilman, P. C., Voeller, K. S., & Torgesen, J. K. (1991). Phonological awareness training and remediation of analytic decoding deficits on a group of severe dyslexics. *Annals of Dyslexia, 41*, 193–206.

Bradley, L. L., & Bryant, P. E. (1983). Categorizing sounds and learning to read: A causal connexion. *Nature, 301*, 419.

Bradley, L. L., & Bryant, P. E. (1985). *Rhyme and reason in reading and spelling.* Ann Arbor: University of Michigan Press.

Brooks, G., Flanagan, N., Henkhuzens, Z., & Hutchison, D. (1998). *What works for slow readers?* Berkshire: NFER.

Bruck, M. (1992). Persistence of dyslexics' phonological awareness deficits. *Developmental Psychology, 28*(5), 874–886.

Bus, A. G., & van Ijzendoorn, M. H. (1999). Phonological awareness and early reading: A meta-analysis of experimental training studies. *Journal of Educational Psychology, 91*(3), 403–414.

Clay, M. (1985). *The Early Detection of Reading Difficulties.* Auckland: Heinemann.

Engelman, S., & Carnine, D. (1982). *Theory of Instruction: Principles and Applications.* New York: Irvington.

Hatcher, P., Hulme, C., & Ellis, A. W. (1994). Ameliorating early reading failure by integrating the teaching of reading and phonological skills: The phonological linkage hypothesis. *Child Development, 65*, 41–57.

Iversen, S., & Tunmer, W. E. (1993). Phonological processing skills and the reading recovery program. *Journal of Educational Psychology, 85*, 112–126.

Langenberg, D. N. (Chair) *et al.* (2000). *Report of the national reading panel, teaching children to read; an evidence-based assessment of the scientific research literature on reading and its implications for reading instruction.* National Institute of Child Health and Human Development: Washington, DC.

Lindamood P., & Lindamood, P. (1998) *The Lindamood Phoneme Sequencing program for reading, spelling and speech. Austin, TX: PRO-ED, Inc.*

Lindamood, C., & Lindamood, P. (1975). *Auditory Discrimination in Depth (ADD) program.* Allen, TX: DLM Teaching Resources.

Lovett, M. W., Ransby, M. J., Hardwick, N., & Johns, M. S. (1989). Can dyslexia be treated? Treatment specific and generalized treatment effects in dyslexic children's response to remediation. *Brain and Language, 37*(1), 90–121.

Lovett, M. W., Lacerenza, L., Borden, S. L., Frijters J. C., Seteinback, K. A., & DePalma M. (2000). Components of effective remediation for developmental reading disabilities: Combining phonological and strategy-based instruction to improve outcomes. *Journal of Educational Psychology, 92*, 263–283.

Lundberg, I., Frost, J., & Peterson, O. (1988). Effects of an extensive program for stimulating phonological awareness in preschool children. *Reading Research Quarterly, 23*, 263–284.

Olson, R. K., Wise, B., Conners, F. A., & Rack, J. P. (1990). Organization, heritability and remediation of component word recognition and language skills in disabled readers. In T. H. Carr, & B. A. Levy (Eds.), *Reading and its development: Component skills approaches* (pp. 261–322). New York: Academic Press.

Olson, R. K., & Wise, B. W. (1992). Reading on the computer with orthographic and speech feedback: An overview of the Colorado Remedial Reading Project. *Reading and Writing: An Interdisciplinary Journal, 4*, 107–144.

Olson, R. K., Wise, B. W., Ring, J., & Johnson, M. (1997). Computer-based remedial training in phoneme awareness and phonological decoding: Effects on the post-training development on word recognition. *Scientific Studies of Reading, 1*, 235–253.

Palincsar, A. S., & Brown, A. L. (1984). Reciprocal teaching of comprehension-fostering and comprehension-monitoring activities. *Cognition and Instruction, 1*, 117–175.

Rack, J. P. (1994). Dyslexia: The phonological deficit hypothesis. In A. Fawcett and R. Nicolson (Eds.) *Dyslexia in children.* (pp. 5–37). Hemel Hempstead: Harvester Wheatsheaf.

Rack, J. P. & Hatcher, J. (2002). *SPELLIT Summary Report.* Staines: The Dyslexia Institute.

Rack, J. P., Snowling, M. J., & Olson, R. K. (1992). The nonword reading deficit in developmental dyslexia: A review. *Reading Research Quarterly, 27*, 29–53.

Snowling, M. J. (1987). *Dyslexia: A cognitive developmental perspective.* Oxford: Blackwell.

Stanovich, K. E. (1988). Explaining the differences between the dyslexic and the garden-variety poor reader: The phonological- core variable- difference model. *Journal of Learning Disabilities*, 21, 590–612.

Topping, K. J. (1995). Paired reading, spelling and writing: The handbook for teachers and parents. London & New York: Cassell.

Topping, K. J. (1986). *Parents as educators: Training parents to teach their children.* London: Croom Helm.

Topping, K. J., & Lindsay, G. A (1992). Paired reading: A review of the literature. *Research Papers in Education, 7*(3), 199–246.

Torgesen, J. K., (2001). 'The theory and practice of intervention: Comparing outcomes from prevention and remediation studies. In A. Fawcett (Ed.), *Dyslexia: Theory and good practice* (pp. 185–200). London: Whurr.

Torgesen, J. K., Alexander, A. W., Wagner, R. K., Voeller, K., Conway, T., & Rose, E. (2001). Intensive remedial instruction for children with severe reading disabilities: Immediate and long-term outcomes from two instructional approaches. *Journal of Learning Disabilities, 34*, 33–58.

Troia, G. A. (1999). Phonological awareness intervention research: A critical review of the experimental methodology. *Reading Research Quarterly, 34*, 28–52.

Truch, S. (1994). Stimulating basic reading processes using auditory discrimination in depth. *Annals of Dyslexia, 44*, 60–80.

Wise, B. W. (1987). Word segmentation in computerized readings instruction (Diss. Abst.). Boulder: University of Colorado.

Wise, B. W., & Olson, R. K. (1995). Computer-based phonological awareness and reading instruction. In C. Hulme, M. Joshi (Ed.), *Cognitive and Linguistic Bases of Reading, Writing and Spelling.* Hillsdale, NJ: Lawrence Erlbaum.

Wise, B. W., Ring, J., & Olson, R. K. (1999). Training phonological awareness with and without attention to articulation. *Journal of Experimental Child Psychology, 72*, 271–304.

Wise, B. W., Ring, J., & Olson, R. K. (2000). Individual differences in gains from computer-assisted remedial reading with more emphasis on phonological analysis or accurate reading in context. *Journal of Experimental Child Psychology, 77*, 197–235.

Wolfendale, S. W., & Topping, K. J. (Eds.) (1996). *Family involvement in literacy: Effective partnerships in education.* London & New York: Cassell.

9

An Introduction to the Theory and Practice of Psychological Testing

Martin Turner

PRINCIPLES OF ASSESSMENT

Why Assessment?

Any enquiry about the nature of the world must, to be useful, make some discrimination. This is larger, that is smaller. This event happened before another one. For over a century now, scientists have made significant advances in the study of behavior, including human behavior, perhaps the most slippery field of endeavor ever selected by the intrepid enquirer.

Even today, there is still a constituency of those who think all measurement of human behavior unnecessary, if not positively offensive. This is the view that it is sufficient to say of two children: "This child can read; that child cannot." However, such a statement already implies some discrimination, even a primitive form of measurement. The lowest level of measurement is the categorical. We have devised two categories: that of children who can read, and that of children who cannot read. We have achieved a level of measurement that is important to every goldfish: this crumb is worthy of eating, this crumb should not be eaten.

Martin Turner, Chartered Educational Psychologist, Brocksett Cottage, Kennel Lane, Windlesham, Surrey GU20 6AA, UK.

The Study of Dyslexia, edited by Turner and Rack.
Kluwer Academic Publishers, New York, 2004.

We may want to progress beyond this and say of two books: this book is harder to read than that book. With respect to our children, we might now say: this child cannot read either book; this child can read the first book; this child can read both the first book and the second book, which is harder than the first. We are now introducing the concept of *rank*. That is to say the books can be ranked in the order of difficulty and the children, too, can be ranked in the order of their reading ability. However, there is considerable scope for uncertainty and even disagreement at this level of measurement, the *ordinal*. After all, we might regard the estimation of the difficulty of reading books as intrinsically difficult, since there may be many books at appropriately the same level of difficulty. The same consideration applies to children. It is not always possible to place children unerringly in rank order of ability, in the same sort of way that a classroom of children can quickly sort itself into a line with the tallest at one end and the shortest at the other. As with height, there may be some disputes toward the middle of the range, where there are perhaps several children of equivalent reading ability or equivalent stature. Nevertheless, this level of measurement appears to be highly satisfactory to parents, who are often told only that their child came fifth in the class or that their child came second from bottom. This relative standing no doubt satisfies the instincts for competition that lurk in every human breast.

We have by now progressed a great deal beyond the goldfish. But nevertheless achievement in reading, if not in growth, may be of considerable importance in our lives. Just as the child who fails to grow may qualify for various arcane treatments, or the child who is excessively tall may feel so out of sorts that he must resort to bullying, so we can agree that it is desirable to bring to bear a more developed level of measurement. Accordingly we will want, if possible, to advance to the *interval* level of measurement. This implies that measurements are being made on the same scale. In terms of reading, we may say that the difference between Betty and Harry, both aged 8, is the same as the difference between Thomas and Sarah, both aged 17. The gap in reading ability within the two pairs is the same because we are able to measure them on a single scale of reading ability that encompasses both less able and more able readers and determines the exact difference between them.

Almost any aspect of human life, which is of importance to somebody, is a candidate for this more serious degree of measurement. The urge to quantify seems to be built into the human brain and can be observed very soon after birth. Almost the first question to follow the identification of the presence of something, is to ask how much of it is there? We come to a river: how broad is it? My neighbor gives me a sack of corn: how much corn is there in it? The doctors tell me my tumor is fatal: how many weeks do I have to live? The answers to all these questions are essentially quantitative. Lots of things, that can be measured, are not measured, but these usually are of little concern to people. As Galileo said long ago, "The book of nature is written in the language of mathematics."

Although education is a subject area in which the introduction of mathematical and objective methods has not proved very popular, the progress of children in

Those who judge a work without any rule stand with regard to others as do those who have a watch with regard to those who have not. One man says: "Two hours ago"; another says: "It is only three quarters of an hour." I look at my watch and tell the first: "You must be bored", and the second: "You hardly feel the time passing", because it is an hour and a half ago. I take no notice of those who tell me that time must hang heavily on my hands and that I am judging it according to my own fancy.
They do not know that I am judging it by my watch.

Blaise Pascal, 1662.

Figure 1.

their acquisition of knowledge and skills is of huge importance to us, as family members, as teachers and as members of the wider society in which these children will grow up to play a part. Accordingly, a great deal of human effort has been expended in devising rational measurements for the aspects of human behavior that we value. By adopting forms of measurement that tell us what we want to know, we are, therefore, parting company with those who are content with primitive or subjective or impressionistic forms of evaluation of children's progress. (The latter point of view is sometimes made to proceed from a consideration of equal opportunities, hostility to competition or the belief that to test children is "divisive.")

The reason for doing so has been perfectly plain at least since it was first expressed by the French mathematician and philosopher, Blaise Pascal (1623–62) in the mid-17th century. He observed (see Figure 1) the remarkable advantage the possession of a watch gave him over various other people when their subjective estimations of the passage of time were compared.

Psychometrics, the Measurement of Abilities

Since the last decade of the 19th century, successive efforts have achieved a measurement of that slippery phenomenon, skilled human performance. Such measurement is increasingly valid, reliable, and fair. It has often proved quite difficult to define what the ability in question might be, even for reading, for instance, but still more for intelligence. Nevertheless, a rough and ready empirical approach to these matters has yielded measurement scales that measure different kinds of reading ability at different stages and different kinds of intelligence at different ages. It matters not that intelligence is largely innately and genetically determined, that is, it unfolds through the tissue of opportunities that life presents, and that reading is largely acquired through teaching. Such measurements must satisfy fundamental criteria, which are required of all forms of measurement.

One does not measure the width and the height of a doorway in different units, inches and centimeters. Measurements must be in the same scale. A window should not be differently measured when different instruments are used. Any measuring instrument should give the same result as any other. The height of some steps must be the same on Tuesday as it is on Thursday. Having escaped from the morass of subjective opinion and impression, we do not want our efforts to founder in a quagmire of inaccurate and unreliable instruments.

All this is a great deal to ask of the measurement of human behavior. Nevertheless scientists have set themselves exactly the same targets in the measurement of human behavior as physicists have set themselves down the centuries in the measurement of physical extent and capacity, goals that have been largely achieved.

We will look in detail at concepts of validity, reliability, and fairness in a later section. For now let us move on to think about different kinds of assessment and different domains of skills.

Different Kinds of Assessment

Let us first consider two forms of assessment that are often opposed to each other and then reconcile them with each other by means of a third.

Normative assessment compares an individual to his or her age group. That is to say, compared with other 10-year-olds, this 10-year-old is advanced in her mathematical reasoning. Compared to other 8-year-olds, who have by and large stopped reversing *b*s and *d*s, this 8-year-old, who continues to do so, is exceptional. Normative assessment, obviously, is by far the most widely practiced and meaningful form of assessment to most people.

A few, though perhaps not many, qualifications need to be borne in mind. Ten-year-olds do not always perform at the same levels in different countries and in different cultural zones. Nor can one assume that a 10-year-old today does much or little of what a 10-year-old did 200 years ago. All these judgments are relative to a specific time and context. Nevertheless, it can be demonstrated convincingly that the performance of all 10-year-olds in the developed or Western world is remarkably similar and independent of their education systems. Since children commence their schooling at different ages, there are noticeable differences in literacy abilities amongst 6- and 7-year olds in, say, Sweden and Cornwall. Two or three years may separate the age of entry into school of such different populations.

Similarly, across time, it is possible to peer at such data as are available and conclude on the evidence that most Durham miners in the 1830s were in fact literate; that by the end of the 19th century most women and virtually all men were able to sign their names (as they had by law to do) in the parish register when they married; and that in 1850, the population of Britain had achieved levels of school attendance and literacy that were generally achieved in the rest of the world a century later. Another qualification on normative assessment must be that the rate of developmental ascent is remarkably steep, especially in the skills of literacy and numeracy, in the earliest years. It is, therefore, more important to compare a child of 6 years and 2 months with other children whose age falls between 6 years 1 month and 6 years 3 months, just as it is acceptable to compare a child of $13\frac{1}{4}$ and $13\frac{3}{4}$ with all 13-year-olds. Important differences cannot be observed between 13-year-olds who are at the beginning and end of their thirteenth year; but differences among young children grasping important skills early in their school careers can be very striking.

An alternative to comparing a child with his or her contemporaries is *criterion-referenced assessment.* This ignores differences amongst people and looks instead at differences within the work. In terms of our earlier example, we now ignore the differences in reading ability and concentrate on the difficulties of the text in books. This is the original model that underlay the development of the National Curriculum in the United Kingdom, a model characterized by a great desire to escape from traditional forms of assessment. It must be said that this experiment has not been a success. Nevertheless criterion-referenced assessment has a number of well-established uses. If you want to evaluate the progress of handicapped preschool children, you may well want a list or inventory of fundamental skills that they should acquire. Thus, toilet training, handling of a knife and fork, tying shoe laces, engaging in shared play may all form behavioral targets which can be refined so as to permit evaluation of small steps toward and between them. Clearly, the age of the child is now less important. Indeed we may wish to use exactly the same inventory with a severely handicapped 10-year-old as with a normal 4-year-old.

However, criterion-referenced assessment, which relates progress purely to set academic objectives, tends to be less useful as the targets become harder to quantify and as the behavior more sophisticated. It is difficult in principle, for instance, to evaluate progress in historical understanding of the French revolution. What are the objective academic benchmarks that might differentiate between the doctoral and an A-level student? This indeed proved to be a difficulty when the National Curriculum attainment targets in history were being devised. The attainment targets for historical understanding at 10 different levels were famously cut up and shuffled and given to eminent historians to arrange in order of difficulty. There were unable to do so.

A particular implementation of criterion referencing, however, dominates the point of view of teachers in relation to reading. This is the concept of the reading "age." A similar concept in North America concerns "grade levels." This is simply the level of proficiency, in English or history or reading, that would be expected of an average child at a given age. Therefore, we expect an average 10-year-old to have a reading age of 10. If he or she does not, but instead has a reading age of 8, we know roughly where the child should be at (but is not) and also where the child is at (and should not be). We also, as teachers, know roughly the ground that lies between these two levels that must be made up. All of these considerations relate an individual to the objective curriculum, such as it is, and provides a framework for the thinking and planning of teachers.

I make these points because age equivalent scores (reading ages, spelling ages, mental ages), while they perform this function tolerably well, also fail the criteria mentioned earlier for an interval scale of measurement. They are therefore a much better guide to curricular progress than they are to the skills of any individual.

Since this is a point of considerable practical importance, let us pursue it a little more closely. In Figure 2, we can see the result of mapping raw scores (i.e., the total number of items correct in each individual case) at the mean for each of 12 different age groups on the Wide Range Achievement Test-Revised (WRAT-3), a 1994 revision of a very widely used test. We are able to do this for the three different subtests involved: Reading, Spelling, and Arithmetic. This enables us to follow the ascent of raw scores on the three tests throughout the age groups. It is clear that only the test of Arithmetic seems to increase at a steady rate between the ages of 5 and 30 years. With the two tests of literacy, reading and spelling, the lines are far from straight. In particular, there is a notably steep ascent in the rate of increase in raw score between the ages of 6 and 8.

This of course is exactly what we expect. The majority of children are settling to learning the skills of literacy in their early years at school. Thereafter, the skill of spelling increases only gradually and rather little as the years go by; the skill of recognizing words, however, increases rather more noticeably into adulthood, presumably as a result of continued reading experience beyond the end of formal schooling. However, at certain ages, the ascent in raw scores in reading

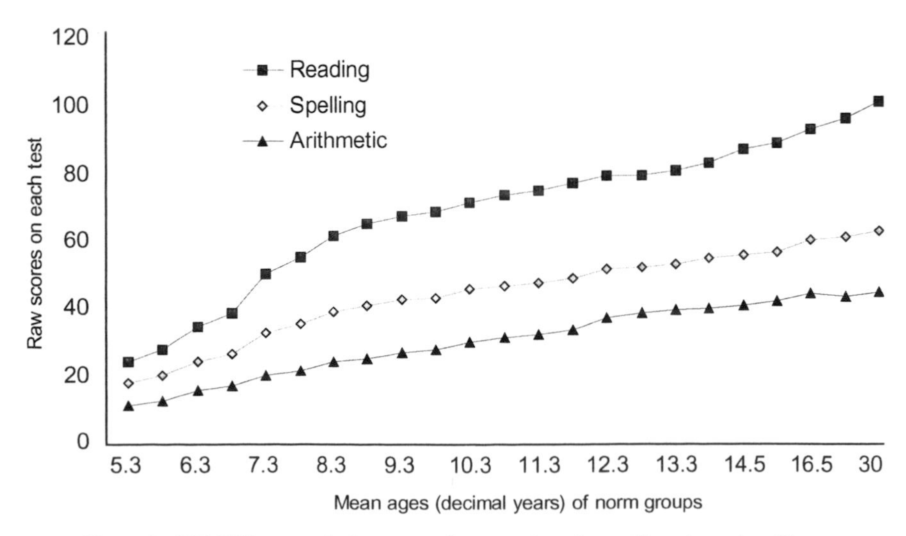

Figure 2. WRAT-R age equivalent scores for tests of reading, arithmetic, and spelling.

seems as flat as those in spelling. Between the ages of 12 and 13, for instance, there is little or no progress in the number of words read.

One thing is clear from this: a year in adolescence is not worth the same as a year in infant school. Moreover, the common unit of measurement used in age equivalent scores, the month, has virtually no constant meaning. A month would appear to be worth a lot at 7, very little at 13. But if there is no stable unit of measurement, then there is no interval scale. Accordingly, the best measurement that such data can yield are rough and ready indications of progress in relation to an expected curriculum. This is why serious measurement will always prefer standard scores of any form. These have a constant meaning. Age equivalent scores do not.

If normative and criterion referenced measurements give us alternative ways of evaluating progress, it takes a third class of assessment to reveal what they have in common. After all, we do not expect normative tests to contain for any age group material irrelevant to that stage; similarly, we expect levels of progress to reflect the average performance of many pupils at that stage in their learning.

In Figure 3, I have chosen overlapping circles to demonstrate the secret affinity that each of these three kinds of assessment in practice has with the other two. The third kind of assessment to be introduced is that of *ipsative* assessment. This is a way of compiling information so as to construct a profile of the individual child. Such ipsative assessment data is necessarily dependent on the assembly of numerous scores on normative tests, just as normative tests depend on curricular

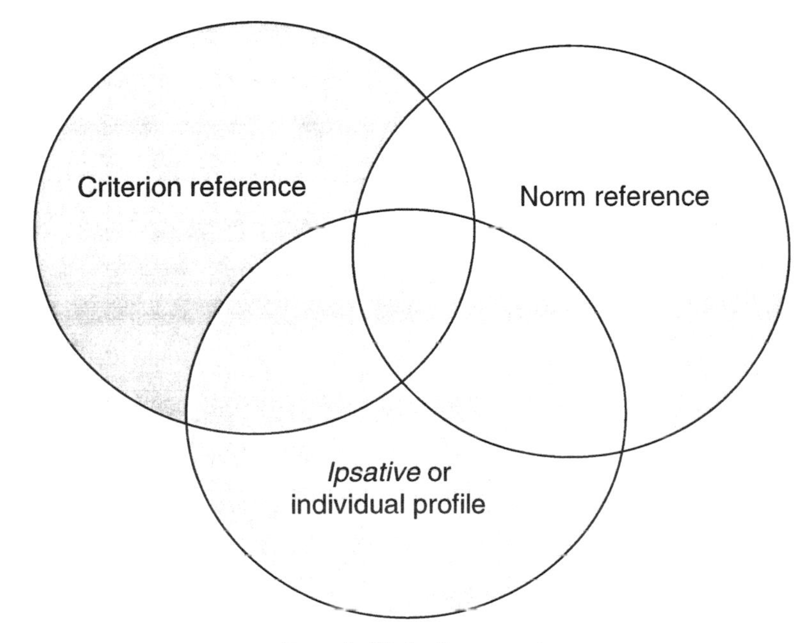

Figure 3. Kinds of assessment.

relevance to the levels or criteria relevant to the pupils' stage of progress in school subjects. It is therefore easiest to think of these three forms of assessment as entailing each other. Indeed, in the specialist teaching of dyslexia and literacy, we use all three kinds of assessment and balance them against each other. For instance, we note a child's progress through a structured reading program (*criterion-referenced assessment*); we assess a child's standing in relation to his or her contemporaries (*normative assessment*); and we evaluate the child's own individual strengths and weaknesses in relation to him- or herself or in relation to his or her own past performances (*ipsative assessment*). All three forms of assessment have a continuing relevance and serve similar purposes. The last of the three, *ipsative assessment*, aims to construct a meaningful psychological profile of an individual, which serves to illustrate the presence of a learning difficulty. It is possible to design an ipsative assessment that has common elements and in which the balance between strengths and weaknesses is entirely standardized. However, an ipsative profile requires a large amount of high quality data in order to be convincing.

Three Domains of Measurement

This brings us to consider next what different kinds of test may be brought to bear on the problem of dyslexia. Recall that the contrast view of dyslexia has been central to our understanding for well over a century. That is to say, some abilities or skills, well developed, form a painful contrast with others, poorly developed. The failure of literacy skills is, in this sense, unexpected: unusual or improbable. This is what requires further investigation and this contrast, or discrepancy, leads us to a scientific understanding of literacy failure in which the causes are internal to the child rather than external (pertaining to the school or the teaching). Such internal causes are usually thought of as central, higher order, and perhaps ultimately genetic in origin.

In order to identify dyslexia, we need to embody our theoretical understanding in terms of three classes of test. Let us examine each of these in turn.

Tests of General Intellectual Ability

There are, to be sure, many different kinds of abstract reasoning and general problem-solving ability. However, one general factor (or *g*) tends to swamp the data on individual differences. A single general intelligence factor typically accounts for three times as much of the variance in a cognitive assessment as all the other factors put together. There is therefore a strong case for assessing general ability as a first step in an assessment.

Ability testing has something of the characteristics of a general cognitive survey and provides a framework within which all other results may be securely understood. The logic of this is very close to the logic of the scientific experiment. It is necessary to control for the most powerful variables. Most people would agree that these, in order of priority, are age and general ability. We have, first, to compare an individual with others of the same age in order to get a fair comparison, and, second, to compare her or him with others of similar intellectual ability. This may not matter greatly about half of the time, since about half of the population is conventionally described as "average" (i.e., with IQs falling between 90 and 110). But if the middle 50% of the population may be so described, that leaves us with individuals of higher or lower ability at least half the time.

IQ is a by-product of a comprehensive assessment of performance on as many as a dozen tests of different abilities and is roughly the arithmetic mean of the results of these subtests. (IQ is properly measured using complex test batteries restricted in use to educational psychologists, but general intellectual ability is often estimated by teachers and schools using similar tests.) However, though it is a composite score depending on many smaller direct observations, IQ has enormous statistical stability and predictive validity. It has been called the most powerful variable to be discovered in the whole of Western social science. This leads us to our second reason for employing it: it is simply too powerful safely to be left out of account. If IQ tests are not given, then every other test will start behaving like an IQ test. For instance, reading tests will tend to measure individual differences in intelligence. This of course is going to distort results of the performances of dyslexic individuals.

It is often doubted, but should not be, that intelligence plays the major role in learning. Such a common sense view is accepted almost universally, but many education professionals are uncomfortable with the measurement of intelligence. It is true that numerous scruples and much controversy accompany the use of any powerful technology. However, we are here solely concerned with the best interests of individuals who may need special educational services, not with other issues, except insofar as they remind us constantly of the importance of handling all test results with the greatest *sensitivity*.

Most scientists and professionals are nowadays in good agreement about the uses and methods of intelligence testing. A committee of experts was formed in 1994 to report on areas of current scientific agreement, and they concluded, in part, as follows:

> The relationship between [intelligence] test scores and school performance seems to be ubiquitous. Wherever it has been studied, children with high scores on tests of intelligence tend to learn more of what is taught in school than their lower-scoring peers ... intelligence tests ... are never the only influence on outcomes, though in the case of school performance they may well be the strongest (Neisser *et al.*, 1996, pp. 82–83).

In addition to furnishing an IQ or general statement of intellectual ability, however, such tests give us important information about the balance or otherwise of cognitive development. It is useful to know if the development of any individual has tended to a preference for verbal or visuo-spatial tasks, simultaneous or successive processing, abstract logical reasoning or direct, intuitive, holistic thought. Moreover, general abilities often contrast, in gifted as in dyslexic individuals, with the lowly skills of information management. (For more on diagnostic tests, see below.)

Tests of Achievement

These are the most familiar and best-established psychological tests in general use. It is comparatively easy to test an individual's reading of words, or comprehension of connected passages, by means of various reading tests. Similarly, assessment of spelling ability, both by means of tests of single word spelling and in free writing, is straightforward. Dyslexic individuals frequently have problems with verbal aspects of arithmetic and these, too, are easily tested.

Finally, for all but the least literate individuals, it is essential to gather information on writing ability using a structured or semi-structured task. All of these so-called basic skills comprise the class of tests known as *achievement tests*.

Diagnostic Tests

Finally, we take the greatest theoretical interest in exploring exactly why an individual of average intelligence has such difficulty learning highly sequenced or memory-intensive basic skills. It soon appears that such individuals have remarkable problems with short-term verbal memory, as can easily be demonstrated by the inability of even 8- or 9-year-old children to learn their times tables or recite the months of the year in order. Children often have similar difficulties learning the alphabet or even the days of the week. We typically elicit samples of such memory by using tests of repetition of digits or of made-up words. Both are inherently meaningless (semantic memory is unimpaired in dyslexics) and what is being tested, therefore, is the ability to hold the verbal sequence in memory. Made-up or *non*words are used so as to limit the familiarity conferred by having a good vocabulary. This skill is an example of what was described earlier as *phonological sensitivity*.

If, in addition to storing and reproducing a sequence of nonwords or numbers, the individual is asked to do something else with them, to manipulate them in some way, then this brings into play the element of attention or concentration known as *executive function*. For instance, an individual may be asked to recite digits or the months of the year in reverse order. This is akin to knowing, not that $5 \times 6 = 30$, but that the next product to be remembered is 6×6; that is, in *keeping place*.

There is agreement nowadays that, in addition to tests of verbal memory and executive function, tests of *speed of processing* are able to tap overall efficiency. These seem to gain in importance as children become older and their skills better automated. The mid- to late-adolescent typically performs routine tasks at enormous speed. By contrast, the older dyslexic pupil has the greatest difficulty in achieving fluency. Tests of number sorting or visual search often discriminate effectively between dyslexic and non-dyslexic pupils.

Figure 4 shows the relationship of these three different domains of test as three concentric circles in a target. Another way of looking at the matter may be seen in Figure 5, in which a "platform" of clinical judgment is sustained by the three pillars of these three different kinds of assessment. At the same time clinical judgment is informed by the progress of general research into dyslexia, while the technical efficiency of the tests on which we rely is informed by progress in the fundamental statistics of psychometric assessment.

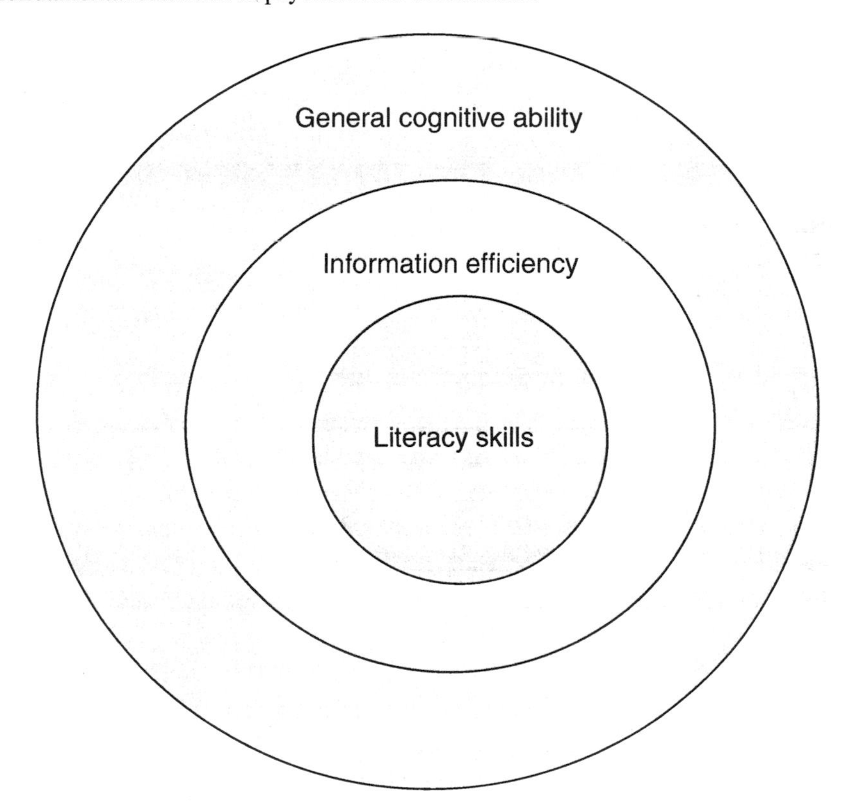

Figure 4. Three classes of test.

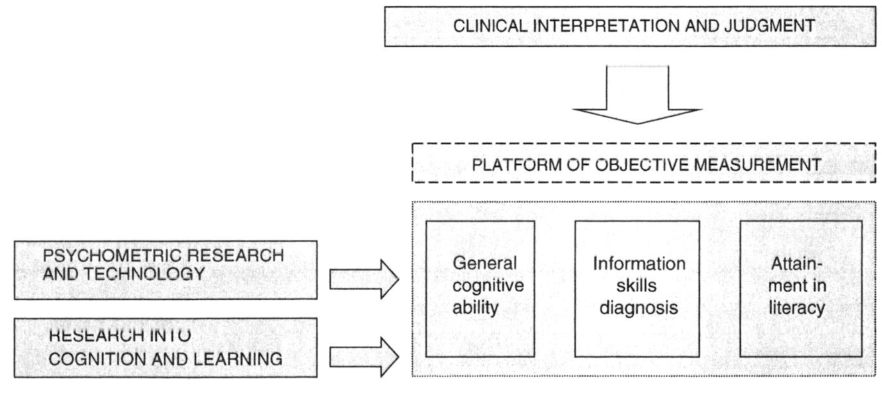

Figure 5.

Summary

We have seen that it is desirable to achieve not merely categorical and ordinal levels of assessment, but the higher, *interval level* of assessment also. However, assessment may relate the individual to his or her age group (*normative assessment*) or—without regard to age—to the relevant body of work to be tackled (*criterion-* or *curriculum-based assessment*). Third, an individual profile may be constructed using the results of standardized tests. This is called *ipsative assessment*.

It is helpful to classify all the tests that are relevant to a dyslexia assessment into three kinds, so that contrasts may appear among them. Accordingly, we use tests of *general intellectual ability* as a framework for interpreting other results in a controlled way; *tests of achievement* in literacy and numeracy skills; and *diagnostic tests* of the efficient management of low level information, for instance, in verbal memory or in tasks that must be performed at speed.

Finally, let us return to the child and the book. We began by considering that books may differ in terms of their intrinsic difficulty: three or four books may be ranked in terms of how difficult we perceive them to be. Similarly, pupils or readers may be ranked in terms of their ability to read these books. Some pupils are manifestly better at reading than others. It is worth mentioning, in conclusion, that the most advanced statistical technology, which now underlies most modern tests, looks at an interaction between these two variables. The ability of pupils may be ranged along one line, which then intersects with the difficulty values of items in a reading test ranged along another line. The point at which the difficulty of the task and the ability of the pupil interact is the goal of the test. Once this point is reliably identified, the test may be discontinued. This has greatly simplified testing and indeed resulted in a new generation of shorter and more efficient tests.

SOME STATISTICS[1]

Introduction

It has been well said—by Snow and Yalow (1982)—that a psychological test is rather like a little white pill. It is not obvious how it works or what it contains. Similarly, a test is sometimes no more than a single sheet of paper. It may look quite like other sheets of paper. How do we know what to make of it?

A small white pill may be produced quite simply from the bark of the willow tree: aspirin. Or it may be the result of decades of costly research: a cure for cancer or AIDS. How can we tell? Not by trying it out, surely. We can swallow a pill and notice little difference. Perhaps a headache improves, but it might have improved anyway. The effects may depend upon certain dosages or a course of

[1] Terms in italic case are explained in the Glossary at the end of this chapter.

treatment being sustained over time. Perhaps one has to have a specific ailment to begin with, to notice any improvement.

The tests we want to use are not difficult to operate. Anybody can obtain one and try it out. But, like the pill, the effect may be uncertain or unknown. Similarly, if the test is administered correctly and given to a suitable candidate, we may discover something about his or her performance that we didn't know before. It is often said that a test is valuable as a structured interview! Giving the test will permit some useful observations.

But of course this is scarcely a reason to employ something as complex and sophisticated as a psychological test. There are much easier ways to achieve interviews and observations. No, we give a test for the astonishingly simple reason that we want to measure something, to compare someone's reading, for instance, with the reading of other children the same age or similar ability. In this case, the main focus of our attention will be on the *interpretation* of the test result.

In order to do this, we need to collect information on the comparison group and it is with the interpretation of the individual in relation to this comparison group that this section will deal.

Test Development and Standardization

But first there must be a result to interpret. Tests are constructed from a pool of *items*—questions or tasks—so that there are enough at the levels of difficulty that the test will be concerned with. This is done by means of *trials* with samples of subjects (test takers), and items are selected or refined in relation to these subjects' performance, drawing on *item statistics* that permit effective *scaling*.[2] Once a test has come to include only items that *discriminate* effectively at the various levels of difficulty, it is ready to be standardized.

This procedure involves teamwork and cooperation among a large number of professionals, both research and field workers. Initially a *sample* is designed so as to obviate the necessity of testing *everyone*. An appropriately stratified sample reflects the total population in its strata; that is, it consists in the appropriate proportions of people of both sexes and the various socioeconomic classes, ethnic groups, drawn from all geographical regions. Given scientific sampling of these variables, quite a small number of individuals will suffice to give an accurate representation of the whole population of which they are members.

The comparison is usually a *universal* one, that is, we are interested in comparing an individual with *all others* of a similar age.

In order for this to be possible, the test must be given to all those in the *standardization sample* in exactly the same way; test administrators are, therefore, carefully trained in the administration of the test and the evaluation of the responses given. It follows, too, that those who later purchase and give the published test must also adhere exactly to the instructions. Test results obtained under

[2] There is a useful and more detailed discussion of score conversion in Cronbach (1990, ch. 4).

some variation of these conditions can bear only an approximate relationship to properly obtained results and cannot be easily interpreted.

Interpretation of Test Results

Many educational tests contain items that are marked right or wrong; the total number correct is the *raw score*. Other types of test, for instance essays, are marked in accordance with a mark scheme, so many for each question, with points given for specific information included or quality of response. Again, the total mark is the raw score.

As is common practice in schools, this raw score may be the end of the story. Teachers and parents are often content to know a child's percentage-correct on a science exam. Other questions follow: where did the child come in rank order; was the child above or below the average for the class; was the exam especially hard or easy? But such questions lead nowhere. In the absence of a properly designed test, there are no satisfactory answers, even for children in the one school, let alone for all children at that age and stage.

The main step in interpreting a raw score, then, on any well-designed test, is to discover its *standard score* equivalent. This is a matter of looking up the raw score in a table, using the individual's age as a coordinate. The value at the intersection of age (in years and months) and raw score is the standard score.

A standard score has the inestimable value of meaning the same thing whenever it occurs. It is also intuitive: it means what you think it means: the individual's performance is compared with all others of the same age.[3] There are many forms of standard score developed for different purposes, but they all assume a normal distribution, and so here we must enquire into some basic and simple, but unfamiliar, statistics.

The Normal Distribution

Many characteristics and traits in nature occur in a *normal distribution*. That is, they occur frequently at the mean (average), but with reducing frequency further from the mean, high or low. Examples are the obvious biological characteristics of height, weight, and strength. Genetically this is convenient, since polygenic traits, traits that are controlled by many genes, typically give rise to normal distributions. In humans, most of the complex traits in which we are interested, such as intelligence and personality, are associated with many genes. However, the main reason for assuming that the abilities we wish to measure are normally distributed, even in the case of literacy skills that may depend upon heritable skills but are not themselves directly heritable, is the great convenience for measurement purposes.

[3] The venerable method, attributed to Wilhelm Stern in 1912, of forming a *quotient* by means of the ratio of two *age equivalent* scores (Mental age ÷ Chronological age × 100) compounds the shortcomings of this type of score and should no longer be used. The scores thus generated are not on a true interval scale and so cannot be compared.

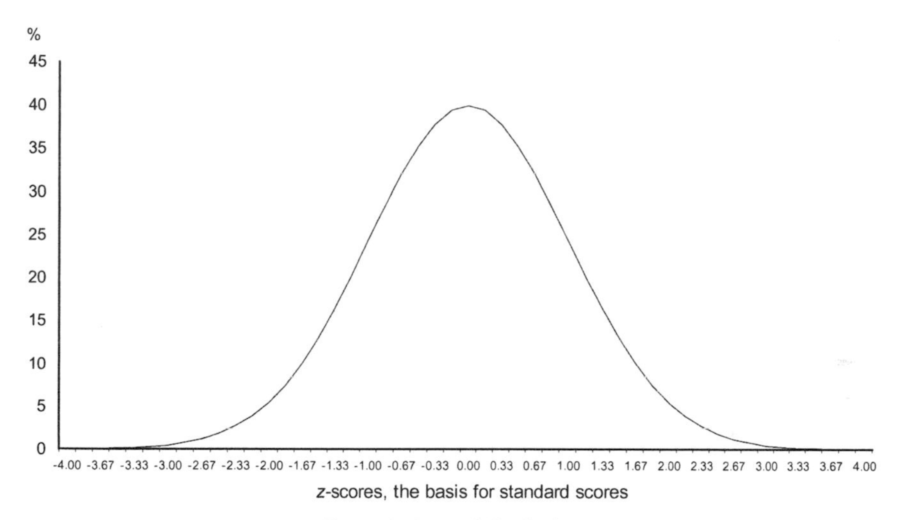

Figure 6. Normal distribution.

Figure 6 shows a simplified normal distribution, with the x-axis marked in standard intervals.

We have already spoken of the arithmetic mean, or average, in connection with exam results and test scores. The mean is perhaps the foremost statistic associated with any distribution and one that causes little trouble. In fact, the normal distribution can be described fully using just *two* statistics, and it is the second one, the *standard deviation*, that is unfamiliar.

The concept of the standard deviation is simple: it is the amount of dispersion (or scatter) among our scores. It is so useful in testing that we should have a look at how it is calculated, though a modern calculator will do it readily enough.

In Table 1, a hypothetical set of scores obtained by candidates (column 1) in a test is given. Each candidate's raw score is given in column 2. The steps for calculating a standard deviation are as follows:

1. Average the candidates' scores.
2. From each obtained score, subtract the group mean. This has been done in column 3, showing half the scores above and half below the mean (minus values).
3. Square each of these deviations to remove the minus signs (column 4).
4. Calculate the mean of each squared deviation from the mean (bottom cell in column 4).
5. Finally take the square root of this value: in this case 3.47.

This relatively simple procedure enables the distribution of these (raw) scores to be described in terms of the mean (10.7) and standard deviation (3.47). In real life we know the parents of these examinees would be pleased to know their

Table 1. The standard deviation

Candidate number	Raw score	Deviation from mean	Squared deviation from mean
1	10	−0.7	0.5
2	8	−2.7	7.5
3	15	4.3	18.2
4	7	−3.7	13.9
5	19	8.3	68.3
6	11	0.3	0.1
7	12	1.3	1.6
8	12	1.3	1.6
9	4	−6.7	45.3
10	9	−1.7	3.0
11	8	−2.7	7.5
12	10	−0.7	0.5
13	14	3.3	10.7
14	12	1.3	1.6
15	10	−0.7	0.5
Mean	10.7	0.0	12.1

offspring's scores *in standard form*. Using these two derived facts, we can in fact determine what these would be. Two further steps are involved:

6. Given the mean and standard deviation for the distribution, from each raw score subtract the mean.
7. Divide the result by the standard deviation.

The result, the most basic of all forms of standard score, the *z-score*, has a mean of 0 and a standard deviation of 1. When we talk about standard deviation units (or Ds), this is what we mean. More on this in a moment.

By convention, the *z*-score is usually converted to a quotient with a mean of 100 and a standard deviation of 15. This is the type of score most often found in psychology and education, the currency of IQs and most reading test scores. The standard quotient, as I shall call it, is produced by means of two final steps:

8. Take the result of Step 7 and multiply by 15.
9. Then add 100 and round to the nearest integer (whole number).

These further steps are shown in Table 2. This gives the candidates' numbers again and their raw scores in columns 1 and 2; column 3 now gives their *rank*, their ordinal position in the class of 15; column 4 gives their relative standing in *z*-score units; finally column 5 displays the results of their performances in familiar standard quotient form, with the achievement of each expressed relative to the performance of the class as a whole.

Clearly, any reported data, for instance for the number of words beginning with *s*- that 12-year-old dyslexic and non-dyslexic boys can generate,[4] can be turned to use as a *de facto* test.

[4] See Frith, Landerl, and Frith (1995).

Table 2. Candidates' standard scores

Candidate number	Raw score	Rank	z-score	Standard quotient
1	10	8	−0.21	97
2	8	12	−0.79	88
3	15	2	1.23	118
4	7	14	−1.07	84
5	19	1	2.38	136
6	11	7	0.08	101
7	12	4	0.36	105
8	12	4	0.36	105
9	4	15	−1.94	71
10	9	11	−0.50	93
11	8	12	−0.79	88
12	10	8	−0.21	97
13	14	3	0.94	114
14	12	4	0.36	105
15	10	8	−0.21	97
Mean	10.7	7.5	0.0	100.0

Note that we are making no assumptions here about the normality of this distribution. If the candidates' scores were normally distributed, then we could derive percentile equivalents from them (see below). We simply take a set of scores, find their mean and standard deviation, and express each one in a form that relates it to all.

Communicating the Scores Derived from Standardized Tests

Useful though this is, our main business is with fully developed psychometric tests. Here the assumptions of normality can be shown to hold. Turn again to Figure 6, the normal distribution. By now it may have become a little clearer that all normal variation can be described in terms of this Gaussian or bell-shaped curve. The demarcations along the horizontal axis are in z-score units, and these can be used to describe equal intervals up and down the scale. It is apparent that very few individuals are ever to be found four standard deviations (SDs) above or below the mean; the tails of the bell curve reach very close to the horizontal axis at these points. Similarly three SDs would be highly exceptional. When we close in on two SDs above (+) or below (−) the mean, we do begin to find a few individuals but still 96% of people fall within these limits. It is only as we move closer to the mean that we find many more individuals. Indeed two thirds of people fall within +/−1 SD; that is, their standardized quotients lie between 85 and 115. By convention, half the population is described as average: +/−0.67 or with standard scores between 90 and 110.

For all serious measurement purposes, the standard quotient is to be preferred. These can be added and subtracted; the difference between 90 and 95 corresponds to that between 115 and 120. At all ages, thanks to sampling that selects a different comparison group of children every few months, and of older pupils every year, each individual can always be compared with all others of the same age in terms of traits that are expected to be normally distributed.

But it also helps to conceptualize the meaning of scores if we say what *percentage* of individuals are found at that level or between two levels, as we have just done. Indeed for every standard score in the normal distribution, there is an associated *percentile* value. This is a great help in communicating the results of tests to teachers and parents who are not, as a rule, familiar with standardized quotients.

It is time to draw together the many ways in which the findings of tests may be expressed in terms of the normal distribution. Table 3 gives standardized quotients, their percentile equivalents, together with two further standard scores commonly to be met with:

1. The Wechsler subtest score, which has a mean of 10 and a standard deviation of 3, and
2. the *T*-score, which has a mean of 50 and a standard deviation of 10.

Because Wechsler subtest scores are somewhat inflexible, I have given them to one decimal place the better to illustrate their variation, but they are never quoted except as integers (whole numbers). Similarly percentile scores, here given to one decimal place, need to be reported as integers, except at the extremes—high and low—of performance, for instance, within the upper or lower 1 or 2% of the population.

The lower half of the normal distribution is given in the left half of Table 3, while the upper half is given on the right. Any of these standard scores may be expressed in terms of any other—or in terms of the percentage of cases that falls *at or below* this level (percentile equivalent).

Summary

This completes our brief survey of the main model that underlies psychological and educational testing: the normal distribution. If we assume that the abilities we are measuring are normally distributed, and even for instruction-sensitive skills this is usually the case, then we are enabled to bring to bear the useful set of statistics associated with the normal distribution.

Any set of results (or scores) may be summarized by two key features: the mean or average; and the standard deviation or measure of dispersion. We have calculated both of these by means of a worked example. In principle, we can now compare any individual with a representative group on some task for which we know the mean and standard deviation.

Additionally, if we use a published test on which scores have been *normalized* during scaling, we can fully quantify the performances of individuals on that test: their status, their progress, and differences between them (subject to a margin of error: see next section). We will always be comparing any one individual with the totality of his or her contemporaries.

Table 3. Scaled scores and their equivalents

z-scores	Standard quotients	T-scores	WISC subtest scores	%iles	z-scores	Standard quotients	T-scores	WISC subtest scores	%iles
−3.67	45	13		0.01	0.07	101	51	10.2	52.7
−3.60	46	14		0.02	0.13	102	51	10.4	55.3
−3.53	47	15		0.02	0.20	103	52	10.6	57.9
−3.47	48	15		0.03	0.27	104	53	10.8	60.5
−3.40	49	16		0.03	0.33	105	53	11.0	63.1
−3.33	50	17	0.0	0.04	0.40	106	54	11.2	65.5
−3.27	51	17	0.2	0.05	0.47	107	55	11.4	68.0
−3.20	52	18	0.4	0.07	0.53	108	55	11.6	70.3
−3.13	53	19	0.6	0.09	0.60	109	56	11.8	72.6
−3.07	54	19	0.8	0.11	0.67	110	57	12.0	74.8
−3.00	55	20	1.0	0.1	0.73	111	57	12.2	76.8
−2.93	56	21	1.2	0.2	0.80	112	58	12.4	78.8
−2.87	57	21	1.4	0.2	0.87	113	59	12.6	80.7
−2.80	58	22	1.6	0.3	0.93	114	59	12.8	82.5
−2.73	59	23	1.8	0.3	1.00	115	60	13.0	84.1
−2.67	60	23	2.0	0.4	1.07	116	61	13.2	85.7
−2.60	61	24	2.2	0.5	1.13	117	61	13.4	87.1
−2.53	62	25	2.4	0.6	1.20	118	62	13.6	88.5
−2.47	63	25	2.6	0.7	1.27	119	63	13.8	89.7
−2.40	64	26	2.8	0.8	1.33	120	63	14.0	90.9
−2.33	65	27	3.0	1.0	1.40	121	64	14.2	91.9
−2.27	66	27	3.2	1.2	1.47	122	65	14.4	92.9
−2.20	67	28	3.4	1.4	1.53	123	65	14.6	93.7
−2.13	68	29	3.6	1.6	1.60	124	66	14.8	94.5
−2.07	69	29	3.8	1.9	1.67	125	67	15.0	95.2
−2.00	70	30	4.0	2.3	1.73	126	67	15.2	95.8
−1.93	71	31	4.2	2.7	1.80	127	68	15.4	96.4
−1.87	72	31	4.4	3.1	1.87	128	69	15.6	96.9

Table 3. *Continued*

z-scores	Standard quotients	*T*-scores	WISC subtest scores	%iles	*z*-scores	Standard quotients	*T*-scores	WISC subtest scores	%iles
−1.80	73	32	4.6	3.6	1.93	129	69	15.8	97.3
−1.73	74	33	4.8	4.2	2.00	130	70	16.0	97.7
−1.67	75	33	5.0	4.8	2.07	131	71	16.2	98.1
−1.60	76	34	5.2	5.5	2.13	132	71	16.4	98.4
−1.53	77	35	5.4	6.3	2.20	133	72	16.6	98.6
−1.47	78	35	5.6	7.1	2.27	134	73	16.8	98.8
−1.40	79	36	5.8	8.1	2.33	135	73	17.0	99.0
−1.33	80	37	6.0	9.1	2.40	136	74	17.2	99.2
−1.27	81	37	6.2	10.3	2.47	137	75	17.4	99.3
−1.20	82	38	6.4	11.5	2.53	138	75	17.6	99.4
−1.13	83	39	6.6	12.9	2.60	139	76	17.8	99.5
−1.07	84	39	6.8	14.3	2.67	140	77	18.0	99.6
−1.00	85	40	7.0	15.9	2.73	141	77	18.2	99.7
−0.93	86	41	7.2	17.5	2.80	142	78	18.4	99.7
−0.87	87	41	7.4	19.3	2.87	143	79	18.6	99.8
−0.80	88	42	7.6	21.2	2.93	144	79	18.8	99.8
−0.73	89	43	7.8	23.2	3.00	145	80	19.0	99.9
−0.67	90	43	8.0	25.2	3.07	146	81	19.2	99.9
−0.60	91	44	8.2	27.4	3.13	147	81	19.4	99.91
−0.53	92	45	8.4	29.7	3.20	148	82	19.6	99.93
−0.47	93	45	8.6	32.0	3.27	149	83	19.8	99.95
−0.40	94	46	8.8	34.5	3.33	150	83		99.96
−0.33	95	47	9.0	36.9	3.40	151	84		99.97
−0.27	96	47	9.2	39.5	3.47	152	85		99.97
−0.20	97	48	9.4	42.1	3.53	153	85		99.98
−0.13	98	49	9.6	44.7	3.60	154	86		99.98
−0.07	99	49	9.8	47.3	3.67	155	87		99.99
0.00	100	50	10.0	50.0					

We have seen that test development is an expensive process involving cooperation and teamwork on a large scale. Moreover, the sophistication of the finished technology is not physically apparent to the user. Most of the technology is "under the bonnet" (perhaps in the technical manual). A test is first designed; then items are written for it and subjected to statistical analysis, scaling and selection; then the test is given live trials; then the final version is standardized on a carefully stratified sample of the general population.

When the trained test administrator, teacher, or psychologist comes to give the test to a live subject, important educational decisions may hang upon the outcome. Even the informal influence of such test results is considerable. It is therefore ethically incumbent upon us to give the test in the standard way enjoined by the detailed instructions given in the manual.

Interpretation proceeds by means of the measurement properties of the test. We draw only the conclusions we are enabled to draw by virtue of the test's reliability and validity (see next section). This will primarily cover the individual's standing in relation to his or her age group, though our findings may be enriched by observations of the individual's strategies and errors. Comparing this performance with others by the same person on contrasting tests enables us to build up a *profile* of abilities and difficulties, but such an edifice of interpretation depends crucially upon conscientious administration of the tests and accurate scoring of the results.

MEASUREMENT IN PRACTICE

Introduction

We have seen that a psychological test is an instrument of considerable power. Much of this power derives from the statistical model and properties that underlie its development. How does this affect day-to-day use of the test? Having said that we wish to make use of normal scientific expectations of measurement, how precise do we expect our measurements to be? With what confidence can we make and report measurements of ability and attainment?

The Concept of Reliability

Absolutely central to all efforts to make meaningful measurements is the concept of reliability.[5] If measurement information is not reliable, it is not worth even collecting. Reliability is the unspoken assumption behind all measurement. A recent

[5] More detailed psychometric accounts for the interested reader may be found in: Bartram (1990), Carmines and Zeller (1979), Salvia and Ysseldyke (1998), and Sattler (1988).

helpful definition of reliability is given in the manual to the Single Word Spelling Test[6]:

> The reliability of a test is a measure of the extent to which a pupil's test scores would vary with repeated testing, assuming that there were no fatigue, learning or lack of motivation.

Though it is sometimes supposed that psychological tests are inherently unreliable, this is true only to a limited extent. Indeed, an entire branch of statistics has arisen to quantify and control the element of error in psychological measurement. Most tests are a great deal more reliable than the readings of height and weight made in a doctor's surgery:

> The reliability coefficients of blood pressure measurements, blood cholesterol level, and diagnosis based on chest X-rays are typically around 0.50.[7]

Reliability is measured on a scale from 0 to 1, with values for a psychological test below 0.80 rendering the test professionally unacceptable. This is the reliability coefficient.

Let us consider for a moment what kinds of reliability we expect. Human beings are unpredictable creatures and measurement is likely to be slippery. Nevertheless we expect, if we test a child's arithmetic capability on Monday, that it will be the same on Wednesday. Moreover, if we have three acceptable tests of arithmetic, we expect, if a child takes all of them, that the results will closely agree. (Conversely, if we teach a child arithmetic skills intensively for a month, we might hope that his score on a test would improve. It is often noted that to teach the structure of syllable division brings large advances in reading ability, so perhaps reading might improve even between Monday and Wednesday!)

We know that attention and motivation can affect the results of cognitive tests. Group tests tend to be less reliable than individual ones, because one can get an individual child's attention and commitment much more effectively in a distraction-free environment with a sympathetic adult and stimulating test materials. If a large class of children is being tested in summer and a motorized grass-cutter repeatedly passes the window, it may be hard for some children not to be distracted. It seems likely, too, that high scores are more likely to be reliable than low ones, since a child can do badly for many reasons (his canary died that morning), but can only do well for one reason: he got sufficient harder items right.

The simplest ways to estimate the reliability of a test follow these examples. If a test is given twice to the same group of people, we will be in a position to evaluate both practice effects (the extent to which the group mean advances with experience of the test) and reliability. Each individual's first score can be correlated against his or her second score. Again, we may give equivalent forms of the test on different occasions. Alternatively, we may give odd numbered items from the test on one occasion and even numbered items on another occasion.

[6] Sacre and Masterson (2000, p. 30).

[7] Jensen (1998, p. 50).

All these methods can help us to establish empirically what the reliability coefficient of a test is and we can build this into our subsequent calculations.

The Concept of Measurement Error

Because human behavior is generally harder to measure than the width of a door or window, we should get into the habit of regarding every test result as having a true score component and an error component. As mentioned, the error component can be established empirically and used to qualify our results.

The best way to do this is to report a band or margin of error around each standard score. The actual margin of error will depend directly on three things:

1. The reliability coefficient of the test.
2. The standard error of measurement (derived from the reliability coefficient above).
3. The desired level of confidence.

The latter is best understood as how anxious one is to be right! If you want to be sure this result could not have been obtained by chance more than one time in a hundred, you will choose a probability of 0.01. (This is sometimes referred to as $p = 0.01$.) A less stringent and more common criterion of sureness is 5% or $p = 0.05$. This means that the result could be obtained by chance no more than one time in twenty. In clinical work, it is often argued that a 10% or even 15% level of confidence is appropriate. However, complete definiteness is not to be had, since reliabilities very close to 1 are not found. While this is a limitation of testing, it is well understood, highly quantified, and a useful discipline, leading us to strive for better technical standards.

Table 4 shows a selection of commonly adopted levels of confidence.

Let us see how an average score of 100 on a reading test with a reliability of 0.94 can be associated with varying levels of confidence or *confidence intervals*. Table 5 shows the confidence intervals for the same four levels of confidence.

Table 4. Levels of confidence

p-value	Percent	One chance in
0.15	15	6.7
0.10	10	10
0.05	5	20
0.01	1	100

Table 5. Confidence intervals

p-value	Lower limit	Upper limit
0.15	96	104
0.10	95	105
0.05	94	106
0.01	91	109

Table 6. Confidence intervals in relation to test reliability

Reliability coefficient	Standard error of measurement	Lower limit	Upper limit
0.99	1.50	98	102
0.90	4.74	92	108
0.85	5.81	90	110
0.80	6.71	89	111
0.75	7.50	88	112

In other words, though we have a highly reliable test, we should report a result at the 95% confidence level as plus or minus six (+/−6) points of standard score. In practice, it is both desirable and manageable to emphasize a margin of error in this way. The concept of a margin of error does not need explaining and a confidence interval, even with a test of excellent quality, conveys a high standard of accuracy.

Let us see, next, how this margin of error widens alarmingly with each decrease in the reliability of the test. In Table 6, an average score of 100 is taken as an example again, together with a 95% level of confidence ($p = 0.05$).

To interpret: with a reliability about as low as is acceptable (0.80), an average score of 100 would have to be reported as falling between 89 and 111, that is +/−11 points. With the usual 95% level of confidence, this means that there is a spread of 22 points within which the true score might actually fall.

This serves to demonstrate how critical to our practical assessment work the reliability of a test actually is.

The Concept of Validity

However, a test should also be valid. That is to say, it should measure what it is supposed to measure. We are not justified in making inferences about a person's geographical knowledge from her or his performance on a test of reading (even if these tests correlate at some level). One is not a valid measure of the other.

All unreliable tests are invalid, but a reliable test too may be invalid. If a ruler systematically exaggerates length by 6 cm, all measurements obtained by means of it will be wrong, even if the error involved is always exactly the same. (A ruler that was always inaccurate by a different amount would be unreliable as well as invalid.)

Just as we can estimate the reliability of a test by how well it correlates with itself, so can we estimate the validity of a test by how well it correlates with the other tests. Validity is a broad concept and there are several approaches to it:

1. *Face validity*. Though the person administering the test may have faith in it, is the test likely to seem reasonable to the person taking it? Does it seem to consist in relevant material?

2. *Content validity*. Would maths teachers used to the National Curriculum accept the content of an old-fashioned test as having a balance of content that adequately reflects programs of study in use in schools?
3. *Construct validity*. Does the test conform to or exemplify what is known or thought about the attribute in question? If it is generally accepted that intelligence may be measured in both *crystallized* (culturally specific) and *fluid* (flexible, transferable) forms,[8] does a new test adequately reflect this theoretical understanding?
4. *Criterion-related validity*. How well does a test relate to a specific aspect of its purpose? If management trainees are selected according to their performance on a test of management potential, does this prove valid in relation to their practical performance during training?
5. *Predictive validity*. This is one example of criterion-related validity. What is the most valid predictor of occupational success, for instance, when an earlier performance is related to a later?
6. *Concurrent validity*. Criterion-related validity can also be estimated by performance on two tests taken at the same time. Does a new test of reading correlate to an acceptable extent with a well-established one?

Obviously these six approaches overlap very considerably, but it is helpful to emphasize now one aspect, now another.

Skilled Cognitive Performance

Recognizable psychological tests have been around for most of the 20th century. Elements of both theory and practice existed from the 1860s, when Francis Galton analyzed trends in anthropometric data[9] he collected from adult volunteers; and a three-question test of mentality was proposed as long ago as 1510. But credit for the first modern test goes to Alfred Binet who, with Simon, developed for the government of France in 1910, a useful test for screening the intelligence of school children.

It is on tests of intelligence or general cognitive ability that most experience has been gained in psychometric testing. During the first half of this century, the general summary of an individual's performance, the IQ or intelligence quotient, was something of a preoccupation. In the latter half, however, there has been more concern with the issue of *sensitivity*, given that IQ has come to be seen as a very general statement about a person's total worth.

We need to consider the place of IQ or any other measure of someone's general abilities, because a general ability factor tends to dominate any data collected in an individual or group assessment of human skilled cognitive performance.

[8] For example, Horn and Cattell (1966).

[9] Reaction times, sensory discrimination, motor coordination.

Factor Analysis

How do we know this? Because of a set of mathematical techniques that nowadays virtually defines the subject of testing: *factor analysis*.

Any large body of performance data can be analyzed with respect to underlying trends. A table of random numbers would show no trends of any kind. But it is a fact that if somebody is good at one task, and gains a high score on a test of it, he or she is likely to be good at another one. This tendency for there to be a positive correlation among tests of skilled cognitive performance can be subjected to factor analytic techniques. First and foremost, a general ability factor, or *g*, emerges from these analyses and this is thought to represent psychometric intelligence.

After decades of controversy, it is also accepted that below the level of *g* there are group factors which, in spite of high correlations with *g* and with each other, have a distinct identity. These have most obviously been described as verbal and visual abilities.[10] For some individuals who show an imbalance in cognitive development or a strong preference for either verbal or visual modes of thought, an analysis of test results at this level may be important.

Figure 7 shows an example of how factor analysis might produce grouping of films at lower and higher levels.

Analyzing Underachievement

This brings us close to the end of this section. But having established that we are interested only in tests that meet high standards of reliability and validity, we need to look at the framework for testing—to be developed more fully in the next section—which will support higher level interpretations of complex data.

We are particularly interested in reading and spelling. As we shall see in the section on reading and spelling (p. 234) anybody can in principle be measured on an absolute scale of reading ability, since on a test with a large range of items ranging from extremely easy to extremely difficult, anybody can obtain a score. In this way, all individuals can be compared with respect to a criterion, say word recognition ability in English. Some people will be happy, others sad, with their reading ability. But how do we know who, if anybody, has a problem. Given scarce resources in terms of the time available to a skilled specialist teacher, who needs our attention?

The common-sense answer to this takes into account a person's age. We expect older individuals to read more effectively than younger ones. We need not be concerned that a 4-year-old is a non-reader; but we should be concerned that an adult is handicapped in this way.

But children of a given age, say 10, are not all alike. Some learn quickly; some learn everything more slowly. Indeed, at the extremes, gifted children

[10] Or Verbal and Performance (Wechsler), Verbal Comprehension and Perceptual Organization (Kaufman), Verbal Conceptualization and Spatial Ability (Bannatyne), depending on your authority.

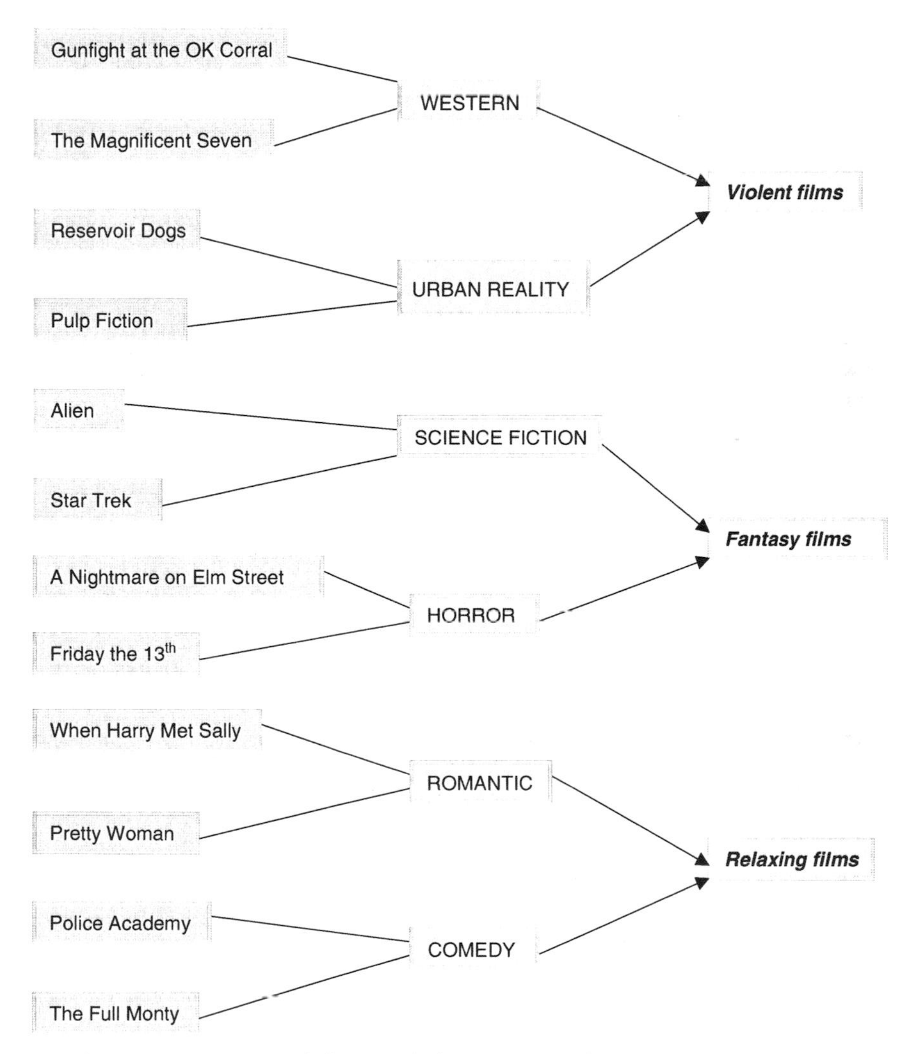

Figure 7. Factor structure in factor analysis: An example (after Cooper, 1999, pp. 36–37).

or those with learning difficulties may be educated for some or all of the time at separate, specialist schools. So the second part of our common-sense answer is that we need to take account of individual differences in ability.

Other variables will doubtless be important: for instance, whether someone has received adequate instruction or has attended school regularly, whether English is their native language or whether they are prevented from concentrating by domestic burdens or other anxieties. But age and ability are our starting points.

This is the logic of the *controlled experiment*. Given a child's age and level of general ability, what do we expect his or her levels of reading and spelling to be?

Such a prediction must necessarily be made within broad bands, but nevertheless this is the statistical basis for evaluating levels of achievement that may be *unexpected*. It is not a definition of dyslexia, but a procedure for investigating literacy-learning difficulties.

Summary

Testing in professional practice is enormously dependent upon the quality of tests used. Foremost among other considerations is the question of validity and, especially, reliability.

Validity is (broadly) correspondence with reality. If a test of some knowledge domain does not measure what it is supposed to measure in a way that meets current standards of how that domain is organized and how we think about it, then the test may not be a valid one.

Assuming validity, the next question is about reliability. Does the test give consistent results? An elastic ruler is of little use. The test manual should report a test–retest reliability coefficient of at least 0.8 (or an internal consistency reliability coefficient above 0.9) before we will want to trust important decisions to its results.

General intellectual ability looms large in any assessment of individual abilities. We know from factor analysis that the general factor in an assessment accounts for three times as much of the variance as all the other factors put together (Brand, 1996).

This gives us a way of discerning those who, on the basis of their age and ability (as well as other considerations) have a level of spelling and reading that is *unexpected* and thus worthy of further investigation and remediation.

COGNITIVE ASSESSMENT

Introduction

We have now seen something of how tests are constructed, how they work, how they are administered, and how results are obtained. We must now consider their use in actual settings, schools or clinics, and how such results may be interpreted.

At this point we may want to branch out from the rather neutral, empirical path of test development and confess that we have a prior agenda. We come to the testing situation with a precise purpose (to investigate an alleged problem in learning), fairly distinct expectations of how children[11] progress in school and, nowadays, positive ideas about the factors that may interfere with their learning.

[11] Children, rather than adults, are referred to in this section since tests for children are used as examples.

While test theory strongly influences how we work with tests as *measuring instruments*, other kinds of theory—about literacy teaching or dyslexia—influence how we look at children.

Using the Three Classes of Test

There is a relatively simple logic to investigating children's learning difficulties, and it is as follows:

1. First we want to sample a child's general learning ability, since this estimate will give us a context in which to interpret particular aspects of learning.
2. Next we want to sample his or her efficiency of information management, especially in terms of memory and processing speed.
3. Third, we want to evaluate performance in key acquired skills: reading, spelling, writing, number.

In a sense all these kinds of investigations require the use of *cognitive* tests, since our present focus of concern is with the learning of basic scholastic skills. Of course, children have other kinds of difficulties, and these may affect learning to some extent, but cognitive tests may not be appropriate to emotional or behavioral difficulties. For instance, in relation to Attention Deficit Hyperactivity Disorder (ADHD), Goldstein and Goldstein (1998) point out that

> Clinic test scores alone or in combination resulted in classification decisions that frequently disagreed with the diagnosis of ADHD based on parent interview and behaviour rating scales (DuPaul, Anastopoulos, Shelton, Guevremont, & Metevia, 1992, p. 328).

These authors note further that

> ... efforts to develop batteries of ADHD measures to assess attentional skills have for the most part been discontinued (p. 329).

Any enquiry, such as a case study, into a child's life should be handled with the greatest sensitivity and respect. We cannot know in advance what factors might be relevant to an enquiry into the learning of basic skills. But we should not claim any expertise, such as clinical training or knowledge, that we do not have. We should always restrict our conclusions, opinions, communications, and feedback to those areas in which we are acquiring professional knowledge and expertise. And in cases where a fuller investigation by other professionals seems warranted, we should not hesitate to recommend such a referral. In this way, we avoid making inflated claims as to our own expertise and satisfy ourselves that the client is pointed toward the best available source of relevant help. This may mean audiology, optometry or ophthalmology, psychiatry, social work, occupational therapy, speech and language therapy, educational psychology, or other disciplines.

Assembly and Presentation of Scores

For any individual child (or adult), we seek to construct an *individual profile*. This is an assembly of test scores drawn from good quality tests. Two tests results from each class—six in all—will provide a minimum number for the construction of a profile. Other tests may be given to supplement, confirm, or disconfirm those first six.

In section 1, we looked at the use of normative tests and saw how, if their results were combined into a profile, they allowed us to compare an individual with him- or herself, his or her development in one area with another. This is sometimes referred to as an *ipsative* analysis.

First we must assemble our test results into a common form, in order to compare them. For preference this is done with *standard scores*, scores that relate an individual to his or her contemporaries in a standard way that always means the same thing at any age (and always means what it appears to mean). But as we have seen in section 2, there are three main types of score in use:

1. *Standard scores* (mean 100, standard deviation 15 points). These can be added and subtracted, multiplied and divided, analyzed and compared.
2. *Percentile equivalents.* These are a linear transformation, using the normal curve, of standard scores but, whereas standard scores are unfamiliar to the lay person (and to most teachers), the percentile builds on the common understanding of the *percentage of individuals who score at or below this level.* Given that the percentage seems to be the world's favorite statistic, this way of reporting a test result conveys its meaning readily. However, percentiles should not be used for evaluating *differences.*
3. *Age equivalent scores* (mental age, reading age). These convey an intuitive meaning, and may be reported, but allow little reliable comparison. To avoid the appearance of decimal form, *these should always be expressed with a colon between the years and months*, thus 8:7 (8 years 7 months).

For ease of reference, all scores in all their possible forms should be assembled in a simple table. An example is given in Table 7.

Several points should be noted about this table.

1. All reliability coefficients refer to internal consistency (Cronbach's alpha) except for the Neale (parallel forms reliability) and the Hedderly Sentence Completion task (Hedderly, 1995).
2. The Matrix Analogies Test (Short Form) (Naglieri, 1985) gives only percentile equivalents of raw scores, but standard scores may be inferred from these using a table such as Table 3 in Section 2.
3. Confidence intervals have been constructed as follows:
   (i) Subtract 100 from the individual's score
   (ii) Multiply the observed score by the reliability

Table 7. Assembly of test results for Joey (8:7)

Name	Joey				
Date of assessment	07-Dec-90				
Date of birth	14-Jul-99				
Age at testing	8 years 7 months				
Test results					
Test name	Raw score	Standard score	Percentile	Conf. interval	Age equivalent
BPVS-II	96	108	70	101–112	9:10
MAT short form	18	103	58	96–109	8:9
PhAB spoonerisms	4	85	16	82–92	
Writing speed	5.4 wpm	82	12	79–87	
WRAT-III reading (blue)	28	93	32	89–98	
WRAT-III spelling (blue)	18	79	8	76–86	
NARA-II Acc Form 1	33	89	23	85–95	7:4
NARA-II Comp Form 1	20	102	55	95–108	8:10
TAAS phoneme del.					6:0

(iii) Add 100 again
(iv) Add and subtract one standard error
(v) Round to whole numbers.

This gives a confidence interval of only one standard error, that is, a 68% confidence interval.

4. Tests that do not report a particular type of score, age equivalent scores in the case of WRAT-III (Wilkinson, 1993) or PhAB Spoonerisms (Frederickson, Frith & Reason, 1997), or standard/percentile scores in the case of Rosner's Test of Auditory Analysis Skills (Rosner, 1993), should not have them invented. It is better to leave blank cells in the table.[12] However, in the next section, a method for local derivation of age equivalent scores from the WRAT-3 norm table *is* described.
5. Hedderly's (1995) test of free writing by means of sentence completion is scored in Joey's case as if he were already 9 years old. This may therefore slightly overstate his lack of productivity.

Note that in steps 3(i) to 3(v) above, the confidence intervals are built around the (hypothetical) true scores. This is why they are not symmetrical. This method takes into account the reliability of the test and is therefore preferable to simply adding and subtracting one standard error of measurement (SE_m) to or from the observed score. The importance of this for interpretation becomes clear in a moment.

[12] Since standard scores have direct percentile equivalents, these may be looked up in either direction, as in the case of the Matrix Analogies Test.

Table 8. Reliabilities and SEms for selected teachers' tests

Test	Age range	Reliability	SE_m
BPVS-II	3:00–15:08	0.86	5.61
MAT short form	5:0–17:11	0.83	6.18
WRAT-III reading	5:0–74:11	0.90	4.74
WRAT-III spelling	5:0–74:11	0.89	4.97
WRAT-III arithmetic	5:0–74:11	0.85	5.81
NARA-II Acc Form 1	6:0–12:11	0.89	4.97
NARA-II Comp Form 1	6:0–12:11	0.82	6.36
NARA-II rate	6:0–12:11	0.66	8.75
PhAB spoonerisms	6:0–14:11	0.89	4.97
Hedderly sentence completion	9:0–adult	0.93	3.97

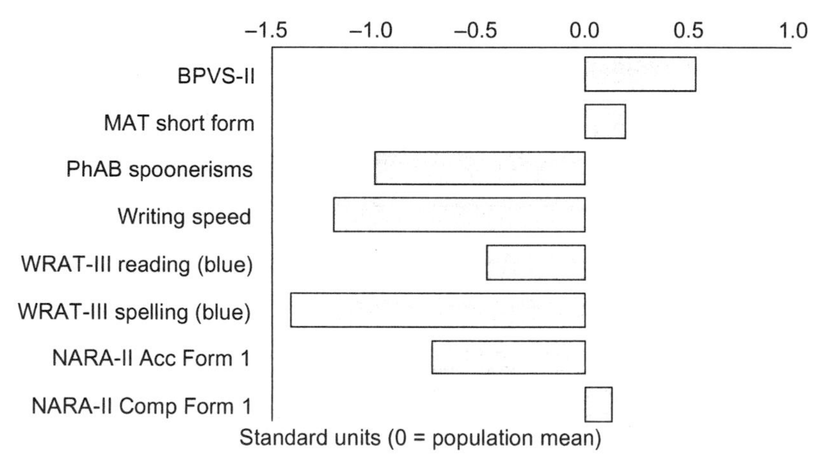

Figure 8. Profile of scores for Joey (8:7).

It is therefore necessary to know the reliability of the test. This information[13] is normally found in the test manual, but for convenience is given in Table 8 for some of the tests commonly used by teachers. In fact, the standard error of measurement (SE_m) is produced by a formula from the reliability coefficient, but both are usually given and may be combined as in steps 3(i) to 3(v) above.

Visual Presentation of a Profile of Scores

Test users do not have equal facility with the data-handling capacities of personal computers nor, indeed, equal access to computers. A profile like the one above may be represented very simply, using an application such as Microsoft Excel and converting the standard scores to *z*-scores, as in Figure 8. Here the standard scores

[13] Often the median reliability coefficient for internal consistency using Cronbach's Alpha or, where items are scored 1 or 0, Kuder-Richardson's formula 20.

(in z-score form) nicely contrast Joey's normal vocabulary (BPVS-II) and nonverbal reasoning (MAT-sf) with his poor phonological (speech sound) processing, reading, spelling, and written language skills. Like many dyslexic children, his reading *comprehension*, though, is mysteriously higher.

Since all these scores represent *deviations from the mean*, it makes sense to represent them rising above or falling below that mean, here 0. A standard score or percentile graph that shows all scores rising from an impossibly low base (percentile 0 or—even more improbably—standard score 0) runs the risk of making the differences between scores disappear in a forest of bars.

However, such a graph, reflecting an individual's standing on all tests in relation to his peer-group, may be disheartening if the person is of uniformly low ability and attainment, since all the bars will fall to the left to some extent. In any case, such a graph can be done perfectly satisfactorily by hand using a standard score scale and marking in only the blocks represented by +/− one standard error of measurement (SE_m). There will thus be short cylinders scattered about instead of long bars. This solves the problem of the origin of the bars. Moreover, a standard score scale from 70 to 100 will seem less depressing than one from 70 to 130.[14]

Finally, only test results expressed in the form of standard scores should be used in a graphical display. All other forms of score are too approximate to be used to convey precise information.

Interpreting Test Scores

We are now very much closer to fulfilling our aim of interpreting the performance of any individual on our tests. Once test results of a uniformly high quality are assembled, and especially once they are seen in a graph, it becomes clear how his or her strengths and weaknesses should be evaluated.

It may be that, conveniently, an individual's *ability* scores seem normal while his *diagnostic* and *attainment* scores seem depressed. In that case, there is clear evidence of a learning difficulty with a cognitive origin. That is, there is

1. an unexpected shortfall in those attainments, such as literacy, that have probably caused the concern in the first place; *and*
2. evidence of poor information processing that might explain the observed underachievement.

However what if the scatter of results does not conform to a convenient pattern? Then we may look at the differences *between* the scores and relate them to the skill domains they represent. But this is precisely where our confidence intervals, by now carefully graphed by hand as short bars on a standard score scale, enable us to make judicious interpretations.

[14] For any individual a scale starting 10 points below his lowest, and ending 10 points above his highest, score should be adequate.

When a difference *between* two standard scores is evaluated, there is double the quantity of measurement error to be taken into account. A *difference* is harder to interpret than either score on its own. As a rough guide, the lower edge of one margin of error should not overlap with the upper edge of the other one. Both margins should be clear of each other before there can be any probability of a *real difference* between the scores. This is because the standard error of a difference contains two sets, not one set, or measurement error.

However, where such differences are marked, and the bands of error do not overlap, the contrast within the profile may be of interest and may deserve comment. This is where the skill domains, rather than the test results, are evaluated. For instance, poor short-term auditory–verbal memory may contrast with good vocabulary, or good vocabulary with poor nonverbal reasoning, or poor accuracy with better comprehension in reading. Every individual, even if he or she falls obligingly into a diagnostic category, will have features that are unique and that have implications for classroom learning.

Summary

In this section, we have drawn together our theoretical knowledge of test scores, reliability, and validity to make reasoned interpretations of performance within a context of child development and school learning, rather than test behavior merely. Ever-mindful of the limitations of our qualifications, we have concentrated on educational tests of cognitive ability and attainment.

It has been necessary to compile test scores in all their forms in a table so as to evaluate them together and look for patterns. Once we have designed a simple graph using bands of error instead of point scores, we have found it still easier to begin to see where true differences may exist. These form the basis of our interpretations in terms of the individual's cognitive development and school learning.

ADAPTING THE MODEL FOR TEACHERS

Introduction

So much for theory. We have seen, from one end to the other, how tests are made and work. Testing remains quite an abstract activity: it is at the high value-added end of the spectrum of educational services and requires a high level of skill. While we test, we have frequent recourse to our model, to maintain our sense of what we are doing and to refresh our purpose.

Nevertheless, no training in testing can rely on theory alone. Teachers in particular need live, close supervision of testing, since their training has led them to intervene rather than to observe, to explain rather than to elicit, to advise actively rather than to adopt an attitude of supportive neutrality, to teach rather than to assess.

In this section, we are going to explore a small group of tests that can be administered by specialist teachers, including teachers receiving specialist training. The choice of tests is guided by two considerations:

1. *Test quality.* There are far too many tests of indifferent quality and splendid antiquity in circulation. No passionate loyalty to a test we have grown up with should prevent us from using another, technically improved one if the opportunity arises. Fewer, better tests will present clear, crisp pictures of what is going on, much preferable to the common fog.
2. *Simplicity.* Modern tests tend to be easier to administer and interpret. Psychometric testing can be quite complex enough without the unnecessary complications that arise from using older tests. Trainees have often brought me the results of tests patiently and laboriously given to long-suffering children, with the request that I interpret the results. Alas, the results are often uninterpretable!

What follows is an introduction to several of this core group of current tests, with some discussion of each. But at best this section will only supplement the practical training in testing that is needed.

A Test of General Intellectual Ability

In spite of recurrent controversies about IQ, we shall continue to need estimates of general ability in order to provide a context for understanding the individual. Two 12-year-olds with a reading age of 8 are not similarly retarded, and will need very dissimilar teaching, if one is intellectually bright and the other is a slow learner. The central importance of genes is being unfolded to us by scientists on a daily basis and suggests that this view of *potential*, the common sense view, may be the right one.

Teachers cannot administer recognized batteries of intelligence tests as such,[15] but they can sample general problem-solving ability in verbal and nonverbal domains so as to arrive at an *estimate* of a child's functioning away from areas of possible impairment. This provides a framework of knowledge about the individual's *general* cognitive development within which to examine *specific* developments that may be rather different. General cognitive ability is one of the most powerful and pervasive explanatory variables in human science and it will tend to swamp any data that is collected. To control (as in an experiment) for IQ is hard enough, and can give rise to all sorts of problems, but not to control for it is to ensure that it controls you!

Let us look, first, at a simple measure of vocabulary, the **British Picture Vocabulary Scale.**[16] This test enables us to observe the growth of

> linguistic *creativity*, the capacity for inventive thinking that is evident in language learning, a type of learning that occurs without benefit of instruction. (Cohen, 1999, p. 62)

[15] See Turner (1999).

[16] Dunn, Whetton, and Burley (1997). Tests will throughout be treated with all the respect accorded to books and listed with full publication details in the references.

The individual to be tested is presented with a set of four pictures and hears a single word spoken. He or she has to point to the picture that goes with the spoken word. This is thus a measure of *vocabulary*, the most central and robust component of measured intelligence, but of *receptive* or *comprehension* vocabulary. The test is *convergent*, in that the child or adolescent must narrow down the possibilities to a single right answer, rather than elaborate on the definition orally, furnishing associations and examples and increasing by volubility the chances of getting a mark. Moreover, by pointing to the preferred picture out of the four presented, he or she is not required to demonstrate any expressive language skills. The test therefore offers a good guide to verbal intelligence that is fair with respect to the shy child or the child for whom English is an acquired language but who understands it better than he or she speaks it.

The recent revision of this test has made administration easier. At the same time, the normative range has diminished somewhat. It now covers individuals from 3 years 0 months to 15 years 8 months. The test taker is administered items in whole sets of 12 items each, from a total of 14 possible sets (and 168 possible items therefore). Suggested starting points are printed in the exemplary record form and the individual must make no more than a single error in a set for this set to be considered a *basal*, that is, a starting point. The individual is then given succeeding sets until he or she makes eight or more errors in a set. This highest set is then considered a *ceiling* set, that is, a finishing point. The raw score for this individual is then the number of the last item in the ceiling set minus the number of errors.[17] A standard score is obtained from the table of norms in the Manual (pp. 40–47) at the intersection of the column with the raw score obtained and the column that includes the subject's age (in ears and completed months). Percentile equivalents for standard scores are provided on the front cover of the record form, while a graph of the normal distribution is provided on the back.

Split-half reliability of 0.86 gives a standard error of measurement (SE_m) of 5.6 standard score points. Confidence intervals, arranged around the hypothetical *true* score, are given in Table 10 on page 34 of the Manual. A Technical Supplement provides norms for 410 pupils aged 3 years 0 months to 8 years 5 months with English as an Additional Language (EAL).[18]

At this stage, a reliable, modern measure of *verbal ability* must suffice for estimation of general ability. IQ—or general intellectual ability as measured by appropriately qualified psychologists—improves, in any case, only a little on verbal ability as a predictor of academic progress. Based on the individual administration of this test, the expectation would be that a pupil's scores on tests of reading and spelling would fall either within the observed score's margin of error

[17] In other words, the individual is credited with any items not administered below the ceiling set.

[18] Year 2 norms are interpolated between Years 1 and 3. In general, the age equivalent scores in months obtained by the EAL sample were about three quarters of their chronological age in months.

(confidence interval) or about half way between this observed score and the population mean (100).[19]

A Test of Efficiency at Processing Linguistic Information

We now turn to the area of diagnostic testing. The most useful recent publication here must be the **Phonological Assessment Battery** (PhAB).[20] This is a battery of eight diagnostic and two comparison) tests (see Table 9) aimed at the heart of the processing difficulty in dyslexia, that of phonological processing.[21] It offers recent national norms for school-aged pupils aged 6 years 0 months to 14 years 11 months,[22] and therefore has more test variety and a wider normative range than any comparable test. No additional materials are required (most stimulus materials are printed as pages of the Manual), so no equipment need be carried around beyond the Manual and record forms.

Administration is clearly described in the Manual and need not be rehearsed here. Internal reliabilities are good. About 50–80 individuals in the standardization sample took each test at each age level. There is no composite score ("phonological ability"), however, and though an average score will serve, interest resides in the profile of an individual's scores across the range of tests.

Although the teacher getting acquainted with this battery will want to administer all tests at first, two are of particular interest. **Spoonerisms** require(s) manipulation, not just awareness, of phonemes and is therefore the most interesting of the tasks, given that a measure of phoneme deletion was not included. Along with **Nonword Reading**, a decoding test of alphabetic skills, it correlates highest with measures of reading accuracy. Though the latter has little relationship with general ability, however, **Spoonerisms** have the highest correlation with IQ, perhaps because of the complexities of response required. In general, though, all PhAB

Table 9. Phonological Assessment Battery—subtests

Alliteration
Rhyme
Spoonerisms
Nonword Reading
Naming Speed—Pictures
Naming Speed—Digits
Fluency—Alliteration
Fluency—Rhyme
Alliteration (with pictures)
Fluency (semantic)

[19] This rule of thumb conservatively assumes a correlation between the two measures, say, vocabulary and reading, of only 0.5.

[20] Frederickson *et al.*, 1997.

[21] The ability to hear, store, reproduce and manipulate the *sounds* of speech, such as the sequence of syllables in *car park* or *barbecue* (rather than *par cark* and *cubeybar*).

[22] The Alliteration Test is normed from 6 years 0 months to 11 years 11 months.

tests relate less closely to general ability than does the Neale reading test[23] with which they were co-standardized. The PhAB has more of a relationship with the *rate* measure (speed of reading) in the Neale. Perhaps rate of articulation, an important variable in phonological skill, is important to both. However, all these tests are of diagnostic and instructional significance and have a developmental importance in literacy attested by varieties of research on both sides of the Atlantic.

Tests of Word Recognition and Spelling

We now move to individually administered tests of reading and spelling. The **Wide Range Achievement Test**, 3rd edition (WRAT-3)[24] offers us simple tests of single word reading and spelling, recently normed from 5 years to 75 years. This is but the first step in assessing literacy skill. We shall see in a moment how to sample more complex aspects of reading. Nevertheless word-level skills are the foundation for later, more complex skills. Guessing is a strategy noticeably absent in skilled readers.

These two tests follow convention. The individual reads single words out loud, with one row exposed at a time, until ten are failed. Similarly, he or she spells single words dictated, one at a time, with a sentence of supporting context for each. The raw score is each word read (and pronounced) or spelled correctly, but don't forget to add 15 points of raw score on each test if the elementary items (letters to be named or written) are omitted.

Obtaining standard scores is straightforward. For each form of the WRAT-3—Blue, Tan, and Red—**standard scores** are given in the Manual (pp. 36–163) opposite each raw score. We are mainly interested in the standardized quotient, for which we can then look up **percentile** equivalents inside the back cover.

Much of the best measure of progress for these short tests—though they are not designed for the purpose—is the *second* column, the **absolute** scale. This is the equivalent of a ruler underlying all the score increments. At last we have here a true interval scale. Progress between 130 and 142, therefore, means exactly the same thing as progress from 250 and 262. The progress between two very different children can be compared in this way.

Tradition in the United Kingdom dictates that we give age equivalents of raw scores, since these seem intuitive and to need no explanation. Also to teachers they signify progress through a learning or curriculum sequence. In the United States, *grade scores* serve much the same purpose. These now coincide helpfully with our National Curriculum years. (We have joined the international standard for describing school years.) There is moreover a ready-made way to report reading, arithmetic, and spelling *ages* (**age equivalent scores**) from the Manual

[23] Neale (1997).
[24] Wilkinson (1993).

without having to devise a special table. This is best done *locally* using the norm tables as follows:

1. Take the grade score at the end of the row and add 5 for the age at which children in the United States are at grade 0 (actually K for kindergarten).
2. Add a decimal place as appropriate for the numbers of steps through the grade represented by this score.
3. For instance, suppose we obtain a score of 26 on the Tan reading tests for a child of 8:4. The relevant norm table appears on p. 61 of the manual.
4. This raw score represents a standard score of 91 (percentile 27). The Grade Score is 2. So the reading age *year* is 7.
5. Four rows contribute to Grade 2, so each point of raw score represents 0.25 (or three months) of a year. Our child's score is third of four rows. So we add three quarters of a year (or nine months) to the 7 years.
6. The reading age is therefore 7 years 9 months (7-9 or 7:9—always use a colon or hyphen to avoid a spurious appearance of decimalization).

If one is interested in age equivalents beyond about, 14 (when in reality progress in such terms is pretty flat), consult p. 177 of the Manual, which, in table 15, gives the median raw score for each age group. One may see that a single point of raw score gives a year or more of progress.

Bear in mind that each measurement has an associated **error of measurement**—roughly from 3 to 6 points of raw score—and that the progress of any pupil must be better than this to show up on a measurement. This is true of any test. WRAT-3 actually tells you the values of errors of measurement (table 12, p. 174). Even at lower ages, the margin of error is 6 months or so.

It should be clear that these short tests are not ideally suited to registering subtle steps of progress in teaching.

UK Infant Norms for WRAT-3 Reading and Spelling

Children in the United States attend school full-time only from the age of 6, after 2 years of Kindergarten. With less formal schooling, their capability in reading and spelling at that age tends to be less than that of comparable British children, though differences have become imperceptible by the end of the primary phase.

US norms therefore tend to flatter British children of infant age. Recently, however, both Reading and Spelling tests were administered to 325 infants in a London Borough[25] concurrently with the similar reading and spelling tests of the recently standardized British Ability Scales, 2nd edition (BAS-II).[26] This has enabled an alternative norm-group to be established for the ages 5 years 0 months to 8 years 6 months.

[25] Olisa and Campbell (1999).
[26] Elliott, Smith, and McCulloch (1996).

Table 10 gives standard scores for WRAT-3 raw score values for tests of reading and spelling for children of infant age currently in British schools. These norms may possibly have more relevance for a UK population than the norms provided in the WRAT-3 Manual.[27]

Passage Reading—the Neale Analysis

Turning next to the reading aloud of connected prose, we have the Neale reading test, already mentioned (Neale, 1997). This requires the primary age school pupil to read aloud passages of increasing difficulty from a realistic-looking book. The test is normed only from 6 years 0 months to 12 years 11 months, and should not be used with children older than this. Reported findings of "RA > 12:10" are merely evidence of half an hour wasted.

Administration is complex and requires the examiner, not only to time each passage, but to prompt the reader after each error or "after several seconds" if the child hesitates or attempts unsuccessfully to decode a word. Passages are timed from the first word spoken to the last and the time recorded in seconds. The *rate* is the total number of words read (below the ceiling passage) divided by the number of seconds, then multiplied by 60 to give words per minute.

For the child who does not start at Level 1, the first read is the *basal passage* if no more than two errors are made; otherwise, the preceding passage is given (and the same criterion applies to this too). Comprehension questions are not asked for any passage on which 16 or more errors are made (20 errors on Level 6); nor is the time taken on this *ceiling passage* included in the total.

The test provides standard scores for accuracy, rate (fluency) and comprehension, percentile and age equivalent scores. All three should be reported, though only standard scores are suitable for analysis.

The test is identical in form to the 1989 Neale, and uses the same record forms, but has been given a large-scale, authoritative new standardization. The two parallel forms of the test (Form 1 and Form 2) were administered to a total of 3,474 pupils in mid-1996.[28] Reliabilities achieved are given in Table 11. It will be seen that reliability for the *rate* measure is unacceptably low, while *accuracy* usually exceeds *comprehension*.

Furthermore, a sub-sample of 303 pupils aged from 6:0 to 14:11 was concurrently administered both the British Ability Scales, 2nd edition (BAS-II; Elliott *et al.*, 1996) and the Phonological Assessment Battery (PhAB). The latter, usable as we have seen by teachers, allows a profile analysis in relation to the scores predicted by a child's concurrent performance on the Neale. The Manual (pp. 35–36) offers an analysis of just such predicted and observed scores for "John" who is "nearly nine." John's difference scores, or departures from expectation,[29] on

[27] A fuller account of the work that led to the development of these norms is in preparation.

[28] This is about twice the number of the standardization sample for the 1989 revision.

[29] Observed minus predicted scores, so that problem areas appear to the left of the axis, as in ipsative graphs.

Table 10. UK norms for children aged 5:0 to 8:6 for WRAT-3 Reading and Spelling

Ages (Y:M)	1	2	3	4	5	6	7	8	9	10	11	12	13	14	15	16	17	18	19	20	21	22	23	24	25
Raw scores on WRAT-3 Reading																									
5:0–5:3	65	68	70	72	74	77	79	81	83	86	88	90	92	95	99	102	106	109	113	116	120	123	127	130	133
5:4–5:6	61	63	66	68	70	73	75	77	80	82	85	88	92	95	97	99	101	104	106	108	110	113	115	117	119
5:7–5:9	56	59	61	64	67	69	72	74	78	81	84	87	89	91	94	96	99	101	103	106	109	111	114	116	119
5:10–5:11	55	58	60	63	65	67	71	74	77	80	82	85	88	90	93	95	98	101	103	106	108	110	113	115	117
6:0–6:3	50	54	57	61	64	67	70	72	75	77	79	82	84	87	89	91	94	96	99	101	104	106	108	11[illegible]	113
6:4–6:6			50	52	55	58	61	63	66	69	72	74	77	80	83	85	88	91	94	96	99	102	105	103	111
6:7–6:9				51	54	57	59	62	65	68	71	73	76	79	82	84	87	90	93	96	99	102	105	107	110
6:10–6:11					52	55	58	61	64	66	69	72	75	78	81	84	87	90	93	96	98	101	104	107	109
7:0–7:3						52	55	57	60	63	65	68	71	74	77	79	82	85	88	91	94	97	99	102	105
7:4–7:7						51	54	57	59	62	65	68	71	73	76	79	81	84	87	90	92	95	98	100	103
7:7–7:9							51	53	56	59	61	64	67	69	72	75	77	80	83	85	88	91	94	96	99
7:10–7:11											51	54	57	60	62	65	68	71	74	76	79	82	85	8[illegible]	90
8:0–8:6												50	53	56	60	63	66	69	72	75	78	81	84	87	90
Raw scores on WRAT-3 Spelling																									
5:0–5:3	80	83	85	87	90	92	95	97	100	102	105	107	109	112	114	117	119	122	124	126	129	133	137	14[illegible]	145
5:4–5:6	66	69	72	75	77	80	83	86	89	92	95	97	100	103	106	110	114	118	121	125	129	131	134	137	142
5:7–5:9	56	59	61	64	67	71	75	79	83	87	91	94	98	102	106	109	112	115	118	121	125	129	133	136	139
5:10–5:11	53	56	60	64	66	70	74	77	81	85	88	92	95	99	103	106	110	114	117	121	125	128	132	135	139
6:0–6:3	53	56	59	63	66	69	72	75	78	81	85	88	92	96	101	105	109	113	117	120	123	126	129	132	135
6:4–6:6	53	56	59	62	65	69	71	74	78	81	84	88	92	95	99	102	106	109	113	116	120	123	126	130	133
6:7–6:9	52	55	59	62	65	68	71	74	77	80	84	88	91	94	97	100	103	107	110	113	116	119	123	126	129
6:10–6:11	52	55	58	61	64	67	71	74	76	79	82	85	88	91	94	97	100	103	106	109	111	114	117	12[illegible]	123
7:0–7:3	52	54	57	60	63	66	69	72	76	78	81	83	86	89	92	95	98	100	103	106	109	112	115	11[illegible]	121
7:4–7:7	51	54	57	60	63	66	69	72	75	77	80	83	86	88	91	93	96	98	101	104	106	109	112	11[illegible]	117
7:7–7:9		53	57	59	62	64	68	71	74	77	80	82	85	88	90	93	96	98	101	103	106	108	111	115	116
7:10–7:11			51	56	60	64	67	70	72	75	78	80	83	85	88	91	93	96	99	101	104	106	109	112	114
8:0–8:6					50	53	56	59	61	64	67	70	73	76	78	81	84	87	90	93	95	98	101	10[illegible]	107

Table 10. *Continued*

Ages (Y:M)	26	27	28	29	30	31	32	33	34	35	36	37	38	39	40	41	42	43	44	45	46
Raw scores on WRAT-3 Reading																					
5:0–5:3	137	140	144	147																	
5:4–5:6	122	124	127	129	132	135	137	140	143	145	148	150									
5:7–5:9	122	124	126	128	131	133	135	138	140	142	144	147	149								
5:10–5:11	120	122	124	127	129	132	134	136	139	141	143	146	149								
6:0–6:3	116	118	120	123	126	128	131	134	137	140	143	146	148	150							
6:4–6:6	114	117	120	123	125	128	130	132	135	138	140	143	146	149							
6:7–6:9	113	116	118	121	124	127	129	132	135	137	140	143	146	148							
6:10–6:11	112	115	118	121	123	126	129	132	134	137	140	142	144	147	149						
7:0–7:3	108	111	114	117	119	122	125	128	131	134	137	139	142	145	148						
7:4–7:7	106	108	111	114	116	119	122	125	127	130	133	135	138	141	143	146	149				
7:7–7:9	102	104	107	110	112	115	118	120	123	126	128	131	134	137	139	142	145	147	150		
7:10–7:11	93	96	99	102	105	109	112	115	118	121	124	127	130	133	136	139	142	145	148		
8:0–8:6	93	96	99	102	104	107	110	113	116	118	121	124	127	129	132	135	138	141	143	146	149
Raw scores on WRAT-3 Spelling																					
5:0–5:3	148																				
5:4–5:6	146	150																			
5:7–5:9	143	146	150																		
5:10–5:11	141	144	146	148																	
6:0–6:3	138	140	144	147																	
6:4–6:6	137	140	143	146	149																
6:7–6:9	132	135	138	142	145	148															
6:10–6:11	126	129	132	135	138	141	143	146	149												
7:0–7:3	123	126	129	132	135	138	141	144	146	149											
7:4–7:7	120	123	125	128	131	133	136	139	141	144	147	149									
7:7–7:9	118	120	123	125	128	130	133	135	138	141	143	146	148								
7:10–7:11	117	120	122	125	127	130	133	135	138	140	143	145	148	150							
8:0–8:6	110	112	115	118	121	124	127	129	132	135	138	141	143	146	149						

Table 11. NARA-II parallel form reliability

	Accuracy	Comprehension	Rate
All ages	0.89	0.82	0.66
NARA-II parallel form reliability for two presentation orders			
Form 1	0.92	0.84	0.65
Form 2	0.87	0.78	0.67
NARA-II Cronbach's Alpha (internal) reliability			
Form 1	0.86	0.94	NA
Form 2	0.85	0.94	NA

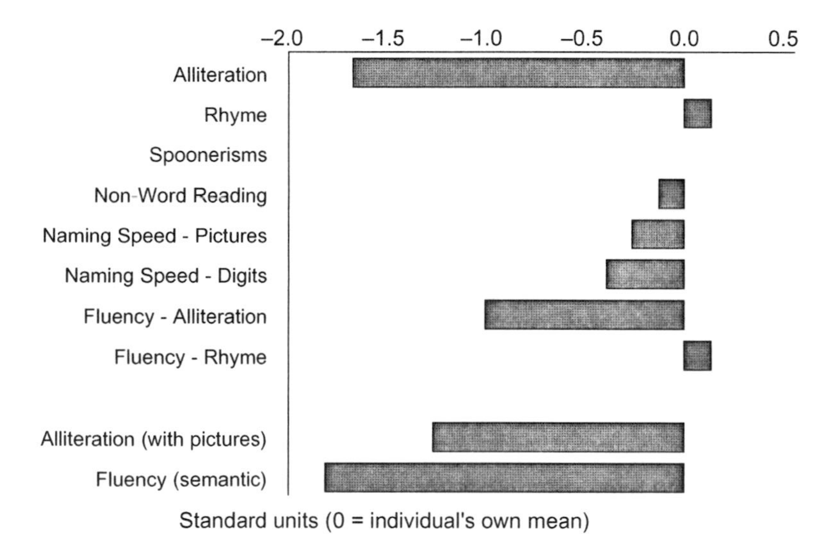

Figure 9. PhAB: Departure from expected scores for John (9).

10 PhAB subtests (in standard or z-score units) may be seen in Figure 9. John's scores on the Neale are given in Table 12. He appears to read quickly but not to understand very well what he is reading. On the PhAB his problems appear to center on alliteration and semantic fluency. The analysis suggested is that

> John's difficulties in comprehension arose from a reticence in answering questions, originating from low levels of self-esteem. This also contributed to his low scores in Alliteration and fluency, rather than these being a reflection of underlying phonological delay or deficit (Manual, p. 36).

Teachers who administer both the PhAB and the NARA-II to a child may find that some such additional *profile analysis* is able to cast new light on the literacy-learning difficulties that this child is having.

Table 12. John's Neale-II Scores

Accuracy	90
Comprehension	80
Rate	100

Interpreting the Educational Psychologist's Report

Thus far, we have adapted a psychological assessment model, originated by psychologists, for use by skilled teachers. Experience has shown that interested teachers, offered a greater degree of training, can perform such educational assessments successfully, providing to families and schools a useful and valued service that, because of the greater number of teachers, might not otherwise be accessible.

However, individual psychological assessments continue to provide influential input to the educational planning for children who have complex learning difficulties. Assessment reports circulate in photocopy and must be assimilated by the teachers responsible for designing and implementing specialist tuition for such children. How to grasp what they are saying and interpret any recommendations?

Of course such psychological assessment reports vary in quality, but the best of them will use precisely the methodology discussed in preceding sections. A firm evidence base of good quality psychometric test results, combined by a relevant developmental and school history, will comprehensively audit the child's areas of strength and weakness, leading to implications for remedial action.

In Britain, two major test batteries are most commonly in use:

1. The second edition of the British Ability Scales (already mentioned) and
2. The Wechsler Intelligence Scale for Children, 3rd edition (WISC-III).[30]

To these major batteries, which largely cover the two domains of *general ability* and *diagnostic* testing, are added tests of attainment in literacy and numeracy, together with any supplementary tests selected. More on these in a moment. It is not proposed to go into detail regarding the subtests that comprise these batteries, not to describe and give examples of the types of items used. An overview will suffice.

The WISC-III

Order has been produced out of the clinical chaos of the historic Wechsler tests, in that twelve subtests have now been grouped together on the basis of their factorial structure.[31] This has become a four-factor test:

1. Verbal Comprehension
2. Perceptual Organization

[30] Wechsler (1992a).

[31] Factor analysis permits a sort of controlled generalization, based on trends, mathematically analyzed, in large bodies of data on skilled test performance.

3. Freedom from Distractibility
4. Processing Speed.

The first two factors survey abilities that are mainly verbal and visuospatial; combined, they yield a Full Scale IQ, the traditional measure of general cognitive ability. The third and fourth, so-called *small* factors, are in effect diagnostic scales, useful in dyslexia assessment. Freedom from Distractibility (ungainly title) is elsewhere called the Working Memory factor (digit span and mental arithmetic), which clearly shows the importance it has in diagnostic work. Processing Speed becomes more useful with older children who remain poor spellers. We have all met the kind of dyslexic child who appears to get easily *congested* with information, so slowly does he or she process it.

Also part of the Wechsler family of tests (but sold separately) are the WORD, WOLD, and WOND tests of achievement (Rust, Golombok, & Trickey, 1993; Rust, 1996a, 1996b).[32] These are all Wechsler Objective Dimensions, but the letters that vary stand for Reading (and spelling), Language (oral), and Number (reasoning and operations). It is the three literacy tests of WORD that are most used:

1. Basic Reading
2. Reading Comprehension
3. Spelling.

The apparatus provided with WISC/WORD enables a *discrepancy analysis* of underachievement in terms of differences between scores expected and scores observed, followed by evaluation of statistical significance and prevalence. The second criterion for dyslexia, evidence of *processing deficits*, is made possible by a similar apparatus for evaluating unusual discrepancies between general ability and working memory/processing speed.

Though it is quite reasonable to stay within the Wechsler system, in fact most psychologists add other tests, especially diagnostic ones, to complement the basic battery.

Since the WISC virtually manufactures "verbal-performance discrepancies," it is important for teachers not to get diverted into a fruitless discussion of a possible imbalance in cognitive development. As the leading expert on the WISC wrote about the previous edition,

> ... one out of three normal children has a significant V-P IQ discrepancy at the 5% level (12 or more points) (Kaufman, 1976b), and the average scaled score range (highest minus lowest scaled score) for the Full Scale is a substantial seven points (±2) (Kaufman, 1976a).[33]

[32] In the United States, these are a single entity called WIAT (Wechsler, 1992b).

[33] Kaufman and Ishikuma (1993, p. 195).

It is wise to seek a pediatric occupational therapy assessment if a child's difficulties suggest *dyspraxia*, a difficulty in the planning and execution of certain movement skills. Overall slowness of performance, and a pencil grip too tight or too slack, are two such signs. But the interpretation of a P > V (Performance greater than Verbal) discrepancy as grounds for possible "visual dyslexia" seems unproductive.

The British Ability Scales

When the first edition of this test appeared in 1980, it moved forward the frontiers of psychometric assessment in Britain and contained many features, such as interval scaling and discrepancy analysis, that have now been taken up by all new tests. BAS-II contains tests of Word Reading, Spelling, and Number Skills that are integral to the battery and offers three factors:

1. Verbal
2. Nonverbal Reasoning
3. Spatial.

This represents a further subdivision of visuospatial ability into purer measures of visualization and the kind of linear, step-by-step nonverbal reasoning that entails the management of sequences in verbal memory. Perhaps for this reason, this middle factor has considerable diagnosticity. Like the WISC, the battery contains diagnostic tests and offers precise discrepancy analysis to isolate specific learning difficulties (SpLD).

Understanding Common Phraseology

Finally, it may be useful to survey some common phraseology used in psychologists' assessment reports. Many phrases and expressions recur that may be taken for granted in the world of special education but which are regarded as jargon by parents and others outside it.

Table 13 gives many common examples, together with an interpretation of each and an indication of the context in which each skill or difficulty may be manifested.

Summary

In this final section, we have drawn many of the threads together in an examination of the practical test choices that face specialist teachers, together with a brief look at the work of psychologists with which they are likely to become familiar.

Test choices are governed by considerations of test quality (which often means modernity) and simplicity (ease of use). Four of the most powerful basic tests available to teachers, three of them in effect batteries (seventeen different

Table 13. Psychological terminology in reports

Skill	Means	May cause difficulties in ...
Short-term auditory memory	Cannot hold information while processing it	Mental arithmetic, multiplication tables, learning by heart, following instructions, spelling, remembering what he/she has heard, attentive listening.
Visual memory	Remembering shapes or patterns.	Checking spelling, look-and-say reading, copying shapes.
Auditory sequencing	Managing sequential order in material heard.	Oral spelling or tables, alphabet and reference skills, following sequence of instructions.
Visual sequencing	Organizing symbols or shapes in order	Spelling (especially irregular words), copying, arithmetical routines, some aspects of CDT.
Visual-motor skill (or hand–eye coordination)	Coordination of vision with movement.	Handwriting, ball skills, PE; clumsiness—may spill food, paint.
Visuo-spatial ability	Perception of objects in space, position, distance, speed, abstract form.	Page layout, aspects of handwriting, relative size, map work, shape work in maths.
Listening or auditory comprehension	Understanding the spoken word.	Following instructions, attending to any verbal material, including stories; distractible, low attention span, easily confused, slow to learn, especially literacy. May depend on following others, looking at gesture, pointing.
Auditory discrimination	Hearing fine differences between sounds.	Even when this is said to be okay there may still be difficulties with sounds of speech—rhyme, sounds within words (segmentation)—that are essential for literacy.
Phonological awareness	Perception of sounds within words.	Sequence of sounds in words; beginnings and endings of syllables; rhyme; alliteration; identification of individual sounds or blends.
Speed of information processing	Mental or clerical speed.	Inefficiency with the management of streams of information, especially written symbols, as in copying.

measures in total) have been briefly reviewed. However, the discussion here, which highlights points of interest, will be no substitute for the beginner for a close reading of the appropriate test manuals.

Finally we have arrived at the sphere of practical action with an examination of the major instruments in use by that other profession, close to the specialist teacher, of educational psychologist (WISC-III, BAS-II). Psychologists and specialist teachers often work together in creative partnership, sharing a common discipline and methodology, of which this chapter has been a close examination.

GLOSSARY OF TERMS USED

Confidence interval: This is the range of values within which the true score on a test is likely to fall, taking account of inevitable measurement error. If the standard error of measurement (SE_m) of a test is known, then the confidence interval will be the observed score plus or minus the SE_m. For instance, an individual obtains a score of 95 on a test for which the standard error of measurement is 4. Then the confidence interval is 95 +/−4 or 91–99. Every observed score should be interpreted with this penumbra of caution.

Discrimination: The ability of an item at a given level of difficulty to divide equally those who succeed and those who fail.

Interpretation: The process whereby the result of a test expressed in numerical form is given its statistical meaning.

Item: The basic component of a test, a question to be answered or a problem to be solved.

Item statistics: The body of statistical knowledge that enables observed scores to be adapted to the frequencies and distributions of a desired statistical model.

Normal distribution: A statistical assumption, made for convenience of the model, that there are fewer individuals at the extremes of any distribution (height, weight, intelligence), but many more clustered around the average. Visually this gives the Bell Curve.

Percentile: This is another way of expressing the standard score. It helps to conceptualize the meaning of scores if we say what *percentage* of individuals are found at that level or between two levels. Indeed, for every standard score in the normal distribution, there is an associated *percentile* value. This is a great help in communicating the results of tests to teachers and parents who are not, as a rule, familiar with standardized quotients. The percentile is the percentage of individuals *below* a given raw score, plus 50% of individuals *with* this raw score.

Reliability: A test should give practically the same result if an individual takes equivalent forms of it or takes the test twice. If a test is unreliable, the results are not worth collecting. Reliability can be empirically determined and precisely quantified, either from the patterns of actual responses to its items (internal reliability) or from the results of administering the test twice (test-retest reliability). Reliability is measured on a scale from 0 to 1, with values for a psychological test below 0.80 rendering the test professionally unacceptable. This is the reliability coefficient.

Quotient: "A result obtained by dividing one quantity by another." *Concise Oxford Dictionary*.

Raw score: This is the total number of items passed (or marks obtained) by the individual taking the test. As yet the score is not transformed or scaled or expressed in a standard form.

Sample: A small number of selected subjects about whose behavior test information is sought, because they reflect accurately a wider population.

Scaling: The use of selected items with appropriate statistical properties to form a scale of measurement.

Score: The result of a test, usually a quantity. A raw score is the number of responses accepted as correct; a scaled or standard score is one converted to standard form to permit the desired comparison.

Standard deviation: A statistic expressing in a universal form the amount of scatter, or dispersion, in a set of scores. In words, it is the root mean squared deviation from the mean.

Standard error of measurement (SE_m): This depends directly upon the reliability coefficient of a test, from which it is derived by a formula, and is normally expressed in the same units used by the test, for instance points of standard score.

Standardization: The process of giving a test to a sample of individuals in a standard way, according to exactly specified rules, with responses judged according to common criteria, so that the information obtained may form the basis of controlled comparison for other, comparable individuals.

Standard score: There are many varieties of standard score with different attractions and conveniences. But all standard scores always express an individual's standing in relation to his or her age group.

***T*-score**: A sensitive form of standard score with a mean of 50 and a standard deviation of 10. The latter makes it easily divisible in the mind.

Trial: Administration of items from a test in the design stage to selected candidates of the intended group to permit their evaluation and refinement.

Wechsler or WISC subtest scaled score: Another common form of standard score used in psychologists' reports, especially those that report results using one or other of David Wechsler's tests (WISC, WAIS, WPPSI). This has a mean of 10 and a standard deviation of 3.

***z*-score**: The pure form of standard score, with a mean of 0 and a standard deviation of 1. All other forms of standard score can be converted to and from this form. It is thus the common language of ability measurement. When someone is described as falling a *standard deviation* above or below the average, or a research effect is expressed in *standard deviation units*, this is the convention that is being used. *z*-scores are also useful as the basis of graphs that convey an immediate intuitive meaning.

REFERENCES

Bartram, D. (1990). Reliability and validity. Chapter 3. In J. R. Beech, and L. Harding (Eds.) *Testing people: A practical guide to psychometrics*. Windsor, Berks: NFER-Nelson.

Brand, C. R. (1996). The importance of intelligence in Western societies. *Journal of Biosocial Science 28*, 387–404.

Carmines, E. G., & Zeller, R. A. (1979). *Reliability and Validity Assessment*. California: Sage.

Cohen, D. B. (1999). *Stranger in the nest: Do parents really shape their child's personality, intelligence or character?* New York: Wiley.

Cooper, C. (1999). *Intelligence and Abilities*. London: Routledge.

Cronbach, L. J. (1990). *Essentials of psychological testing* (5th ed.). New York: HarperCollins.

Dunn, Ll. M., Dunn, L. M., Whetton, C., & Burley, J. (1997). *British picture vocabulary scale* (2nd ed.) (BPVS-II). Windsor, Berks: NFER-Nelson.

DuPaul, G. J., Anastopoulos, A. D., Shelton, T. L., Guevremont, D. C., & Metevia, L. (1992). Multimethod assessment of attention-deficit hyperactivity disorder: The diagnostic utility of clinic-based tests. *Journal of Clinical Child Psychology, 21*, 394–402.

Elliott, C. D., Smith, P., & McCulloch, K. (1996). *British ability scales second edition* (BAS II). Windsor, Berkshire: NFER-Nelson.

Frederickson, N., Frith, U., & Reason, R. (1997). *Phonological assessment battery* (PhAB). Windsor, Berkshire: NFER-Nelson.

Frith, U., Landerl, K., & Frith, C. (1995). Dyslexia and verbal fluency: More evidence for a phonological deficit. *Dyslexia, 1*(1), 2–11.

Goldstein, S., & Goldstein, M. (1998). Managing attention deficit hyperactivity disorder in children: A guide for practitioners, (2nd ed.). New York: Wiley.

Hedderly, R. G. (1995). The assessment of SpLD pupils for examination arrangements [pp. 12–16]; and *Sentence Completion Test* [pp. 19–21]. *Dyslexia Review*, *7*(2).

Horn, J. L., & Cattell, R. B. (1966). Refinement and test of the theory of fluid and crystallized intelligence. *Journal of Educational Psychology, 57*, 253–270.

Jensen, A. R. *The G factor*. (1998). New York: Praeger.

Kaufman, A. S. (1976a). Do normal children have "flat" ability profiles? *Psychology in the Schools, 13*, 284–285.

Kaufman, A. S. (1976b). Verbal–Performance IQ discrepancies on the WISC-R. *Journal of Consulting and Clinical Psychology, 44*, 739–744.

Kaufman, A. S., & Ishikuma, T. (1993). Intellectual and achievement testing. In T. H. Ollendick, & M. Hersen (Eds.) *Handbook of child and adolescent assessment*. Needham Height, Massachusetts: Allyn and Bacon.

Naglieri, J. A. (1985). *Matrix analogies test—Short form (MAT)*. Columbus, Ohio: Charles Merrill.

Neale, M. D. (1997). *Neale analysis of reading ability—Revised (NARA-II*; Second Revised British Edition: Standardization by C. Whetton, L. Caspall, and K. McCulloch) Windsor, Berkshire: NFER-Nelson.

Neisser, U., Boodoo, G., Bouchard, T. J., Boykin, A. W., Brody, N., Ceci, S. J. *et al.* (1996). Intelligence: Knowns and unknowns. *American Psychologist, 51*, 77–101.

Olica, J., & Campbell, S. (1999). The literacy assessment profile: A dyslexia screening approach for teachers. *Dyslexia Review, 11*(1), 12–15.

Pascal, B. (1623–1662), (1966). *Pensées* (no. 534/5, p. 217) In A. J. Krailsheimer (Ed. and trans.). Harmondsworth: Middlesex Penguin Books.

Rosner, J. (1993). Test of auditory analysis skills. In *Helping children overcome learning difficulties* (3rd ed.). New York: Walker.

Rust, J. (1996a). *Manual.* Wechsler Objective Language Dimensions (WOLD), London: The Psychological Corporation (Harcourt Brace and Company).

Rust, J. (1996b). *Manual.* Wechsler Objective Numerical Dimensions (WOND), London: The Psychological Corporation (Harcourt Brace and Company).

Rust, J., Golombok, S., & Trickey, G. (1993). *WORD: Wechsler objective reading dimension*. Sidcup, Kent: Harcourt Brace Jovanovitch, The Psychological Corporation.

Sacre, L., & Masterson, J. (2000). *Single word spelling test* (SWST). Windsor, Berkshire: NFER-Nelson.

Salvia, J., & Ysseldyke, J. E. (1998). *Assessment in special and remedial education* (7th ed.). Boston: Houghton Mifflin.

Sattler, J. M. (1988). *Assessment of children* (3rd ed.). San Diego, California: Jerome M. Sattler.

Snow, R. E., & Yalow, E. (1982). Education and intelligence. In R. Sternberg (Ed.) *Handbook of human intelligence* (pp. 493–585). Cambridge: Cambridge University Press.

Turner, M. (1999). A note on test restriction. *Dyslexia Review, 11*(1), 10–11.

Wechsler, D. (1992a). *Wechsler intelligence scale for children—Third edition UK* (WISC-III). New York: Harcourt Brace Jovanovitch, The Psychological Corporation.

Wechsler, D. (1992a). *Wechsler individual achievement test* (WIAT). San Antonio, Texas: The Psychological Corporation.

Wilkinson, G. S. (1993). *The wide range achievement test*—3rd Edition (WRAT-3). Wilmington, Delaware: Wide Range.

10

Dyslexia and Self-Esteem

Jacky Ridsdale

INTRODUCTION

In our society, the association between bad spelling and stupidity is so strong that it is almost taken for granted. A misspelled public notice, for example, is a trigger for laughter and derision; it is commonplace for jokes and cartoons to be based on that association. The implications of this for the thoughts and feelings of bad spellers in our society are clear. It is no surprise then if we find that those who have poor literacy skills also have more widespread feelings of intellectual and social limitation.

This chapter will be exploring this relationship in more detail: looking at theories of self-concept and self-esteem, the relationship between these and academic attainment, the relationship between dyslexia, self-esteem and behavior difficulties and what sort of interventions may enhance the self-esteem and adjustment of dyslexic students.

GLOBAL SELF-ESTEEM AND DYSLEXIA

The issue of whether suffering from dyslexia inevitably lessens your core, or global, sense of who you are—or whether it just makes you feel bad about your

---

Jacky Ridsdale, The Dyslexia Institute, Broom Hall, 8–10 Broom hall Road, Sheffield S10 2DR, UK.

The Study of Dyslexia, edited by Turner and Rack.
Kluwer Academic Publishers, New York, 2004.

school performance—will be addressed in this section. The related question of whether how good you feel about how you do in a subject actually determines to some extent how well you *actually* do will also be addressed. This is a very complex area of theory and research. There is no absolute consensus here that can be easily summarized, but there are findings that can illuminate our thinking. One of the first ideas that is necessary to consider in more depth is how we develop our sense of self at all. Are we born as a "self"? Or is this something we become—and how does that happen?

This section will thus address topics of:

- The concept of self
- Dyslexia and academic/global self-esteem
- Reciprocity between attainment and self-esteem.

The Concept of Self

When we are born, we do not know who we "are." We do not know even that we "are." We start life as an almost indistinguishable part of our mother's body and have no realization of ourselves as an independent being, no "self." Some authors (Kohut, 1977; Maslow, 1954; Rogers, 1961) would say that the "making of a self" is *the* fundamental human passion and need. It is certainly a vital part of human development.

Fundamental self-concept, our sense of self, could be said to be made up of two types of self, the existential and the categorical. The fact that we are able to introspect, to think, about ourselves, implies that there *must* be two types, a sort of "thinker about," an "I," which we can classify as our *existential* self (subjective self or self as agent; Harré, 1979) and one "thought about," a "me" (James, 1890, 1983) or categorical (objective) self whom we "present" to others (Schlenker, 1980); the latter is to some extent created by our existential self. If we want to change someone's view of oneself, we have to change the way the "existential" self thinks and feels about the "categorical" self. To do this, we may have to create new categories altogether, or at least modify existing ones. The self, both types, develops over time through the interactions between the individual and the social world that surrounds that individual.

The idea of self-concept as it has been introduced would suggest that this is a simple and well-established entity. This is not, however, the case. Some (Mruk, 1999) would suggest that the self is a "real" thing. That it is possibly the "fundamental structure of our existence," a stable, central, holistic identity. By contrast the post-modernist view espoused by Katz (1996) would suggest that "self" is no more than a narrative.

If we deliberately abandon attempts to define self fully and adopt a pragmatic and modifiable way of thinking about the self as a hypothetical construct, this allows us to study people in terms of their self-concept without getting stymied by a philosophical debate as to whether it exists. Four possible ways in which hypothetically constructed models of self-concept may be classified are as follows: nomothetic, multifaceted/hierarchical, taxonomic, and compensatory (Burden, 1998; Byrne, 1984). Burden concludes that the model of self-concept best supported by empirical evidence is the multifaceted/hierarchical model of Shavelson and Bolus (1982). This model suggests a general self-concept at the apex of a pyramid, which is then split down into academic and nonacademic self-concept. On the next level down, these two are split again: academic self-concept into subject areas such as literacy and numeracy, nonacademic into areas such as emotional, physical, and social. Theoretically there is no particular level at which these splits need cease to occur. This model, flexibly interpreted, seems able to encompass the other three and will be adopted throughout the chapter.

Dyslexia, Academic/Global Self-Esteem

The Shavelson and Bolus (1982) model of self-concept, as discussed, postulates a pyramid structure for self-concept with a sizeable apex representing global self-concept. The next layer down, structurally and in terms of central importance, is divided into two—academic and nonacademic, both of which are then divided into narrower areas.

If we accept that self-esteem is an intrinsic, isomorphic dimension of self-concept, that is, it can be "mapped" onto it, then this implies that academic self-*esteem* can also be differentiated from global self-*esteem* and into specific academic areas. Research has been devoted to the relationship between academic self-concept, academic attainment, academic self-esteem, and global self-esteem. Much of it is difficult to evaluate comparatively because of inconsistency in nomenclature (Bear, Minke, Griffin, & Deemer, 1997; Burden, 1998). The tenor of the work makes it clear, however, that what is under consideration in all cases is how someone thinks and feels about himself or herself. It is beyond the scope of this chapter to explore with rigor the derivation in each case of either the terminology used or the measures of self. Given this caveat, the almost universal finding is that there *is* a relationship between self-esteem and academic attainment, (Bandura, 1989; Burns, 1982; Byrne, 1984; Chapman, 1988; Harter, 1983).

Clearly, academic attainment encompasses much more than literacy. Our interest, however, is primarily in those people with a dyslexic difficulty, unfortunately yet another area where nomenclature is confusing. Learning Disability (LD), Specific Reading Retardation (SRR), Specific Learning Difficulties (SpLD), are but a few examples of alternative terms used for dyslexia in the research. This chapter will use the terms dyslexia or dyslexic at all times for the sake of clarity and will assume the definition of dyslexia offered by the Dyslexia Institute (2002; www.dyslexia-inst.org.uk).

Factors that emerge as important in determining the relationship between dyslexia and global, or "apex" self-esteem include gender, developmental stage, past academic achievement, and parental attitude. In addition, several hypothetical constructs relating to motivation seem vital in the dynamic of this relationship.

These are discussed in a later section. Most research outcomes tend to support the following observations:

1. Dyslexia influences academic, rather than global, self-esteem.
2. The relationship between academic achievement and self-esteem is not one way.
3. In this relationship, it is academic achievement that is the dominant feature.
4. There is an interactive relationship between dyslexia, poor self-esteem, and emotional and behavioral difficulties.
5. Appropriate structured literacy input can have an impact on both achievement and self-esteem.

Chapman (1988) reviews much work in the area of dyslexia and self-esteem and concludes that dyslexic children tend to have lower general and not just academic, self-concepts than their peers. He suggests, however, that by and large this does not indicate self-esteem so low as to be dysfunctional. He also suggests that dyslexic students may establish markedly lower than normal academic self-esteem levels by School Year Three and that these remain relatively stable throughout school. On a more positive note, dyslexic students are often able to compensate for their loss of self-esteem through acknowledged success in nonacademic or extracurricular activities. Specialized literacy support assistance also impacts positively on the academic self-esteem of dyslexic pupils (Elbaum & Vaughn, 1999).

The work of Marsh (Marsh, 1992; Marsh, Byrne, & Shavelson, 1988) is in contrast to that of Chapman. Rather than looking "up" to global self-esteem, this looks "down" the pyramid model to more narrow divisions of self-concept and their related self-esteem. Marsh suggests that "academic self-concept" *per se* may be a misleading term because further subject specificity is necessary in order to gain valid insight. He delineated eight subject areas. This work implies that in looking at the academic self-esteem of dyslexic students, we should perhaps be considering it in relation to literacy achievement specifically.

As the vast majority of the school curriculum is literacy based, however, a case could therefore be made that as literacy skills are so pervasive, failure here and concomitant low "literacy" self-esteem, will impact all, or most, other areas of academic self-esteem. In addition, reference should be made to the point at the beginning of the chapter regarding the almost universally held, although not always overtly expressed, belief, within our culture, that people who cannot read and spell adequately are lacking in intelligence and education. These two factors suggest that the assertion that dyslexia impacts on only a narrow area of subject-specific academic self-esteem needs, at the very least, much qualification.

Another point of relevance here concerns the measures of self-esteem commonly used. The misleading range of terminology in this field has been mentioned. In addition, it should be noted that we are seeking to access what may be private, painful, and only half-acknowledged self-beliefs through, for the most part, the blunt and insensitive approach of pencil and paper questionnaires (Marsh, Craven, & Debus, 1999). In fact, Watson (2002) would go so far as to suggest that questionnaires, or even the use of words at all, may be not nearly sufficiently sensitive to access the internal schemata that drive and permeate so much of our behavior. To maintain, in the face of these reservations about methodology, terminology, and validity, in addition to the above argument about cultural significance, the premise that dyslexia impacts mainly on academic, rather than global self-esteem, seems contentious.

It is certainly the experience of many practitioners that the sense of failure and concomitant low self-worth endured by some dyslexic individuals ranges far beyond the purely academic. Smith and Nagle (1995) found dyslexic children tended to perceive themselves as less competent than their peers not only in academic skills and intelligence, but also in behavior and social acceptance. Recent unpublished research (Ruddock, 2000), unpublished research paper, Dyslexia Institute found that dyslexic adults reported exceptionally low self-esteem and harrowing recollections of failure at school. Maughn (1995) would suggest that it is because these adults have remained reading disabled that they have such low self-esteem. She feels that childhood reading problems, if confined to childhood, need not have an impact on adult self-esteem. The corollary of this is that if the literacy problem is not confined to childhood, the effect can be devastating. As Gross (1997, p. 1) puts it: "This kind of chronic disappointment can translate psychologically into a generalized sense of diminished value and potential, where children's inner sense of adequacy is seriously compromised."

The Relationship between Attainment and Self-Esteem

The relationship generally between academic attainment and self-esteem is so well established that even the Code of Practice (2001) acknowledges that poor attainment can lower self-esteem. One issue still under consideration, however, is whether the relationship between self-esteem and academic achievement does in fact work both ways, that is, whether how you feel and think about yourself impacts on your progress in any academic field, one example being, acquiring fluent literacy skills.

Numerous studies have examined this relationship in general terms. Most have found it indeed to be reciprocal but with the causal preponderance falling on the attainment side. This means that there *is* a two-way relationship, but that academic attainment tends to be the most important factor (Hamachek, 1995; Marsh & Yeung, 1997; Muijs, 1997). Many factors, including age and gender, parental attitude, actual attainment, need for approval, and social or sporting skills, impact on this relationship.

Chapman, Lambourne, and Silva (1990) evaluated the parental role in generating academic self-esteem and found it to be minimal in comparison to actual past academic achievement. This is very much in contrast to earlier work (Coopersmith, 1967; Song & Hattie, 1984), who attribute much influence on academic self-concept to family characteristics and parenting style. Clearly different parents attach different worth to academic attainment and it may be of significance that the Coopersmith study, for example, was undertaken with boys from families socioeconomically defined as middle class.

Marsh *et al.* (1999) argued that as the children matured, the relationship between academic attainment and self-esteem both lessened and became more subject specific. Burden (1998) found this relationship to be strongest at the extremes of the achievement range, weakest in the middle.

An interesting complicating factor in this relationship is that of either defensive or compensatory self-concept. Alvarez and Adelman (1986) and Lobel and Teiber (1994) found, for example, that the nature of the impact of failure on self-esteem was related to the individual's need for approval. In "approval needy" individuals, failure could ultimately lead to a defensive, self-protective, apparent overestimate of ability to succeed, an inflated self-concept. This is a hypothetical construct, described as a "self-serving bias" (Bentall, Kinderman, & Kaney, 1994).

With reference to dyslexic children, Kloomok and Cosden (1994) and Hagborg (1996) found that some dyslexic children maintained healthy self-esteem in the face of academic failure by apparently perceiving themselves as successful in other areas, for example, sport, or perceiving themselves as enjoying high levels of social support. These are examples of compensatory self-concept, self-protective measures. Other dyslexic children (Feick & Rhodewalt, 1997) were found to have proactively protected their self-esteem by claiming "self-handicap" before undertaking tasks where failure was a likely outcome, that is, adopting a stance of predicting failure as a certain outcome because of insuperable difficulties within themselves.

These cognitive activities—either recognizing one's strengths in areas beyond the academic, holding an inflated estimate of one's ability, or stoutly predicting failure—represent three different ways of dealing with the painful impact of experiencing real failure: compensating, denying, or self-handicapping. It is clear that whereas compensation, of the type described above, is adaptive and beneficial, in contrast, self-serving denial, or self-handicapping, are not, in that neither of these latter stances is likely to promote further effort at what has proved to be a daunting task.

The above research suggests that there *is* reciprocity in the relationship between attainment and self-esteem. It also suggests that in their attempts to deal with the discomfort of failure, some dyslexic students develop adaptive psychological defenses. Others may maladaptively distort their internal constructions of reality. In addition, they may develop more overtly dysfunctional behaviors. In order to present at least a "front" of high self-regard, they may become noncompliant with a system that invalidates them on a daily basis. There is evidence to support the existence of a reciprocal relationship not only between dyslexia and poor self-esteem, but also a third factor, maladaptive behavior.

So it would seem that there is evidence that dyslexia could lower global self-esteem and could have an impact on the child or adult's perceptions of themselves in a negative way. This relationship is not universal, however, and the findings from research in themselves generate further questions. We need to know, for example, what it is that makes some dyslexic individuals resilient in dealing with their lack of success, while others show emotional and behavioral difficulty.

DYSLEXIA AND EMOTIONAL AND BEHAVIORAL DIFFICULTIES

This section will attempt to answer questions as to how dyslexia and emotional difficulties seem to be related and explore just how serious those difficulties can be. It will cover the following topics:

- dyslexia, low self-esteem, and emotional and behavioral difficulties
- dyslexia and conduct disorder
- dyslexia, anxiety, withdrawal, and depression
- dyslexia and delinquency.

Dyslexia, Low Self-Esteem, and Emotional and Behavioral Difficulties

The relationship between low self-esteem and emotional and behavioral difficulties is well established (DfE, 9/94, Lund, 1986; Maines and Robinson, 1995; Margerison, 1996; Rosenberg, Schooler, & Schoenbach, 1989). The link is complex, dynamic, and reciprocal, and not easily disentangled. When we throw academic failure, in the form of dyslexia, into this already volatile mixture, we add a potent catalyst. Given the previous discussion of the relationship between attainment and self-esteem, we can hypothesize that dyslexia may often be found to be the prime factor in generating both poor self-esteem and, emotional and behavioral difficulties. It can, however, also enter an already troubled relationship between behavior and poor self-esteem and make matters considerably worse. The literature in this area suggests that, largely through the agency of poor self-esteem, dyslexia may be associated with both difficult, noncompliant and withdrawn, neurotic, depressive behavior. In addition, there may be comorbidity between dyslexia and delinquency. Generally, in the relationship between dyslexia and maladaptive behavior, dyslexia is causally implicated.

Dyslexia and Conduct Disorder

Much of the research in the area of dyslexia and emotional and behavioral difficulties have focused on difficult and challenging noncompliant behaviors rather than on internalizing neurotic behaviors (Beitchman & Young, 1997). The reasons for this bias in research interest are not hard to surmise given the large class sizes faced by most teachers and the disruption to those classes a disaffected, challenging, and noncompliant child can achieve. Bender and Golden (1988), Frick *et al.* (1991); Heiervang, Lund, Stevenson, & Hugdahl (2001); Hinshaw (1992), Maughan, Pickles, Hagell, Rutter, & Yule (1996); Rutter and Yule (1970); Willcutt and Pennington (2000), have all found evidence of a relationship between dyslexia and difficult, externalizing behavior, although Fergusson and Lynskey (1997) found conduct problems at first apparently associated with dyslexia actually to be predicated by early onset conduct disorders, which had in turn in some cases hampered literacy acquisition.

Haager and Vaughn (1995) found a tendency for dyslexic children to be rated by class peers as having lowered social status, implying less than average pro-social skills. Brown and Heath (1997) found that even when their peers rated them as being of normal social status, they could be viewed by their teachers as having poor social skills and more behavioral problems. There is clearly evidence to suggest that many dyslexic children find relating to their peers, their teachers, and the day-to-day demands of the school system in a positive way difficult. Some respond with challenge and nonconformity, others with withdrawal, anxiety, and even depression.

Dyslexia, Anxiety, Withdrawal, and Depression

Nabuzoka and Smith (1993) found dyslexic children less likely to be judged popular by peers and in addition, tending to be shy, help seeking, and victims of bullying. Given the above catalog of social and emotional difficulty, it is perhaps no surprise that some dyslexic children have been found to be mildly depressed as well as low in self-esteem (Palladino, Poli, Masi, & Marcheschi, 2000). Dyslexic children have in fact been found by some researchers to have increased levels of anxiety, concentration difficulties, and difficulty in problem solving, in addition to social immaturity and depression (Fisher, Allen, & Kose, 1996; Jorm, Share, Matthews, & MacLean, 1986; Klasen, 1972; Livingston, 1990; Richman, Stevenson, & Graham, 1982; Stanley, Dai, & Nolan, 1997; Willcutt & Pennington, 2000). Miles (1996) suggests that an important characteristic of some dyslexic children's inner worlds, their thoughts, and feelings is that they feel frightened, of failing, of being different, of words, of social gaffes. This supposition is certainly borne out by the retrospective reporting on their school days of some dyslexic adults (Berg, 1989) from which the following is a quotation. The individual was asked to read aloud:

> I was the worst in the class. I used to sit and shake, petrified, I would just say please God, please God don't ask me.

Prior, Smart, Sanson, and Oberklaid (1999) found behavioral maladjustment, of both an externalizing and an internalizing type, to be associated with specific learning difficulties. One can hypothesize, for the reasons delineated in the previous discussion of the relationship between academic success, or lack of it, and self-esteem levels, that it is highly likely that low self-esteem is a crucial mediating factor.

What is clear is that a considerable body of research indicates a relationship between dyslexia, other SpLD, developmental disorders, and emotional and behavioral difficulties (Frick *et al.*, 1991; Maughn, 1995; Tomblin, Zhang, Buckwalter, & Catts, 2000). The exact nature of this relationship remains uncertain. This is unsurprising as the relationship between emotional and behavioral difficulties and almost any other precipitating, causal, or preventative factor is also uncertain.

There is evidence, however, that suggests that not all dyslexic children are equally vulnerable to emotional and behavioral difficulty. Gilbert (1992) suggests that there are heritable characteristics that promote higher or lower self-esteem. Some dyslexic children may therefore be protected by their genetic propensity. Two further mitigating factors may be found in general ability and socioeconomic status (SES) of parents. Hales (1987), for example, found that over time dyslexic children of higher intelligence are at less risk of behavior disorder than their less intelligent peers. Hales proposes that, in addition to the generally protective influence of having higher than average intelligence, they face a "more sympathetic world," with regard to their dyslexia, whereas those of lower IQ may find their specific difficulties in literacy simply ascribed to a lack of intelligence, a frustrating and demoralizing situation. Strehlow, Kluge, Moller, and Haffner (1992) found raised education levels and SES of parents to be protective factors long term, enhancing the prospect of a college education for dyslexic students. All the previously mentioned factors that promote compensatory self-esteem will also be protective, as will secure attachment to primary caregivers (Hagborg, 1996). For the most vulnerable and least protected and resilient dyslexic youngsters, however, the future may not hold a college education but, instead, the possibility of increasing levels of behavioral difficulty, truancy (O'Keeffe, 1994), delinquency, a criminal record and, in some cases, incarceration.

Dyslexia and Delinquency

Delinquency will be taken to mean being apprehended by law enforcement officers for behavior that is socially deviant to the extent of involving law breaking and criminal activity.

Once again the picture is complex and unclear. One common source of confusion is the unwarranted popular assumption that illiteracy and dyslexia are synonymous. The prevalence of illiteracy in the delinquent population (Hogenson, 1974) seems to have led to a journalistic and unsubstantiated assumption that there are huge numbers of dyslexic criminals and that dyslexia is a major predisposing factor in crime. This lack of clarity is pointed out by Turner, Sercombe, and Cuffe-Fuller (2000). In Turner's study, it was found that the average IQ of the 97 young male inmates studied was 77, strikingly low, but in line with previous findings *vis-à-vis* young offenders (Herrnstein & Murray, 1994). This finding suggests that general, rather than specific, learning difficulties are what are *most* prevalent in the population of young offenders, although those of lower ability may of course also be dyslexic in addition to having general difficulty in learning.

Hoge, Andrews, and Leschield (1996) established that the main risk factors in delinquency were dysfunctional family relationships and parenting problems. It is of note though that they found good school achievement to be a protective factor. This shield, however, is one that would clearly be lacking in an under-attaining dyslexic child who was, perhaps, already at risk of becoming delinquent by virtue of other predisposing factors.

Dodds *et al.* (1993) indicate the stressful emotions of guilt, anger, and denial that parents of dyslexic children may feel. These stress factors could well be thought capable of contributing to parenting problems and may in some circumstances militate against any protective factors of high SES, intelligence, and positive early attachment experiences. There is indeed evidence, despite the above caveat, about not treating illiteracy and dyslexia as synonymous, that there *is* a raised incidence of dyslexia among the delinquent population (Apiafi, 2001; Brunner, 1991; Daderman & af-Klinteberg, 1997; Strehlo, 1994; Sunderland & Klein, 1998; Thorstad, 1999). Turner *et al.* (2000) found in their study of young men in an institution for young offenders that 18% met stringent criteria for dyslexia more extensive than that of simple discrepancy between ability, age, and literacy level. It should be noted that this is four or five times the incidence expected randomly. Brunner (1991) deduced specific reading failure to be a cause and not just a correlate of delinquency, through the agency of frustration. Turner *et al.* (2000, p. 4) summarize the most pragmatic current view as follows: "Other factors, known to dispose towards crime may be of greater importance than dyslexia, though the latter is known to cause frustration and, if untreated, anger and alienation in the longer term."

What is clear from all the above evidence is that the long-term emotional and behavioral sequelae of dyslexia can, in some circumstances, be devastating. They can also be self-perpetuating—Rosenberg *et al.* (1989) found that not only does low self-esteem foster delinquency, but that delinquency in turn raises self-esteem. What therefore can we offer the disaffected dyslexic young person that is more powerful in its analgesic impact on their wounded sense of themselves as a "contender" than the balm offered to them by being a major player in a subcultural alternative society? In such an alternative society, audacity is all and literacy of little account.

Research supports the idea not only that suffering from dyslexia can lead to emotional and behavioral difficulty, but those negative consequences can be so serious that the final outcome may be delinquency. The question then arises. Can we do anything effective to improve this corrosive situation? If so, what?

INTERVENTION

Carefully structured approaches to improving literacy are crucial and central to any amelioration of the core skill-acquisition deficit and reducing this deficit *per se* will in many cases result in a swift improvement in self-esteem (Elbaum & Vaughn, 1999). This may also improve any concomitant behavioral difficulties. This will *not* however always be straightforward. Self-esteem is a complex construct, slippery, elusive, and obdurate. It would, therefore, be a grave mistake to assume that intervening to raise the self-esteem of the dyslexic individual was easy. It is not at all a case of giving someone some extra literacy support and telling him or her they are doing well. Inappropriate or insensitive intervention can be worse than none at all and can in fact make it even harder for the next protagonist to achieve a positive outcome who is offering teaching combined with soothing input to someone with a longstanding dyslexic problem.

This section will address topics of

- Externalizing, internalizing and coping
- Raising self-esteem
- Effective literacy skill enhancement
- Fostering metacognition
- Goal orientation
- Praise
- Peer group support
- Modifying attribution
- Modifying locus of control and self-efficacy
- Reducing learned helplessness.

Externalizing, Internalizing, and Coping

The success of any intervention strategy will depend not only on that intervention itself, but also on the characteristics of the individual in receipt of such intervention.

To introduce this section, I would like briefly to refer to my own practitioner experience. Most of the dyslexic children and young people that I meet through assessment respond roughly in one of three ways during feedback and discussion, as follows. I will often say something on the lines of "These tests and things we have been doing tell me that you have quite a lot of difficulty with reading and spelling. Has that sometimes made you think to yourself, 'I must be stupid or something'?"

The answers I get are:

1. "Yes I know I'm really thick," accompanied by tears, or,
2. "No, I'm good at spelling actually," accompanied by a challenging glare, or,
3. "Oh no, I know it's not that I'm thick, I'm just not very good at spelling. It doesn't matter that much really. It's just a nuisance."

accompanied by a rueful smile. It would of course be a gross oversimplification to equate these responses totally with, respectively,

1. internalizing, neurotic, and depressed behaviors
2. externalizing, disruptive, and potentially delinquent behaviors, and
3. adaptive, realistic behaviors which reflect well-balanced locus of control and protective compensatory self-esteem processes (Bear *et al.*, 1997).

From a practical point of view, however, we could perhaps simplify the challenge of how effectively to raise self-esteem in all the differing youngsters we see if we were at least to try and profile them in some way. One way to do this could be in terms of these three types of emotional response to prolonged failure: externalizing, internalizing, coping. These could be said to reflect the three ways of protecting self-esteem mentioned earlier: denying, self-handicapping, and compensating (Alvarez & Adelman, 1986; Bentall *et al.*, 1994; Lobel & Teiber, 1994).

Each dyslexic individual brings to the assessment and subsequent teaching situation a unique personal history and a unique internal world. Ideally we should tailor our support to each individual so that it provides a unique fit between need and provision. Pragmatically this is not possible. It is however useless to take a "one size fits all" approach. It is immediately obvious, for example, that what may raise the self-esteem and motivate an anxious, not especially bright, dyslexic 6-year-old girl, whose parents are of high SES, will not however raise the self-esteem or motivate an angry, alienated, highly intelligent, dyslexic 14-year-old boy from a deprived social background, nor yet a dyslexic 8-year-old boy of average ability who lives on and participates actively in a family farm, shines at football, and currently finds accuracy in spelling somewhat surplus to his requirements!

In planning and implementing approaches to raising the self-esteem and enhancing the motivation of those suffering from dyslexia, we need, therefore, to consider crucial core factors such as literacy levels, age, gender, ability, home and school experience. We could perhaps enhance our practice if in addition we considered emotional and behavioral factors. If, in order to do so, we adopt the "three types" classification illustrated above, this might make that task more manageable.

Raising Self-Esteem

So far the term self-esteem has been used throughout this chapter, at face value, without any examination of what the term really means. Both self-concept and self-esteem are complex, controversial, hypothetical constructs subject to much theoretical study.

At the simplest level, Robert Brooks (1999) defines self-esteem as follows: "Self-esteem may be seen as based upon the feelings and thoughts that individuals possess about their sense of competence and worth [...]."

Mruk (1999a) in an extensive exploration of this elusive phenomenon concludes that the many definitions of self-esteem available can be grouped in three different ways. First, self-esteem incorporates ideas of worth and competence. Second, it involves both cognition and affect, and, third, self-esteem has properties of both stability and openness to change. In order to raise self-esteem, therefore, we must not only be cognizant of the dyslexic individual's thoughts and feeling about his or her worth and competence, but also have some awareness of how open these are to change.

Although clearly related, these two ideas of worth and competence are in fact separable. In terms of competence, White (1963) describes self-esteem as having "its taproot in the experience of efficacy." He associates it with the very earliest development of the infant and the experience of mastery. He suggests that the infant's experience of influencing his environment by, for example, grasping, generates within him feelings of efficacy, even if the effect achieved was unintentional. Ultimately, as the infant becomes the child and then interacts purposefully and successfully on his or her environment, these feelings become a "more general cumulative sense of competence."

The implication of this is that although most development of self takes place in the social arena, mediated by significant others, a sense of mastery, efficacy, and competence can to a certain extent develop in direct interaction between the self and the environment without social mediation. Sensitive teaching must acknowledge this and promote self-directed success and active learning where feasible.

By contrast, the concept of worth can only really be understood from a social perspective. How "good" or "bad" we are can only have meaning in a cultural context. The way in which we derive the worth dimension of our self-esteem must therefore be comparative. We must have some socially derived idea of how we "should" or "shouldn't" be and hence the notion (Lawrence, 1988), that our self-esteem reflects how far we feel our actual self-image matches up to our ideal self—how we "should" be. Lawrence suggests the gap between these two—self-image and ideal-self—constitutes a measure of our self-esteem—the bigger the gap, the lower our self-esteem.

Self-esteem as a sense of worth is nicely captured by Baumeister, Smart, and Boden (1996) who define it as a "global evaluation of oneself."

Mruk (1999) describes self-esteem as being "lived." By this he means it is something we *experience*. He suggests that some theorists in this area tend to stress self-esteem as something we think about ourselves, the way we evaluate and perceive ourselves. These descriptions relate to cognition. Others stress that our self-esteem constitutes the way we feel about ourselves, the emotional correlate. These descriptions relate to affect. Reviews of the literature (Branden, 1994; Coopersmith, 1967) make it clear that there cannot be a clear dichotomy between cognition and affect in this area. Experiential introspection on our own global evaluations of our selves confirms this. We cannot wholly separate how we feel about ourselves from what we think we are like. Within that relationship between cognition and affect, one dimension can, however, be emphasized more than the other. The message for the teacher is clear. Empty reassurance is valueless, but promoting both feelings and thoughts of self-worth in their dyslexic student must be a priority. The dyslexic student can only, however, be helped to feel better about his or her worth if they have some real objective evidence on which to base this feeling. Some students will be more open to reassurance than others. Some will have become adamant in their self-perceptions as, although potentially malleable, self-esteem can prove very difficult to mould.

By its very nature as a developmental phenomenon, that is, something which grows and changes as we grow and change from an infant to an adult, self-esteem must be seen as a dynamic concept, something open to change. Efforts made to change a person's self-esteem for the better clearly imply that it is open to change. Paradoxically, however, such efforts also imply certain stability. Why would we have to make strenuous efforts to change something if it were not somewhat resistant? Mruk (1999) clarifies this issue by maintaining that self-esteem is both stable and open. By the time we are adults, we have a fairly steady evaluative view of ourselves but the level of this can still be raised or lowered by circumstance and our interpretation of our experiences. It is this possibility of change that the teacher must pursue.

There is a whole other body of research evaluating the measures commonly used to assess an individual's self-esteem (Byrne, 1996). An overwhelming finding is that agreement as to what exactly is being measured, and how accurately, is hard to attain. Many authors would argue that self-esteem equates with mental health and is on a par with physical health, that is, you can never have too much of it (White, 2002). Others suggest that it is indeed possible to have too much (Emler, 2001) and that this can lead to over-confidence and engagement in risky antisocial behavior.

The combination of the complexity of self-esteem processes and the difficulty of being sure that their measurement is meaningful may be what truly underlies these apparently contradictory views. This is implicitly acknowledged by Stanley *et al.* (1997) who refer to "self-reported" self-esteem. It is conceptually possible that a person may report and indeed express declaratively to themselves, that they have high self-esteem whereas their behavior and emotional state suggests that this is in fact a defensive stance and far from an accurate reflection of reality (Watson, 2002).

White's (2002) view, equating self-esteem with mental health, could be the most productive for adoption by the practitioner, with perhaps the inclusion of the term realistic as a qualifier. Chang (2001) found that "clarity" in terms of self-concept and self-esteem, knowing themselves well and realistically, helped to protect students from becoming depressed in the face of stressful life experiences. What we need to try and aim for therefore is to help our dyslexic clients develop positive, but clear and realistic, views of themselves as worthwhile individuals so that they may face with equanimity the task of improving their literacy.

If practitioners hope to raise self-esteem in the dyslexic learner, they are hoping to change the way someone thinks and feels, both about what they can and can't do and also about how "valuable" they are. In order to be effective in making such changes, one needs to be cognizant of all the research delineating the complex psychological processes involved in generating, maintaining, and adjusting self-esteem. Consideration is also needed of the approaches to raising self-esteem that have been attempted by others. Areas of promise here include: skill enhancement, fostering metacognition, modifying goal orientation, using contingent and meaningful praise, increasing peer group support, promoting adaptive attributions, encouraging realistic loci of control, increasing self-efficacy, and reducing learned helplessness.

Any success in these approaches will rely on an accurate analysis of the dyslexic young person's current position in these areas and in addition accuracy in determining what is motivating to them; what rewards them and reinforces their behaviors and who their significant others are.

Effective Literacy Skill Enhancement

This cannot be considered in isolation, as it is likely that it is within this arena of specifically enhancing literacy skill that the other approaches to raising self-esteem in dyslexic learners will be incorporated and are interwoven. In addition, Bear *et al.* (1997) and Crow *et al.* (1999) suggest that there is little theoretical or empirical support for the idea that teachers should focus on attempting to raise general, global self-esteem *per se* in the hope of impacting on academic problems. Approaches to raising self-esteem in dyslexic learners are therefore best attempted within the core approach of teaching literacy and not separately from this.

Harter (1999) notes that, in general, if poor self-esteem arises from a disparity between aspiration and accomplishment, to improve it one can either reduce aspiration or enhance accomplishment. To be functionally literate is an essential accomplishment in our society; therefore, in this context, there is no choice but to try and enhance it. Its pivotal cultural role means that its "perceived importance" (Harter, 1986) is largely immutable; if the dyslexic learner cannot enhance his literacy, it will be exceptionally difficult to enhance the damage to his global self-esteem that has arisen from this deficit. Many older dyslexics are able to reflect on their experiences and express the view that skilled instruction in literacy, that actually raised their attainment, enhanced their self-esteem (Palfreman-Kay, 2001; Riddick, Farmer, & Sterling, 1997). It is, perhaps, through the autobiographical reflection of adults on their dyslexia and the experiences that have either promoted or damaged their self-esteem that we can best access the long-term impact of skilled teaching on the person as a whole. The complexity of this relationship makes meaningful short-term objective analysis using the sort of self-esteem measures currently available very difficult. There is, however, some published material. The 1999 National Summit on Research in Learning Disabilities included three papers, which indicate the central role of actual teaching and skill enhancement and development, in improving both literacy skills and self-esteem (Elbaum & Vaughn, 1999; Gersten & Baker, 1999; Swanson, 1999). Recent research (Hatcher & Rack, 2001), found dyslexic children receiving multisensory specialist teaching on the "SPELL IT" project tended to gain in self-esteem alongside making progress in literacy. Fink (1998) looked at dyslexic adults who were highly successful both professionally and personally and found that intervention in the form of systematic phonic instruction was part of their exceptionally positive long-term life experiences.

A point worth reiteration here is just how damaging the effects of an inappropriate remediation approach, one that fails to improve literacy, can be, in that this may not only impede the child's academic progress, but may also reinforce feelings of inadequacy and hopelessness (Kline, 1986).

The Dyslexia Institute, amongst others, offers its clients carefully structured learning plans based around multisensory approaches such as the Dyslexia Institute Literacy Programme (DILP, Walker & Brooks, 2000). Specialists there believe this sort of instruction to be intrinsic to any intervention, but would recognize that the positive impact of a good literacy teaching tool on self-esteem must be maximized by tailoring it to the needs of the individual, not just cognitively, but as a whole. One of the tenets of the DILP approach is the necessity for fostering a metacognitive approach to learning. Research suggests (Palladino *et al.*, 2000) that this has implications not just for skill acquisition but also for self-esteem, breaking out of the trap of learned helplessness, gaining mastery and control.

Fostering Metacognition

Flavell (1976) delineates two components of meta-cognition: knowing, in a conscious declarative way, about how we think and learn and also, being able to take effective control of these cognitive processes. Palladino *et al.* (2000) found evidence to support the hypothesis that dyslexic pupils may not spontaneously develop general meta-cognitive skills, as their repeated experience of failure in literacy tends to negate their approaches to taking strategic control of learning tasks. If effective learning depends on the development of meta-cognitive skills and this development does not occur spontaneously in dyslexic pupils, then fostering it through teaching must be considered crucial. We must explicitly teach dyslexic learners to reflect on their own thinking processes; to come out from behind their shield of helplessness and once again commit themselves to taking control of their own learning (Edgar, 2001; Goldup & Ostler, 2000; Hunter-Carsch & Hughes, 2001; Walker & Brooks, 2000).

Encouraging those who have repeatedly failed at a learning task to try again, albeit with better teaching tools and more recognition and control of their own cognitive strengths and weaknesses, is risky for both teacher and learner. Some early experience of success is essential. One way to ensure success is through careful management of task demand; another is through the modification of goal orientation.

Goal Orientation

Cury, Biddle, Sarrazin, and Farnose (1997) divided goal orientation, how you approach the challenge of a task, into ego-involved or task-involved. An ego-involved stance would imply that your success or failure in the task is going to impact highly on how you feel about yourself. A task-involved stance, however, implies focus purely on mastering the task in hand. Failure in this second condition will be a much less threatening outcome.

Cury *et al.* (1997) found that pupils with ego-involved goal orientation who also perceived themselves to be of low ability in the relevant area engaged in self-handicapping behaviors such as prevarication, nonengagement, and displacement activities, when approaching a learning task. Those with task-involved orientation, by contrast, showed persistence and commitment to mastery even in the face of early failure. What this means in practice for the teacher of the dyslexic learner is that he or she must present literacy acquisition tasks to the student in such a way that these are perceived as "unthreatening," manageable, almost game-like, while not being perceived as trivial, patronizing, or a waste of time—no mean feat! It is through this sort of skill in teaching that the motivation of the dyslexic learner can be rekindled and their view of themselves be rendered more accurate and less damning. They can learn to recognize their strengths in some areas, rather than allowing their literacy failure to dominate their whole perception of themselves.

As Gross (1997, p. 4) puts it: "Children with learning problems need help in designing broad and generous self-definitions as learners which contain disparate and often contradictory elements."

They need help to extricate themselves from the "Catch 22" in which they are enmeshed. Our first task is then to help dyslexic students stop worrying so much about how bad they are at literacy. They can then stop expending all their energy on preventing themselves from the hurt of failing at it again. Our next task is to find a new way to teach them, one that leads them into a new perception of themselves. This new perception could shift them into the "coping" category I suggested earlier, the "I'm bad at spelling but so what!" position, which could deter them from any effort! Either an "Actually it just might be worth having another go at this, it looks as if I might be able to do it now" or "Now what strategy can I try with this?" position, from which they can safely risk making a real effort at trying again at literacy, would be preferable. Anecdotal reports from teachers suggest that, in fact, slightly anxious "internalizers" are easier to teach, more highly motivated and make better progress than either "externalizers" or "copers." This reflects previous findings on introversion and academic progress (Eysenck, 1970). This should not, however, deflect the ethical practitioner from the goal of raising self-esteem alongside improving literacy!

In order to attain a positive shift in perception and motivation, the teacher needs to take great care in task presentation, reducing uncertainty of outcome, and promoting mastery orientation in the student. In addition, great sensitivity in the use of praise and feedback is necessary.

Praise

The rewarding quality of praise is relative, not absolute. Whereas in one situation simple praise such as "very good" is rewarding and positive, in another it can be embarrassing, demeaning, and aversive. Factors within the learning situation itself and also within the learner, his peers, and the teacher, all contribute to the effect of praise. Thompson (1994) found praise that was informative was the most rewarding. It helped to prevent self-handicapping behaviors and promote appropriately internal attributions for success. Contingent, informative praise identifies specific accomplishment, helps the students recognize their cumulative progress and acknowledges their efforts. Noncontingent, less effective praise, by contrast, is bland, global, encourages comparison with peers and indicates success having been due to luck or ability beyond the student's locus of control. Teachers can and should modify their use of praise to reflect the above observations. What they cannot do, however, is make an assumption that they are a "significant other" to the student. This means they can never be certain that their praise is valuable.

Significant others impact on the development of self; their role perhaps needs some explanation. In a young child's life, his/her significant others are those people who control access to basic needs such as food, warmth, safety, and love. In addition, they have great power over the child's higher needs, the development of cognition and personality. What the child perceives these people think and feel about him/her generates his or her sense of self, and self-worth. The child finds out who he or she is largely through these people's responses and reactions. When this interaction between the child and significant others is healthy, where praise and warmth are endemic, both secure attachment and confident autonomy are developed so that a firm, basically optimistic, differentiated and whole self-identity is achieved, alongside the capacity to establish and maintain positive relationships. This security constitutes one of the protective factors, promoting resilience against emotional and behavioral difficulty, that can be available to the dyslexic child.

For the very young child, the significant others are usually parents or other major caregivers, possibly siblings and members of the extended family. The people who constitute our significant others change over time. As we enter adulthood, an element of choice enters this relationship and we may then select those whose opinions we value as our significant others.

Prior to adulthood and choice, an important arena in which our self-esteem is molded is that of school. There is great potential for both erosion and enhancement of self-esteem here, and it is here that our group of significant others expands beyond the family, and here that tasks that overtly test our competence and worth are part of our daily life. For the young child, the teacher can assume that they constitute a significant other. This is not necessarily the case, however, for the teacher of an adolescent, who cannot, therefore, be certain as to how valued their praise is by the student. Older pupils may find the positive regard of their peer group more important than that of their teacher (Robinson, 1995). This situation can, however, be harnessed productively. The peer group can be a valuable source of support and influence.

Peer Group Support

Steinhausen and Metzke (2001) found peer group acceptance to be protective against both internalizing and externalizing emotional and behavioral difficulties in adolescent youngsters. Robinson found that although parental support remains an important source of self-esteem at this age, approval from a generalized peer group becomes the strongest predictor of general self-esteem. The implication of these two observations is that the peer group is potentially a valuable source of support.

Utilizing the support of the peer group to promote the successful inclusion of those with Special Educational Needs is a well-established idea (Newton, Taylor, & Wilson, 1996). Creative use of this source of support for the dyslexic learner warrants pursuit. This is particularly so, perhaps, for the adolescent dyslexic student. In this case, as discussed, the teacher may be a less than significant other to him and may, therefore, have less chance of successfully influencing him than someone whose high regard matters to him more.

Attempts have been made to use peer group support to promote both skill acquisition and self-esteem, with varying degrees of success (Bagley & Mallick, 1996; Baum, Renzulli, & Heebert, 1995; Carabine & Downton, 2000; Ogusthorpe, 1984; Westhues, Clarke, Watton, & Claire-Smith, 2001). Their researches covered one-to-one mentoring, peer tutoring, and nonintensive group support. All were found to raise self-esteem, reading achievement and problem-solving skills to some extent in dyslexic students. A rather different approach involved investigating the effects of using dyslexic students to tutor normal peers. This had similarly positive results. It is likely that the peer group is currently an under-used and potentially vital resource, which teachers might wish to consider in their attempts to meet the academic and emotional need of their dyslexic students. Another approach involves focusing on the way dyslexic students conceptualize their experiences of success and failure in school and whether it is possible to modify these attributions.

Modifying Attributions

In order to consider whether it is actually possible to modify the way dyslexic students conceptualize their failures, the attributions they make in the face of their pervasive lack of success in literacy acquisition, it is necessary to have an understanding of Attribution Theory.

When things happen to us, we seek to understand why. This is particularly so when we do badly in some endeavor. We try to work out what the reason is for the result we experience and in doing so we make *attributions*; we attribute the cause of our failure in various ways. This mechanism of attribution is central to any discussion of self-concept and attainment. Dweck (2000) would suggest, in fact, that in order to be valid, any measurement of self-esteem must incorporate consideration of the individual's attributions, particularly those relating to intelligence and self-efficacy.

If we fail at a task, it is extremely important to us to make attributions as to why this happened (Dodds, 1994). Four major types of attribution likely to be made in an achievement situation are postulated by Dweck (2000). These are: task, ability, luck, and effort. What this means is that if, for example, we do badly in a physics exam, we make sense of this failure, come to terms with it, get over it, etc., by telling ourselves why it happened; by attributing the failure to *something*. We may, for example, say to ourselves "the questions were incredibly hard" (task), or "I'm hopeless at anything with numbers" (ability), or "they picked all the topics I hadn't revised" (luck), or "I didn't do enough revision" (effort). We do not make the same attributions in every case and, as a moment's introspection on the last time you did badly at something will tell you, our attributions are complex and interwoven and not as clear-cut as the example implies.

The attributions we tend to make in the face of failure are indicative of how we perceive ourselves. If we tend to opt for the ability attribution, we are demonstrating that we perceive ourselves as in some way "stupid." This is an internal and stable attribution. It is debilitating and likely to hamper response to further challenge. Who wants to try something they are sure they will fail at? If, however, we have opted mainly for task or luck, we have made external and flexible attributions. We might well have another go at the challenge; we might be luckier; it could be easier. If we opt for effort, we have made an internal but flexible attribution. We could face a challenge again. It is within our power to succeed if we have a mind to.

We also make attributions in the face of success. Here the outcomes of the attributions mean that if we attribute our success to ability and effort, that is, internally, we will have more confidence to attempt challenges in the future than if we attribute our success to luck and ease of task. In some cases, we may show a "self-serving bias" whereby responsibility for failure is always attributed externally and success internally (Bentall *et al.*, 1994).

Research has been carried out into the relationship between attribution and academic attainment in dyslexic students. Using the Dweck (2000) model, Dodds (1994), found poor spellers more likely than good spellers to attribute failure in a spelling test to low ability. Butkowsky and Willows (1980) found poor readers tended to attribute any success in reading to external factors, as well as attributing failure to low ability. Other work, however, suggests a more complex picture. Durante (1993) found that not all the dyslexic subjects in her group made maladaptive, self-denigrating attributions and that attributions of this type were more closely related to comorbid behavior disorders. This relationship between maladaptive attribution and behavior difficulty was also found by Eslea (1999).

It is clear that the "healthiest" attribution style must be the one most rooted in reality. The least healthy attribution style is that where failure is always attributed to lack of ability and success to luck or task—the opposite of self-serving, in fact, self-denigrating. There is some evidence (Yasutake, Bryan, & Dohrn, 1996) that promoting healthier attributions through cognitive reprocessing training can have a positive effect on self-perception. The question then becomes, is it a reasonable goal on the part of the teacher to try and promote healthier attributions of failure and success in the dyslexic learner and if so how can it be done?

All the intervention approaches already mentioned, should help contribute to the outcome of giving control, success, and dignity back to the dyslexic learner and removing guilt and helplessness. A key component in all of these and in addition in promoting accurate attributions is that of skilled and detailed individual assessment. The need to assess the student in terms of their emotional state and their self-esteem, which can determine their responses to teaching, has already been explored. In addition, there is a core need to establish the student's cognitive profile through psychometric assessment.

The experience of colleagues and myself is that the response of dyslexic students of all ages to the detailed information available from a psychometric assessment is almost universally positive. It is common for clients to express overwhelming relief: "I thought I was thick," "so that's why I did so badly at school," "I knew there was something" are typical remarks and these indicate a shift in attribution of their failure. As Turner (1997) puts it, the face-to-face debriefing after an assessment is "cathartic" (p. 118) and as such of great importance—it is the first step in including the pupils in their own learning remediation. The need for accuracy and sensitivity on the part of the practitioner at this point cannot be overestimated. Accurate, detailed information about their cognitive strengths and weaknesses, presented in a sensitive and comprehensible manner, allows the dyslexic learner to attribute his or her literacy failure accurately.

The observations of Kline (1986), although expressed some years ago, still have relevance here, with reference to both assessment and subsequent educational intervention. Kline stresses the importance of involving the dyslexic child, even the very youngest, in discussions about testing, results, and remediation plans. Although Kline takes a psychoanalytic perspective, discussing issues of defense mechanisms, his recommended strategy of placing the dyslexic learner, however young, in an active central position in terms of intervention also has relevance from the perspective of attribution analysis. The pupils will then perhaps begin to attain success and subsequently attribute that success to their own agency. Practitioners might intuitively already give detailed feedback to older children. The observations of Kline, taken together with that of Chapman (1988), that children's academic self-perceptions are becoming stable by year 3, suggests that it may be good practice to offer this feedback to much younger children.

Eccles, Wigfield, Harold, and Blumenfeld (1993) indicate that from the age of 5, children may perceive differences in their competence in different domains. They will, therefore, also begin to make attributions, and if they lack the necessary information to be accurate in this they will surely begin to attribute their relative failure in literacy maladaptively, the "I am thick" mode previously noted. Detailed assessment feedback may prevent this and promote development of realistic attribution.

In the case where maladaptive attributions are already established, Craven, Marsh, and Debus (1991) found some success in attribution-retraining interventions. These involved giving internally focused feedback that emphasized children's strengths in particular subject area ("Look how well you did in that tracking task") and also modeling the making of internal attributions for success ("You must feel good about the way you can learn letters").

Although it is a complex area, the teacher of any dyslexic student might find it beneficial to consider the attributions of success and failure being made by that student in the day-to-day teaching situation and make appropriate attempts to mould these attributions through careful, sensitive, and supportive feedback on task performance.

Closely related to Attribution Theory is the personality concept of Locus of Control (Rotter, 1966) and modifying Locus of Control is another way in which the teacher may enhance the self-esteem of the dyslexic student.

Modifying Locus of Control and Perceptions of Self-Efficacy

In a similar way to the consideration of modification of attributions, a discussion of possible modification of locus of control and associated self-efficacy needs to be preceded by a description of these concepts.

If for the most part we attribute the outcomes of what we do, our successes and failures, to internal factors, over which we have influence, we can be said to have an internal locus of control orientation. If, on the other hand, we generally attribute outcomes to external factors, we have an external locus of control orientation. If we generally have an external orientation, we have an associated general feeling of powerlessness. We feel unable to take charge of our lives and at the mercy of chance. We may then abdicate all responsibility and abandon effort.

Looking back at Attribution Theory, however, makes it clear that some internal attributions are paradoxically also unlikely to contribute to our sense of power over our lives. If, for example, we attribute failure to lack of ability, that is an *internal* attribution, but, as we may view it as inflexible, we may feel it is beyond our control— "I was born that way"—and it functions as an *external* locus of control orientation. It is "external" to that over which we perceive ourselves as having "control." The concepts of attribution and locus of control, although related, are thus not interchangeable. It is possible to make an internal attribution that is external to our locus of control. Recent research confirms the links between attainment and locus of control (Hawkes, 1995; Kliewer & Sandler, 1992). Hagborg (1999), however, found dyslexic students with self-concepts he terms "inadequate" did not differ in their locus of control from their counterparts with "adequate" self-concept.

The relationship between locus of control orientation, attribution style, self-esteem, and experience of success and failure is clearly a complex one. The way in which we attribute our successes and failures and the extent to which we feel in control of our lives, both contribute to our perceptions of self-efficacy, our judgments of our capabilities to do given tasks.

Self-efficacy is an individual's belief in his or her potential to succeed in a particular situation.

Referring back to the pyramid, multifaceted, model of self-concept, we can see that one could postulate various levels of self-efficacy perception, from the global, "Actually I am likely to be competent in anything you ask me to do," to the academic, "I'm pretty good in all my subjects," to subject specific, "I'm good at English but in the bottom set for Maths," right down to very situation-specific areas such as "I'm really good on planets, you can test me on the order!"

Much of the research in this area has been at the subject-specific level. Some of it has led authors to generalize from their findings and hypothesize more global impact from subject-specific results.

Rankin, Bruning, and Timme (1994) found a strong predictive relationship between lower self-efficacy beliefs for spelling and actual spelling performance, regardless of spelling ability otherwise measured. At a more global level, Panagos and DuBois (1999) found that dyslexic adolescents' self-efficacy beliefs were a strong and potentially limiting, mediating influence on their choice of possible career. A fascinating study by Hartley (1986) showed that low self-efficacy is so strong a mediator that it can lead to denial of demonstrable success. A classic piece of work in this area, illustrating the clear relationship between low self-efficacy and poor test outcome, is that of Bandura (1989). In this study, children who perceived themselves as "bad" at maths subsequently attained low scores on a maths test, despite having been assessed by teachers as having skills in this area that were in fact equal to those who did well on the test. In 1997, Bandura revisited self-efficacy to postulate that it impacts on success or failure of outcome through its impact on task engagement. We do not try hard, or persist, in tasks where we feel we lack competence. We then fail, once again, to do well at these tasks and our perception of poor competence is, once again, validated. There are clear implications for intervention in this area, but as Bandura (1989, p. 8) warns, these must be addressed with sophistication: "No amount of reiteration that I can fly will persuade me that I have the efficacy to get myself airborne."

The strength of a long-held belief in lack of efficacy will take more than "warm fuzzies" being told nonspecifically how nice/clever/good you are, or even some experience of success, to shift. In fact, as teachers will testify, some dyslexic children find unexpected success in literacy at first quite aversive, as is all cognitive dissonance. This surprise achievement rattles their well-established beliefs and puts them in a position of risk, in that it means they are once again "in the race" rather than "wounded onlookers," a long-held and defensible position of learned helplessness, with which they had become at least reasonably comfortable.

Reducing Learned Helplessness

Learned helplessness could be said to be the ultimate outcome of a combination of internal attributions for failure, external locus of control and consistently poor perceptions of self-efficacy. The question is whether teachers are in a position to shift the dyslexic learner who has reached the state of learned helplessness into a more open and willing approach to skill acquisition. The dynamics of learned helplessness are as follows.

What we do not attempt, we cannot fail at. What we attempt only half-heartedly causes us little pain if we fail at it. It is not surprising, therefore, that if we keep failing at something despite our best efforts, we either no longer attempt it or do so only half-heartedly. "A sense of incompetence ... could give rise to work avoidance" (Seifert & O'Keefe, 2001). It is not, however, failure in itself which leads to maladaptive work behaviors and self-beliefs, but the context in which that failure occurs. Failure *per se* can be useful, "honorable and constructive" (Holt, 1969). Given a supportive context, in which failure is regarded simply as a neutral type of feedback, failure can then be informative, rather than painful. The key features here are the goal orientation of the learner and his view on the nature of ability. Strage (1997), for example, found those students who regarded intelligence as a fixed rather than an incremental entity and, therefore, beyond their control, tended to greater learned helplessness. One could with justification substitute "ability to read and spell" for intelligence and theorize that dyslexics would be similarly affected. A child in this position is at once able to explain his failure, "It's just how I was born" and is also able to exempt himself from further effort, "What's the point, I can't change how I am." Butkowsky and Willows (1980) found dyslexic children who reacted in exactly this way and the more they failed, the more "helpless" they became, and the lower their self-esteem. This finding was situation specific, but raises many issues. For example, the value of the label dyslexia to those experiencing unexpected failure in literacy is well established (Miles, 1996). The finding of Butowsky and Willows suggests, however, that it is a label to be used and explained with care. This caution reflects a possible negative counterbalance to the generally enabling nature of cognitive assessment. Accuracy in attribution of failure may be generated from careful assessment. Care must be taken, however, to ensure that the dyslexic young person recognizes that, although they are now absolved of "guilt," their dyslexic difficulties are intrinsic and not due to laziness; although their dignity is returned to them, as they find they are not lacking in basic ability, they are *not* able, as a result, to abdicate all responsibility for their learning. The reverse is true. They have to make conscious effort to overcome and compensate for their areas of weakness and the effects these have had on their academic development. The teacher has a crucial role here in helping to convince the individual who has learnt to be helpless both that he or she is in fact capable of becoming competent and successful and that this will take hard work.

The practitioner who takes on the task of trying to promote literacy competence in someone who has repeatedly tried and failed to develop such expertise is engaging in an undertaking far beyond that of simply teaching skills. If they are to be successful in this undertaking, their first step must be to generate some sense of competence and control. They must help their pupil enhance their self-efficacy and believe that their difficulties *are* limited, modifiable, and within their locus of control. Seifert and O'Keefe (2001, p. 91) express it very well: "The psychological environment constructed by the teacher should foster confidence and autonomy which are critical for developing self-regulated, adaptive learners. Students' perceptions of teachers as being nurturing and supportive of learning are strongly connected to students' sense of competence and control. It is in this context that teaching is, first and foremost, an exercise in human interaction."

SUMMARY

This chapter has looked at the complex relationship between self-esteem and academic attainment, with particular reference to unexpected failure in literacy acquisition—dyslexia. The reciprocity of this relationship has been explored, as have the relationships between dyslexia, poor self-esteem, and emotional and behavioral difficulties. Research into intervention has been discussed and practical techniques for the teaching practitioner suggested.

One key feature of this area is that all the core constructs are hypothetical. They are elusive and hard to define. Paradoxically they also have the superficial appearance of being just the reverse—"Not being able to read and write properly is bound to make you feel stupid"—"Feeling stupid is bound to make you feel bad about yourself"—"Telling someone they are not stupid will make them feel better." These sorts of statements have a beguiling simplicity. Psychological constructs are, however, seldom truly simple. One aim of the chapter has been to delve beneath the surface of this area so as to shed light on the underlying complexities and thus derive at approaches that meet effectively the diverse needs of those with dyslexia and concomitant damage to their self-concept; approaches which are not only pragmatic, but sophisticated and have their roots in evidence-based psychology.

REFERENCES

Alvarez, V., & Adelman, H. S. (1986). Overstatements of self-evaluations by students with psychoeducational problems. *Journal of Learning Disabilities, 19*, 567–571.

Apiafi, J. (2001). The dyslexia institute partnership with the Nottinghamshire probation service, the PALS project—Positive action through learning support. *Dyslexia Review, 12*, 2, 10–12.

Bagley, C., & Mallick, K. (1996). Towards achievement of reading skill potential through peer tutoring in mainstreamed 13 year olds. *Disability and Society, 11*, 83–89.

Bandura, A. (1989). Perceived self-efficacy in the exercise of personal agency. *The Psychologist*, 411–424.

Bandura, A. (1997). *Self-efficacy: The exercise of control*. New York: Freeman.

Baum, S. M., Renzulli, J. S., & Heebert, T. P. (1995). Reversing underachievement: Creative productivity as a systematic intervention. *Gifted Children Quarterly, 39*, 224–235.

Baumeister, R., Smart, L., & Boden, J. (1996). Relation of threatened egotism to violence and aggression: The dark side of self esteem. *Psychological Review, 103*, 5–33.

Bear, G. G., Minke, K. M., Griffin, S. M., & Deemer, S. A. (1997). *Self concept in children's needs II: Development problems and alternatives*. Bethesda, MD: National Association of School Psychologists.

Beitchman, J. H., & Young, A. R. (1997). Learning disorders with a special emphasis on reading disorders: A review of the past 10 years. *Journal of the American Academy of Child and Adolescent Psychiatry, 36*, 1020–1032.

Bender, W. N., & Golden, L. B. (1988). Adaptive behaviour of learning disabled and non-learning. *Disabled Children Learning Disability Quarterly, 11*, 55–61.

Bentall, R. P., Kinderman, P., & Kaney, S. (1994). Cognitive processes and delusional beliefs: Attributions and the self. *Behaviour Research and Therapy, 32*, 331–341.

Berg, B. L. (1989). *Qualitative research methods for the social sciences*. London: Allyn & Bacon.

Branden (1994). *The six pillars of self-esteem*. New York: Bantam.

Brooks, R. (1999). The self-esteem teacher—Seeds of self-esteem. In *Attribution theory and self-esteem* (Chapter 4) Ohio: Treehaus Communications, Inc.

Brown, A., & Heath, N. (1997). *Social competence in peer-accepted children with and without learning disabilities*. Poster presentation at the 30th annual national convention of the National Association of School Psychologists, Orlando, FL.

Brunner, M. S. (1991). *Reduced recidivism and increased employment opportunity through research-based reading instruction*. Washington DC: National Institute of Justice, US Department of Justice.

Burden, R. (1998). Assessing children's perceptions of themselves as learners and problem solvers. The construction of the Myself-as-a-Learner Scale (MALS). *School Psychology International, 19*, 291–305.

Burns, R. B. (1982). *Self-concept development and education*. Hold, Rinehart and Winston. Austin, TX.

Butkowsky, I. S., & Willows, D. M. (1980). Cognitive-motivational characteristics of children varying in reading ability: Evidence for learned helplessness in poor readers. *Journal of Educational Psychology, 72*, 3, 408–422.

Byrne B. M. (1984). The general/academic self-concept nomological network: A review of construct validation research. *Review of Educational Research, 54*, 427–456.

Byrne, B. M. (1996). Measuring self-concept across the lifespan: Issues and instrumentation. Washington DC: American Psychological Association.

Carabine, B., & Downton, R. (2000). Specific learning difficulties and peer support. *Educational Psychology in Practice, 16*, 487–494.

Chang E. C. (2001). Life stress and depressed mood among adolescents: Examining cognitive-affective mediation model. *Journal of Social and Clinical Psychology, 20*, 416–429.

Chapman, J. W. (1988). Learning disabled children's self-concepts. *Review of Educational Research, 58*, 347–371.

Chapman, J. W., Lambourne, R., & Silva, P. A. (1990). Some antecedents of academic self-concept: A longitudinal study. *British Journal of Educational Psychology, 60*, 142–152.

Code of Practice. (2001). *Special educational needs code of practice*. Nottinghamshire: Department for Education and Skills, DfES Publications.

Coopersmith, S. (1967). *The antecedents of self-esteem*. Freeman: San Francisco.

Craven, R. G., Marsh, H. W., & Debus, R. L. (1991). Effects of internally focussed feedback and attributional feedback on enhancement of academic self-concept. *Journal of Educational Psychology, 83*, 17–27.

Crow, R. E., Hoops, S. M., & Williams (1999). *Improving students reading achievements through their use of self-esteem lessons*. Saint Xavier University: Master's Action Research Project.

Cury, F., Biddle, S., Sarrazin, P., & Farnose, J. P. (1997). Achievement goals and perceived ability predict investment in learning a sport task. *British Journal of Educational Psychology, 67*, 293–309.

Daderman, A., & af-Klinteberg, B. (1997). Personality dimensions characterising severely conduct disordered male juvenile delinquents. *Reports from the Department of Psychology—Stockholm, 831*, 1–21.

DfE (1994). *The education of children with EBD (Circular 9/94)* London: HMSO.

Dodds, J. (1994). Spelling skills and causal attributions in children. *Educational Psychology in Practice, 10*(2), 111–119.

Dodds, P. S. (1993). Beyond the rainbow: A guide for parents of children with dyslexia and other learning difficulties (2nd ed.). Eric Document Reproduction Service.

Durante, J. E. (1993). Attributions for achievement outcomes among behavioural subgroups of children with learning disabilities. *Journal of Special Education, 27*(3), 306–320.

Dweck, C. S. (2000). *Self theories: Their role in motivation, personality & development*. Philadelphia: Psychology Press.

Eccles, J., Wigfield, A., Harold, R. D., & Blumenfeld, P. (1993). Age and gender differences in children's self-and-task perceptions during elementary school. *Child Development, 64*, 830–847.

Edgar, B. (2001). Effective learning in secondary school: Teaching students with dyslexia to develop thinking and study skills. In *Dyslexia and effective learning* (Chapter 6). London: Whurr.

Elbaum, B., & Vaughn, S. (1999). *Can school based interventions enhance the self concept of students with learning disabilities? Two decades of research in learning disabilities: Reading comprehension, expressive writing, problem solving, self-concept. Keys to successful learning: A national summit on research in learning disabilities*. New York: National Centre for Learning Disabilities.

Emler, N. (2001). *Self-esteem: The costs and courses of low self worth*. Published for Joseph Rowntree Foundation by YPS.

Eslea, M. (1999). Attributional styles in boys with severe behaviour problems: A possible reason for lack of progress on a positive behaviour programme. *British Journal of Educational Psychology, 69*, 33–45.

Eysenck, H. J. (1970). *Readings in extraversion–introversion*, Vol. 3. London: Staples Press.

Feick, D. L., & Rhodewalt, F. (1997). The double-edged sword of self-handicapping: Discounting, augmentation and the protection and enhancement of self-esteem. *Motivation and Emotion, 21*, 147–163.

Fergusson, D. M., & Lynskey, M. T. (1997). Early reading difficulties and later conduct problems. *Journal of Child Psychology and Psychiatry, 38*, 899–907.

Fink, R. P. (1998). Literacy development in successful men and women with dyslexia. *Annals of Dyslexia, 48*, 311–346.

Fisher, B. L., Allen, R., & Kose, G. (1996). The relationship between anxiety and problem solving skills in children with and without learning disabilities. *Journal of Learning Disabilities, 29*, 439–446.

Flavell, J. H. (1976). Metacognitive aspects of problem solving. In L. B. Resnick (Ed.), *The nature of intelligence*. Hillsdale, NJ: Erlbaum.

Frick, P. J., Kamphaus, R. W., Lahey, B. B., Loeber, R., Christ, M. A. G., Hart, E. L. *et al.* (1991). Academic underachievement and the disruptive behaviour disorders. *Journal of Consulting and Clinical Psychology, 59*, 289–294.

Gersten, R., & Baker, S. (1999). *Reading comprehension instruction for students with learning disabilities. Two decades of research in learning disabilities: Reading comprehension, expressive writing, problem solving, self-concept. Keys to successful learning: A national summit on research in learning sisabilities*. New York: National Centre for Learning Disabilities.

Gersten, R., & Baker, S. (1999). *Teaching expressive writing to students with learning disabilities. Two decades of research in learning disabilities: Reading comprehension, expressive writing, problem solving, self-concept. Keys to successful learning: A national summit on research in learning disabilities*. New York: National Centre for Learning Disabilities.

Gilbert, P. (1992). *Depression: The Evolution of Powerlessness*. Hove: Lawrence Erlbaum.

Goldup, W., & Ostler, C. (2000). The dyslexic child at school and home. In J. Townend, & M. Turner (Eds.), *Dyslexia in practice: A guide for teachers* (pp. 311–340). London: Kluwer Academic/Plenum Publishers.

Gross, A. H. (1997). *Defining the self as a learner for children with learning difficulties. The World National Centre for Children with Learning Disabilities.* www.ldonline.org.

Haager, D., & Vaughn, S. (1995). Parent, teacher, peer and self-reports of the social competence of students with learning disabilities. *Journal of Reading Disabilities, 28*, 205–215.

Hagborg, W. J. (1996). Self concept and middle school students with learning disabilities: A comparison of scholastic competence subgroups. *Learning Disability Quarterly, 19*, 117–126.

Hagborg, W. J. (1999). Scholastic competence subgroups among high school students with learning disabilities. *Learning Disability Quarterly, 22*, 3–10.

Hales, G. (1987). *Personality Aspects of Dyslexia.* Milton Keynes UK: Open University.

Hamachek, D. (1995). Self concept and school achievement—Interaction dynamics and a tool for assessing the self-concept component. *Journal of Counselling and Development, 73*(4), 419–425.

Harré, R. (1979). *Social being: A theory for social psychology.* Oxford: Basil Blackwell.

Harter, S. (1983). Developmental Perspectives on the self-system. In P. Musson (Ed.), *Carmichael's manual on child Psychology, Vol. 4 Social and Personality development* (pp. 274–386). New York: Wiley.

Harter, S. (1986). Processes underlying the construction, maintenance and enhancement of the self-concept in children. In J. Suls, A. G. Greenwald (Eds.), *Psychological perspectives on the self.* (pp. 3, 137–181). Hillsdale, NJ: Erlbaum.

Harter, S. (1999). *The construction of the self.* London: The Guildford Press.

Hartley, R. (1986). Imagine you're clever. *Journal of Child Psychology and Psychiatry.* May 1986.

Hatcher, J., & Rack, J. (2001). *Spell-it: An overview of the project. Dyslexia Review* (pp. 14, 1, 5–10) Staines: The Dyslexia Institute.

Hawkes, B. B. (1995, November 8–10). Locus of control in early childhood education: Where did we come from? Where are we now? Where might we go from here? Paper presented at the annual conference of the Mid-South Educational Research Association. Biloxi, MS.

Heiervang, E., Lund, A., Stevenson, J., & Hugdahl, K. (2001). Behaviour problems in children with dyslexia. *Nordic Journal of Psychiatry, 55*, 251–258.

Herrnstein, R. J., & Murray, C. (1994). The Bell Curve: Intelligence and class structure. In *American Life.* Free Press (Macmillan).

Hinshaw, S. P. (1992). Externalising behaviour problems and academic underachievement in childhood and adolescence: Causal relationships and underlying mechanisms. *Psychological Bulletin, 111*, 127–135.

Hoge, R. D., Andrews, D. A., & Leschied, A. W. (1996). An Investigation of risk and protective factors in a sample of youthful offenders. *Journal of Child Psychology and Psychiatry, 37*(4), 419–424.

Hogenson, D. L. (1974). Reading failure and juvenile delinquency. *Orton Society Annual Bulletin,* 1974, p. 164–169.

Holt, J. (1969). *How children fail.* Penguin Books. Harmondsworth: Middlesex.

Hunter-Carsch, M., & Hughes, M. (2001). *Spelling support in secondary education. Dyslexia and effective learning in secondary and tertiary education* (Chapter 5). London: Whurr.

James, W. (1983/1890). *The principles of psychology.* Cambridge, MA: Harvard University Press.

Jorm, A. F., Share, D. L., Matthews, R., & MacLean, R. (1986). Behaviour problems in specific retarded and general reading backward children: A longitudinal study. *Journal of Child Psychology and Psychiatry, 27*, 33–45.

Katz, I. (1996). *The construction of racial identity in children of mixed parentage.* Cromwell Press, Trowbridge, Wiltshire, UK.

Klasen, E. (1972). *The syndrome of specific dyslexia.* Lancaster: Medical and Technical Publishing Co. Ltd.

Kliewer, W., & Sandler, I. N. (1992). Locus of control and self-esteem as moderators of stressor-symptom relations in children and adolescents. *Journal of Abnormal Child Psychology, 20*(4), 393–413.

Kline, C. (1986). The dyslexia emotional dyad: Implications for diagnosis and treatment. *Canadian Journal of Psychiatry, 31*, 517–520.

Kohut, H. (1977). *The restoration of the self.* New York: International Universities Press.

Kloomok, S., & Cosden, M. (1994). Self concept in children with learning disabilities: The relationship between global self-concept, academic discounting, nonacademic self concept and perceived social support. *Learning Disability Quarterly, 17*, 140–153.

Lawrence, D. (1988). *Enhancing self-esteem in the classroom*. London: Chapman.

Livingstone, R. (1990). Psychiatric comorbidity with reading disability: A clinical study. *Advances in Learning Disabilities: A Research Annual, 6*, 143–155.

Lobel, & Teiber (1994). Effects of self-esteem and need for approval on affective and cognitive reactions – Defensive and true self-esteem. *Personality and Individual Differences, 16*(2).

Lund, R. (1986). The self-esteem of children with EBD. *Maladjustment and Therapeutic Education, 5*.

Maines, B., & Robinson, G. (1995). You Can.... You KNOW You Can! Lame Duck Enterprises. Bristol, UK.

Margerison, A. (1996). Self-esteem: Its effect on the development and learning of children with EBD. *Support for Learning, 11*, 176–180.

Marsh, H. W. (1992). Content specificity of relations between academic achievement and academic self-concept. *Journal of Educational Psychology, 84*, 35–42.

Marsh, H. W., & Yeung, A. S. (1997). Causal effects of academic self concept on academic achievement: Structural equation models of longitudinal data. *Journal of Educational Psychology, 89*, 41–54.

Marsh, H. W., & Yeung, A. S. (1999). The liability of psychological ratings: The chameleon effect in global self esteem. *Personality and Social Psychology Bulletin, 25*(1), 49–64.

Marsh, H. W., Byrne, B. M., & Shavelson, R. J. (1988). A multifaceted academic self concept: Its hierarchical structure and its relation to academic achievement. *Journal of Educational Psychology, 80*, 366–380.

Marsh, H. W., Craven, R., & Debus, R. (1999). Separation of competency and affect components of multiple dimensions of academic self concept: A developmental perspective. *Merill-Palmer Quarterly, 45*, 567–601.

Maughn, B. (1995). Annotation: Long-term outcomes of developmental reading problems. *Journal of Child Psychology and Psychiatry, 36*, 357–371.

Maughn, B., Pickles, A., Hagell, A., Rutter, M., & Yule, W. (1996). Reading problems and antisocial behaviour: Developmental trends in comorbidity. *Journal of Child Psychology and Psychiatry, 37*, 405–418.

Maslow, A. H. (1954). *Motivation and personality*. New York: Harper.

Miles, T. (1996). The inner life of the dyslexic child. In V. P. Varma (Ed.), *The Inner life of children with special needs* (pp. 112–123). London: Whurr Publishers Ltd.

Mruk, C. (1999a). A phenomenological theory of self-esteem. In *Self-esteem. Research, theory and practice*. (Chapter 5, pp. 155–194). London: Free Association Books.

Mruk, C. (1999b). Major self esteem theories and programs. In *Self-esteem. Research, theory and practice* (Chapter 4, pp. 115–154). London: Free Association Books.

Mruk, C. (1999c). Self-esteem research findings. In *self-esteem. Research, theory and practice* (Chapter 3, pp. 69–113). London: Free Association Books.

Mruk, C. (1999c). The meaning and structure of self-esteem. In *Self-esteem. Research, theory and practice* (Chapter 1, pp. 1–32). London: Free Association Books.

Muijs, R. D. (1997). Predictors of academic achievement and academic self concept: A longitudinal perspective. Symposium: Self perception and performance. *British Journal of Educational Psychology, 67*, 263–277.

Nabuzoka, D., & Smith, P. K. (1993). Sociometric status and social behaviour of children with and without learning difficulties. *Journal of Child Psychology and Psychiatry and Allied Disciplines, 34*, 1435–1448.

Newton, C., Taylor, G., & Wilson, D. (1996). Circles of friends: An inclusive approach to meeting emotional and behavioural needs. *Educational Psychology in Practice*, Volome II pp. 41–48.

Ogusthorpe, R. T. (1984). *Handicapped children as tutors*. Research Report, Brigham Young University.

O'Keeffe, D. J. (1994). Truancy in English Secondary Schools: A Report prepared for the DFE: The Truancy Research Project, 1991–1992.

Palladino, P., Poli, P., Masi, G., & Marcheschi, M. (2000). The relation between metacognition and depressive symptoms in pre-adolescents with learning disabilities: Data in support of Borkows model. *Learning Disability Research and Practice, 15*, 142–148.

Panagos, R. J., & Dubois, D. L. (1999). Career self-efficacy development and students with learning disabilities. *Learning Disabilities Research and Practice, 14*, 25–34.

Palfreman, K. J. (2001). Students views of learning support. In M. Hunter-Carsch, & M. Herrington (Eds.), *Dyslexia & Effective Learning in Secondary and Tertiary Education* (Chapter 16). London: Whurr.

Prior, M., Smart, D., Sanson, A., & Oberklaid, F. (1999). Relationships between learning difficulties and psychological problems in preadolescent children from a longitudinal sample. *Journal of the American Academy of Child and Adolescent Psychiatry, 38*, 429–436.

Rankin, J. L, Bruning, R. H., & Timme V. L. (1994). The development of beliefs about spelling and their relationship to spelling performance. *Applied Cognitive Psychology, 8*(3), 213–232.

Richman, N., Stevenson, J., & Graham, P. J. (1982). *Pre-school to school: A behavioural study.* London: Academic Press.

Riddick, B., Farmer, M., & Sterling, C. (1997). *Students and dyslexia—Growing up with a specific learning difficulty.* London, England: Whurr Publishers.

Robinson, N. S. (1995). Evaluating the nature of perceived support and its relation to perceived self-worth in adolescents. *Journal of Research on Adolescents, 5*, 253–280.

Rogers, C. R. (1961). *On becoming a person.* London: Constable.

Rosenberg, M., Schooler, C., & Schoenbach, C. (1989). Self-esteem and adolescent problems: Modeling reciprocal effects. *American Sociological Review, 54*, 1004–1018.

Rotter, J. B. (1966). Generalized expectancies for internal versus external control of reinforcement. *Psychology Monograph, 80*, 609.

Rutter, M., & Yule, W. (1970). Reading retardation and anti-social behaviour. In M. Rutter, J. Tizard, & K. Whitmore (Eds.), *Education, health and behaviour* (pp. 240–255). London: Longmans.

Schlenker, B. R. (1980). *Impression management: The self-concept, social identity and interpersonal relations.* Monteray, CA: Brooks/Cole.

Seifert, T. L., & O'Keefe, B. A. (2001). The relationship of work avoidance and learning goals to perceived competence, externality and meaning. *British Journal of Educational Psychology, 71*, 81–92.

Shavelson, R. J., & Bolus, R. (1982). Self-concept: The interplay of theory and methods. *Journal of Educational Psychology, 74*, 3–17.

Smith, D. S., & Nagle, R. J. (1995). Self perceptions and social comparisons among children with learning disabilities. *Journal of Learning Disabilities, 28*, 364–371.

Song, I., & Hattie, J. (1984). Home environment, self concept and academic achievement: A causal modelling approach. *Journal of Educational Psychology, 76*, 1269–1281.

Stanley, P. D., Dai, Y., & Nolan, R. F. (1997). Differences in depression and self-esteem reported by learning disabled and behaviour disordered middle school students. *Journal of Adolescence, 20*, 219–222.

Steinhausen, H. C., & Metzke, C. W. (2001). Risk, compensatory, vulnerability and protective factors influencing health in adolescence. *Journal of Youth and Adolescence, 30*, 259–267.

Strage, A. (1997). Agency, communion and achievement motivation. *Adolescence, 32*, 126, 299–312.

Strehlow, U. (1994). Katamnestic studies on dyslexia. *Acta-Paedopsychiatrica:- International Journal of Child and Adolescent Psychiatry, 56*, 219–228.

Strehlow, U., Kluge, R., Moller, H., & Haffner, J. (1992). Long-term course of developmental dyslexia beyond school age—A follow-up study. *Zeitschrift fur Kinder-und-Jugendpsychiatrie und Psychotherapi, e20*, 254–265.

Sunderland, H., & Klein, C. (1998). *Dyslexia in prisons.* London Language and Literacy Unit.

Swanson, H. L. (1999). *Intervention research for adolescents with learning disabilities. Two Decades of research in learning disabilities: Reading comprehension, expressive writing, problem solving, self-concept. Keys to successful learning: A national summit on research in learning disabilities.* New York: National Centre for Learning Disabilities.

Thompson, T. (1994). Self worth protection: Review and implications for the classroom. *Journal of Educational Review, 46*, 259–274.

Thorstad, I. (1999). *An investigation into the incidence of dyslexia at HMYOI, Onley.* Project Research, Birmingham's Regional College.

Tomblin, J. B., Zhang, X., Buckwalter, P., & Catts, H. (2000). The association of reading disability, behavioural disorders and language impairment among second-grade children. *Journal of Child Psychology and Psychiatry, 41*(4), 473–482.

Townend, J., & Turner, M. (2000). *Dyslexia in practice—A guide for teachers.* New York: Plenum.

Turner, M. (1997). *Psychological assessment of dyslexia.* London: Whurr.

Turner, M., Sercombe, L., & Cuffe-Fuller, A. (2000). *Dyslexia and Crime Dyslexia Review 12*, 1, 4–5.

Walker, J., & Brooks, L., (Eds.) (2000). *Dyslexia Institute literacy programme.* London: James and James.

Watson, P. (2002). Why Psychiatry has failed. *New Statesman*, 1st July 2002.

Westhues, A., Clarke, L., Watton, J., & Claire-Smith, S. S. (2001). Building positive relationships: An evaluation and outcome, big sister program. *The Journal of Primary Prevention, 21*, 477–493.

Willcutt, E. G., & Pennington, B. F. (2000). Psychiatric comorbidity in children and adolescents with reading disability. *Journal Child Psychology and Psychiatry, 41*(8), 1039–1048.

White, M. (1963). Ego and reality in psychoanalytic theory: A proposal regarding independent ego energies. *Psychological Issues, 3*, 125–150.

White, M. (2002). 50 *Activities for raising self esteem.* Cambridge: Pearson.

Yasutake, D., Bryan, T., & Dohrn, E. (1996). The effects of combining peer tutoring and attribution training on students' perceived self-competence. *Remedial and Special Education, 17*(2), 83–92.

Index